PREDICTIVE APPROACHES IN DRUG DISCOVERY AND DEVELOPMENT

Wiley Series on Technologies for the Pharmaceutical Industry

Sean Ekins, Series Editor

A complete list of the titles in this series appears at the end of this book.

PREDICTIVE APPROACHES IN DRUG DISCOVERY AND DEVELOPMENT

Biomarkers and *In Vitro/In Vivo* Correlations

Edited by

J. ANDREW WILLIAMS, Ph.D.
JEFFREY R. KOUP, Ph.D.
RICHARD LALONDE, Ph.D.
DAVID D. CHRIST, Ph.D.

WILEY

A JOHN WILEY & SONS, INC., PUBLICATION

Published by John Wiley & Sons, Inc., Hoboken, New Jersey
Published simultaneously in Canada

Library of Congress Cataloging-in-Publication Data:

Predictive approaches in drug discovery and development : biomarkers and in vitro/in vivo correlations / edited by J. Andrew Williams ... [et al.].
p. ; cm.—(Wiley series on technologies for the pharmaceutical industry)
Includes bibliographical references and index.
ISBN 978-0-470-17083-0 (cloth)
I. Williams, J. Andrew. II. Series: Wiley series on technologies for the pharmaceutical industry.
[DNLM: 1. Biomarkers, Pharmacological. 2. Drug Discovery. 3. Drug Evaluation, Preclinical. 4. Technology, Pharmaceutical. QV 744]
615.19–dc23

2011040419

Printed in the United States of America

10 9 8 7 6 5 4 3 2 1

To Becky and Mary Ann, without whose total support we would never have started and could not have finished this effort. While we like to think we know what we're getting into, their continuing faith and support of our decisions are amazing.

J. Andrew Williams
David D. Christ

"Essentially all models are wrong, but some are useful"

George E.P. Box

"For nothing ought to be posited without a reason given, unless it is self evident, or known by experience..."

William of Ockham
(known best as Occam's Razor)

CONTENTS

PREFACE

In the race to discover approvable new drugs faster and with fewer resources, two key elements have emerged that can enhance the drug pipeline and accelerate development, namely, biomarkers and *in vitro/in vivo* correlations (IVIVCs). At the early stage of the race, identifying the concepts and practices that link *in vitro* data with projected *in vivo* performance can lead to the identification of more robust clinical candidates and the more intelligent selection of new leads. Recognizing the limitations of IVIVCs and using IVIVC appropriately are critical to new drug discovery. As clinical trials are conceived, the identification of easily measured, rugged, and reliable markers of disease and the effects drugs have on disease are critical in defining appropriate patients and demonstrating efficacy as early as possible. Biomarkers, defined as an objectively measured indicator of physiological or pathophysiological function, or an indicator of pharmacological response, are important elements in translating basic pharmacology and drug effects into clinical utility and regulatory acceptance. Because of their power, understanding and applying biomarkers is really an expectation for the new drug development paradigm. This book provides a critical compilation of the most important aspects of these two topics from an international perspective.

Everyone involved in the process of new drug discovery, development and regulation should find the concepts and examples described herein useful, both for evaluating the merits of starting programs with these tools and for making decisions based on data from these approaches. Expertise in these two areas is no longer just the province of the pharmaceutical industry and regulatory agencies, but as more academic and government programs become involved in "drug discovery," more scientists, regardless of location, will need familiarity with these topics. The chapters in this book were written so that all scientific interests could find value; everyone from the technical staff to senior management.

Many of the concepts and strategies behind developing and applying biomarkers and IVIVC are complementary, and much of this book's value is contained in these reinforcing themes.

The expert authors responsible for each chapter come from a wide background in the pharmaceutical industry, worldwide regulatory agencies, and academia. While each chapter contains a core of basic information, the chapters also contain each author's perspective and opinion. We hope you will find this important aspect of the book most valuable since it provides the context for much of the science in these rapidly evolving areas.

ACKNOWLEDGMENTS

We are indebted to the chapter authors for their commitment, perseverance, and patience. All are excellent scientists, experts in their field, with overbooked calendars, and we sincerely appreciate the time they dedicated to their chapters. They provided great material to us, and if anything is not clear, we will take editorial responsibility. Thank you.

We would also like to acknowledge the patient guidance and unwavering support of Dr. Sean Ekins, Series Editor for the Wiley Series on Technology for the Pharmaceutical Industry. Sean is a friend and colleague, and his experience and advice throughout our editing efforts have been sustenance. On many levels, this book could not have been completed without Sean.

Jonathan Rose, Amanda Amanullah, and the staff at Wiley have been terrific. We appreciate their expertise, and patience, and the final volume is a product of their support.

J. Andrew Williams, Richard Lalonde,
Jeffrey R. Koup, and David D. Christ

San Diego, CA; Groton, CT; Vonore, TN; and Newark, DE

CONTRIBUTORS

Diane Alessi Fenton, MI, USA

Simon Authier 445 Armand Frappier, Laval, QC H7V 4B3, Canada

Kenneth Bachmann CeutiCare, LLC., 300 Madison Ave, Suite 270, Toledo, OH 43604, USA

Kimberly Brannen Reproductive Toxicology, Charles Rivers Lab., 587 Dunn Circle, Sparks, NV 89431, USA

Bruce D. Car Pharmaceutical Candidate Optimization, Bristol-Myers Squibb Co., Princeton, NJ 08543, USA

Michael J. Curtis Cardiovascular Division, School of Medicine, Rayne Institute, St Thomas' Hospital, London SE17EH, United Kingdom

Rex Denton Discovery Toxicology, Bristol-Meyers Squibb Co., Princeton, NJ 08543, USA

Joachim Drevs Cancer Hosptial UniSantis, An den Heilquellen 2, 79111 Freiburg, Germany

Sean Ekins Collaborations in Chemistry, 5616 Hilltop Needmore Road, Fuquay Varina, NC 27526, USA; Collaborative Drug Discovery, 1633 Bayshore Highway, Suite 342, Burlingame, CA 94010, USA; Department of Pharmaceutical Sciences, University of Maryland, 20 Penn Street, Baltimore, MD 21201, USA; Department of Pharmacology, University of Medicine and Dentistry of New Jersey, Robert Wood Johnson Medical School, 675 Hoes Lane, Piscataway, NJ 08854, USA

Ulrike Fiedler Experimetal Biomarker Research, ProQinase GmbH, Breisacherstr., 117, 79106 Freiburg, Germany

Joseph C. Fleishaker CORTEX, 4320 Forest Park Blvd, St Louis, MO 63108, USA

Scott Fountain Pfizer Inc., 10646 Science Center Drive, San Diego, CA 92121, USA

Ian Hudson Licensing Division, Medicines and Healthcare Products Regulatory Agency, 151 Buckingham Palace Road, London Sw1W 9SZ, United Kingdom

Akihiro Ishiguro Pharmaceuticals and Medical Devices Agency (PMDA), Shin-Kasumigaseki Building, Chiyoda-ku, Tokyo 100-0013, Japan

Lawrence Lesko Office of Clinical Pharmacology, FDA, MD, USA

Jiunn H. Lin 3D BioOptima, Jiangsu Wuzhong Life Sciences Park, 1338 Wuzhong Blvd., Suzhou Jiangsu 215104, China; 2 Willet Drive, Ambler, PA 19002, USA

Rajanikanth Madabushi Cardio-Renal at Office of Clinical Pharmacology, Baltimore, MD, USA

Ralph Paul Maguire Novartis Institutes for BioMedical Research, Forum 1, Novartis Campus, CH-4056, Basel, Switzerland

Marco Mascini Dipartimento di Chimica, Università di Firenze, Via della Lastruccia 3, 50019 Sesto Fiorentino, Italy

Harumasa Nakamura Pharmaceuticals and Medical Devices Agency (PMDA), Shin-Kasumigaseki Building, Chiyoda-ku, Tokyo 100-0013, Japan

R. Scott Obach Pfizer Inc., MS 8118D-2008, Eastern Point Road, Groton, CT-06340, USA

Karin Potthoff University Hospital, Im Neuenheimer Feld 400, 69120 Heidelberg, Germany

Michael K. Pugsley Johnson & Johnson PR&D, Global Preclinical Toxicology/Pathology, Raritan, NJ 00869, USA

William H. Schaefer Merck Laboratories, P.O. Box 4, WP81-205, West Point, PA 19486, USA

Gabriella Szekely-Klepser Drug Safety Evaluation, Allergan Inc., 2525 Dupont Drive, Irvine, CA 92612-1599, USA

Sara Tombelli Dipartimento di Chimica, Università degli Studi di Firenze, Via della Lastruccia 3, 50019 Sesto Fiorentino, Italy

Satoshi Toyoshima Pharmaceuticals and Medical Devices Agency (PMDA), Shin-Kasumigaseki Building, Chiyoda-ku, Tokyo 100-0013, Japan

Eric Troncy Faculté de médecine vétérinaire, Université de Montréal, 1500 des Vétérinaires, C.P. 5000, Saint-Hyacinthe, QC J2S 7C6, Canada

Yoshiaki Uyama Regulatory Science Research Division, Office of Regulatory Science, Pharmaceuticals and Medical Devices Agency (PMDA), Shin-Kasumigaseki Building, Chiyoda-ku, Tokyo 100-0013, Japan

Ruth A. VanBogelen Biomarker Science Group, Manchester, MI, USA

Janet Woodcock Center for Drug Evaluation and Research, FDA, MD, USA

Qiuwei Xu Merck Laboratories, P.O. Box 4, WP81-205, West Point, PA 19486, USA

PART I

BIOMARKERS IN DRUG DISCOVERY

1

THE IMPORTANCE OF BIOMARKERS IN TRANSLATIONAL MEDICINE

JOSEPH C. FLEISHAKER

1.1 INTRODUCTION

The new millennium was to have ushered in a bright new era of drug discovery. The unraveling of the human genome would provide a host of new therapeutic gene targets to treat debilitating diseases (1). The rest of the "omics" (proteomics, metabonomics, and transcriptomics) would provide additional insights on these targets and methods to assess drug effects early in the development process (2, 3). New therapeutic modalities (sRNAi, therapeutic proteins, and vaccines) would allow us to treat diseases, such as Alzheimer's disease, that up until now have eluded our best efforts. This was an engaging vision of the future.

What the new millennium has brought so far is steadily decreasing R&D productivity in the pharmaceutical industry. In 2007, only 16 new chemical entities were approved, compared to the 27 approved in 2000 by the U.S. Food and Drug Administration. The success rate for drugs in phase II proof of concept (POC) testing is at 20% or less (4). At the same time, the cost of bringing a new medicine to the market is approaching US$1.7 billion (5). There have also been several high profile withdrawals of products from the market for safety concerns, most notably rofecoxib (VIOXX® Tablets). This is hardly the vision conjured by mapping the human genome.

The key to addressing these issues and realizing the bright future for drug development is to assess, as early as possible, the properties (good and bad) of a potential target for intervention in a disease process and therapeutic modalities against that target. On the basis of these data, one must make a decision

Predictive Approaches in Drug Discovery and Development: Biomarkers and In Vitro/In Vivo Correlations, First Edition. Edited by J. Andrew Williams, Jeffrey R. Koup, Richard Lalonde, and David D. Christ.

whether to devote resources (private or public) to the development of that particular agent. The challenge is to do this with limited resources and with less than a 100% certain answer. By making early decisions on compounds and targets, we can then assess more targets/treatments for potential benefit and devote our limited resources to those that show the most promise. Traditional drug development paradigms have relied on large and prolonged studies to make go/no go decisions on new therapeutics. For example, a definitive answer on the utility of a disease-modifying agent for rheumatoid arthritis requires the assessment of the progression of joint narrowing and erosion by radiography (6). For Alzheimer's disease, long-term studies are necessary to establish a disease-modifying effect (7). How then do we get an answer within 3 months (or less) in 100 patients (or less) that an investigational treatment for these treatments is likely to be of therapeutic benefit and warrant the resources necessary for continued development?

Translational medicine has been proposed as the answer to the above question, and biomarkers are critical to the successful translation of findings in pharmacological studies in animals to therapeutic benefit in humans. The purpose of this chapter is to examine the integral role that biomarkers play in translational medicine and the development of new medicines. We examine successful applications of biomarkers to speed drug development and discuss examples where the lack of biomarkers has led to repeated failure in drug development. Finally, we discuss some future directions in biomarker research that can enhance drug development.

1.2 TRANSLATIONAL MEDICINE AND BIOMARKERS—SOME USEFUL DEFINITIONS

In any discussion on biomarkers, it is important that it is clear exactly what is being discussed. For example, the question, "Is your company working on biomarkers?" can be difficult to answer. Is the questioner referring to biomarkers for use in translational medicine and early decision making during drug development? Or rather, does the question really relate to a company's development of diagnostic tests to use when a drug is approved? Thus, the various definitions of translational medicine and biomarkers should be clearly understood in order to promote advancement in these areas.

Littman et al. (8) state that "The question of how to define translational research remains unresolved and controversial." They also provide a table (Box 1.1) that describes the areas that define translational research. The FDA Critical Path Initiative (9) describes translational research as being concerned with "moving basic discoveries from concept to clinical evaluation." The interesting part of this definition is that it is unidirectional from test tube to animal to human. Equally important is the back translation of clinical observations that may elucidate important insights into human disease, which drive further basic research aimed at new therapies (10).

Box 1.1 GOALS AND AREAS DEFINING TRANSLATION RESEARCH

Goals
The establishment of guidelines for drug development or for the identification and validation of clinically valid biomarkers.
Experimental nonhuman and nonclinical studies conducted with the intent of developing principles for discovery of new therapeutic strategies.
Clinical investigations that provide the biological foundation for the development of improved therapies.
Any clinical trial initiated with the above goals.
Basic science studies that define the biological effects of therapeutics in humans.

Source: Reproduced with permission from Littman BH, Di Mario L, Plebani M, Marincola M, *Clinical Science*, 2007;112:217–227 (8). This table was adapted from Mankoff SP, Brander C, Ferrone S and Marincola FM (2004), *J Transl Med* 2, 14, published by BioMed Central Ltd (9).

The NIH Biomarkers Definition Working Group (11) defined a biomarker as "a characteristic that is objectively measured and evaluated as an indicator of normal biological processes, pathogenic processes, or pharmacological responses to a therapeutic intervention." This is a relatively broad definition of a biomarker, which would include widely disparate methodologies such as FDG-PET, cognitive test batteries, gene expression, protein expression, and biochemical measures under the realm of biomarkers. The same group identified several uses for biomarkers, including diagnosis of disease, a tool for staging disease, and indicator of disease prognosis, or for prediction and monitoring of a clinical response to treatment. Translating these uses to drug development, biomarkers can be used to select which patients should be treated or to monitor beneficial and harmful effects of a medication. Implicit (but often forgotten) in the use of biomarkers in drug development is that they should be decision making; data obtained should affect either the conduct of a protocol or a development program.

In any discussion of biomarkers, one must differentiate between biomarkers and surrogate markers. The NIH group also defined a subset of biomarkers, the surrogate endpoint, as "a biomarker that is intended to substitute for a clinical endpoint. A surrogate endpoint is expected to predict clinical benefit (or harm or lack of benefit or harm) based on epidemiologic, therapeutic, pathophysiologic or other scientific evidence. In this sense, substitute is generally considered to mean substitute in a regulatory sense for a clinical endpoint." Classical surrogate endpoints are arterial blood pressure reduction as a surrogate for reduced stroke and cardiovascular mortality, LDL-cholesterol reduction for reduced cardiovascular mortality, and prolonged QT interval as a reflection of risk of sudden cardiac death due to torsades de pointes. In this chapter, we deal with biomarkers in the broadest sense of their use, and do not focus on the development of biomarkers as potential surrogate endpoints.

In developing and using biomarkers, one can use various classification systems. One is related to the type of information that the biomarker provides. According to this, a biomarker can be classified as a target, mechanism, or outcome biomarker (12). A target biomarker measures the interaction of a drug with a target receptor. A common example is measuring the binding of an atypical antipsychotic drug to D2 receptors in the brain using positron emission tomography of a ^{11}C-labeled ligand. A mechanism biomarker measures a physiological, biochemical, genomic, or behavioral change that occurs downstream from the target. Examples would be glucose lowering for a diabetes drug, decreased target phosphorylation after a kinase inhibitor, or sedation after the administration of a benzodiazepine. Outcome biomarkers are those that relate to the efficacy/toxicity of a compound, such as viral load as a function of survival benefit for anti-HIV therapy.

One can also consider the linkage between biomarker effects and clinical outcome (13). For example, mydriasis may be an excellent indication of the activity of a norepinephrine reuptake inhibitor (mechanistic biomarker), but it is not necessarily an indicator of potential efficacy in depression (14). On the other hand, occupancy at the D2 receptor, as measured by PET for an antipsychotic (target biomarker), is very closely related to efficacy for this class of compound (15). Thus, the linkage with outcome, as well as the type of biomarker, should be considered when assessing the ultimate utility of a biomarker.

The terms *validation* and *qualification* in relation to biomarker development also cause confusion. Wagner (16) defines validation as "The fit-for-purpose process of assessing the assay and its measurement performance characteristics, determining the range of conditions under which the assay will give reproducible and accurate data." Qualification is defined by Wagner as "The fit-for-purpose evidentiary process linking a biomarker with biological processes and clinical endpoints." The key phrase in both definitions is "fit-for-purpose." The rigor around validation and qualification should be dependent on the use of a biomarker and the decision that it will drive. The rigor around the validation and qualification of a biomarker used to assess whether a compound continues in development may be much less than that for a biomarker used to determine whether a particular patient should be treated with a particular compound. Fit-for-purpose thus means that the assay and its relevance to therapy are sufficient to drive the decision for which they are being developed.

1.3 BIOMARKERS: THE ROSETTA STONE OF TRANSLATIONAL MEDICINE

The term *translational medicine* suggests that we are translating "something" in animals to "something" in humans. During drug development, this would be translation of activity in an animal model of disease to activity in the human disease with great fidelity. Unfortunately, this is not a common occurrence in drug development. Perel et al. (17) systematically reviewed the concordance between animal and human data for six disease areas. Table 1.1 describes the areas reviewed, the number of animal studies reviewed, and the methodological aspects of these studies. Three of the interventions showed concordance in outcomes between animal and human studies (thrombolysis for acute ischemic stroke, bisphonates for osteoporosis, and antenatal corticosteroids), and three did not. In the case of antifibrinolytics for hemorrhage, animal models yielded no reliable data, while clinical trails showed clear benefit. In general, the study designs of the animal studies were poor, generally lacking in randomized treatment allocation or blinding of the allocation or the assessor. Thus, there is substantial room for positive bias in the assessment of the results in these animal studies. There was, however, no correlation between the quality of the experiments and the concordance between animal and human studies.

Irrespective of methodological considerations, there are often differences between human disease and disease models in animals (18). If one considers acute ischemic stroke, many drugs have been studied in animals and humans, and only one, tissue plasminogen-activating factor, has been found to be efficacious and is in clinical use. In the neurological trauma arena, many animal studies are conducted in healthy animals, free of the comorbidities (diabetes, high blood pressure, etc.) that would be present in an elderly patient with an acute ischemic stroke. In addition, genetic homogeneity with a rat strain does not reflect the genetic heterogeneity in the human patient population. Outcome measures in rodents (infarct size) do not reflect relevant outcome measures in humans (functional disability). Animal models, where therapeutic interventions

TABLE 1.1 Quality of Animal Studies Used to Predict Efficacy in Several Disease Indications

Intervention	Random Allocation to Group	Adequate Allocation Concealment	Blinded Assessment of Outcome
Corticosteroids for traumatic head injury ($n = 17$)	2 (12)	3 (18)	3 (18)
Antifibrinolytic agents ($n = 8$)	3 (38)	0	4 (50)
Thrombolysis for acute ischemic stroke ($n = 113$)	43 (38)	23 (20)	24 (21)
Tirilazad for acute ischemic stroke ($n = 18$)	12 (67)	1 (8)	13 (72)
Antenatal corticosteroids ($n = 56$)	14 (25)	0	3 (5)
Bisphonates ($n = 16$)	5 (31)	0	0

Values are number of studies (percentages of total).

Source: Adapted with permission from BMJ Publishing Group Ltd. Comparison of treatment effects between animal experiments and clinical trials: systematic review. Perel P, Roberts I, Sena E et al. *BMJ*, Volume 334, p 197, Copyright 2007 (17).

may be administered before or shortly after a neurological insult, may not be reflective of therapy in humans that will not begin for hours after neurological insult. While these examples are specific to the area of stroke and neurotrauma, similar results are seen across multiple therapeutic areas.

Disease models in animals themselves rarely reflect the human disease in total. Disease models generally reflect some aspect of the human disease. For example, some depression models in animals reflect the learned helplessness typical of human depression, while other models address the cognitive deficits seen in human depression (19). Separate transgenic mouse models in Alzheimers disease have been developed to address abnormalities in β-amyloid protein, tau-protein, and pre-senilin (20), which are commonly found together in the human disease. In many cases then, animal models are set up to reflect certain pathways in human disease, rather than the disease per se. Both disease models and pharmacology models in animals may be used in translation to humans, as shown in Figure 1.1, but both pathways require biomarkers for successful translation (21).

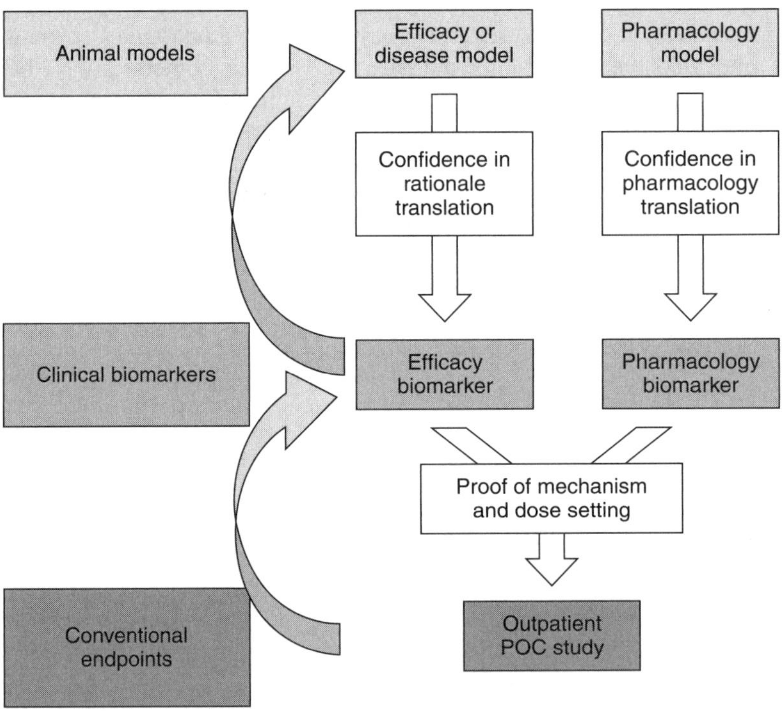

FIGURE 1.1 Linked animal models and clinical biomarkers can be used to confirm translation of preclinical efficacy and pharmacology to clinical effects. Clinical measures are used to set dose range and optimize the design of outpatient studies. *Source:* Reprinted from *Drug Discovery Today*, Volume 12, Sultana SR, Roblin D, O'Connell D, pp. 419–425, Copyright 2007, with permission from Elsevier.

The development of translatable biomarkers is still an evolving field, but there are several examples available. Cardiac troponin has been shown to be indicative of cardiac injury in both animals and humans, so that there is high confidence that the increases in cardiac troponin in animals seen in preclinical drug testing would also be seen in humans (22). As such it is a valuable screening tool. Imaging techniques such as PET for receptor occupancy or fMRI have been useful in the development of antischizophrenic compounds (21). Other soluble biomarkers, such as cyclic GMP, can reflect the pharmacology of agents such as the neuroendopeptidase inhibitor and PDE-5 inhibitors in both animals and humans (21).

Target and mechanism biomarkers that would be translatable from animals to humans are absolutely essential to answer key questions during the early development process. These questions are as follows:

1. Does the drug hit the intended target in humans?
2. Does the drug exhibit the intended pharmacology in humans?
3. What is the relationship between pharmacokinetics and pharmacodynamics in humans?
4. What doses/drug concentrations are appropriate for initial studies in patients to more fully explore the efficacy of the compound? Can these be achieved within the tolerable dose range for the compound in humans?

For novel compounds, positive answers to all of these questions are needed to assure that we adequately test the hypothesis that modulating the target mechanism in humans has beneficial effects on a disease process. While this conclusion is intuitive, large scale development programs have been conducted in the absence of this information.

1.4 DRUG DEVELOPMENT WITHOUT BIOMARKERS—AN EMPTY EXPERIENCE

Tirilazad mesylate (Tirilazad, Fig. 1.2) is a 21-aminsteroid compound that was developed as a free radical scavenger and antioxidant for the treatment of acute neurological trauma (23). Tirilazad was studied in the treatment of head injury, ischemic stroke, spinal cord injury, and aneurismal subarachnoid hemorrhage and was approved in several countries for the treatment of subarachnoid hemorrhage.

Tirilazad was designed to prevent lipid peroxidation following the generation of free radicals due to the initial tissue damage following a neurological insult. A variety of treatment paradigms in preclinical models were utilized for tirilazad, ranging from single-dose administration following head trauma in mice to administration for 6 days in a canine model of subarachnoid hemorrhage (23). These paradigms were designed to cover the time of penumbral neurological damage that could occur after the initial insult. All of these studies had several characteristics in common. Neurological outcome measures (motor scores,

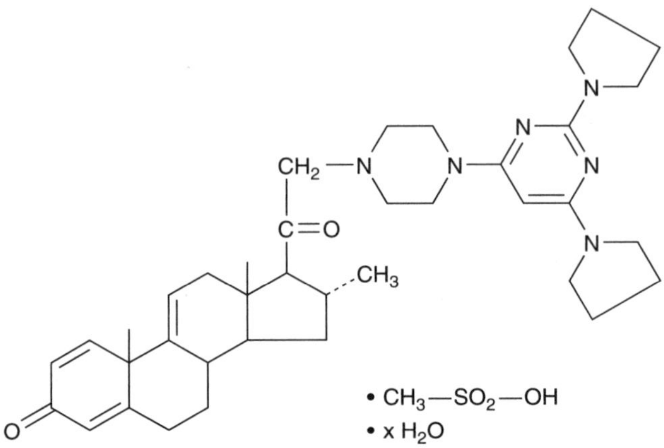

FIGURE 1.2 Structure of tirilazad mesylate.

evoked potentials) or local morphologic/biochemical measures (infarct size, middle cerebral artery vasospasm, lipid peroxide levels, etc.) were the key endpoints for these studies. With the exception of attempts to evaluate the sparing of antioxidant vitamins (vitamins C and E) peripherally by tirilazad (24), neither circulating biomarkers nor circulating or brain levels of tirilazad were measured as part of these studies. Dosing was based on body weight (mg/kg), and exposure was not compared across animal species.

On the basis of the data available in animals and humans, how would we answer the questions outlined in the previous section?

1. Does the drug hit the intended target in humans? We do not know. No assessments of brain uptake of tirilazad were performed in humans.
2. Does the drug exhibit the intended pharmacology in humans? We do not know. Tirilazad elicited no overt pharmacology in early clinical trials.
3. What is the relationship between pharmacokinetics and pharmacodynamics in humans? We do not know. No biomarkers were available to measure tirilazad activity, and there was no correlation between tirilazad dose or exposure and efficacy in humans.
4. What doses/target drug concentrations are appropriate for initial studies in patients to more fully explore the efficacy of the compound? Can these be achieved within the tolerable dose range for the compound in humans? We do not know. The only extrapolation that could be made between animals and humans was based on dose/body weight, not exposure.

Studies of tirilazad in the treatment of head trauma, ischemic stroke, and spinal cord injury failed to show efficacy, and some studies showed worsening of outcome relative to placebo (25–28). Initial studies of tirilazad for the treatment of aneurysmal subarachnoid hemorrhage at a dose of 6 mg/kg/day showed some benefit in men, but not in women (29, 30). On the basis of pharmacokinetic data, premenopausal women showed higher clearance and lower plasma concentrations of tirilazad (31); two additional large studies were conducted in female SAH patients at a dose of 15 mg/kg/day (32, 33). Results from these studies did not show a general benefit of tirilazad in women.

After several thousand patients were treated with tirilazad, what was learned? There remain two possibilities. Either tirilazad is ineffective for the treatment of neurological trauma in humans or the trials that were conducted were sufficiently flawed (wrong dose, imbalance in groups in normal medical care, wrong patient groups, etc.) that the effects of tirilazad could not be seen in these patient groups (23). The available data do not allow a determination of which hypothesis is correct.

1.5 BIOMARKER TRANSLATION SUCCESS STORIES

While the lack of a translatable biomarker impedes the development of new medicines and reduces the probability of ultimate success, the availability of these biomarkers allows early assessment of therapeutic potential and can speed clinical development. The latter situation is described in two case studies that illustrate the power of translatable biomarkers in drug development.

1.5.1 Sunitinib

Various receptor tyrosine kinases (RTKs) and their receptors are overexpressed in different tumor types and contribute to tumor growth and survival. For example, vascular endothelial growth factor (VEGF) receptors are important in melanoma, platelet derived growth factor (PDGF) receptors are key in gliomas, stem cell factor receptors (KIT) are overexpressed in gastrointestinal stromal tumors (GIST), and Fms-like tyrosine kinase-3 (FLT3) receptor is deregulated in acute myelogenous leukemia (AML) (34). Sunitinib (SU11248, SUTENT® capsules) (Fig. 1.3) was designed as a potent inhibitor of these receptor kinase receptors. *In vitro* and *in vivo* measures (in mouse xenograft models) of VEGFR2, PDGF2, and FLT3 along with plasma concentration determinations in animals allowed robust PK/PD analysis that suggested that plasma concentrations in the range of 50–100 ng/ml were effective in various tumor types. The knowledge of overexpression in various tumor types and the PK/PD relationships based on markers of receptor inhibition allowed rapid identification of the doses that would be effective in phase I studies in humans and selection of indications and patients for early clinical evaluations in oncology patients. The initial three indications studied, AML, GIST, and renal cell carcinoma, were selected because sunitinib was active against the kinase targets that are overexpressed in these tumors.

On the basis of the observed *in vitro* and *in vivo* inhibition of FLT3 by sunitinib, a phase I single-dose, dose-escalation study was conducted in AML patients (35) with FLT3 inhibition as the primary endpoint. Twenty-nine patients received

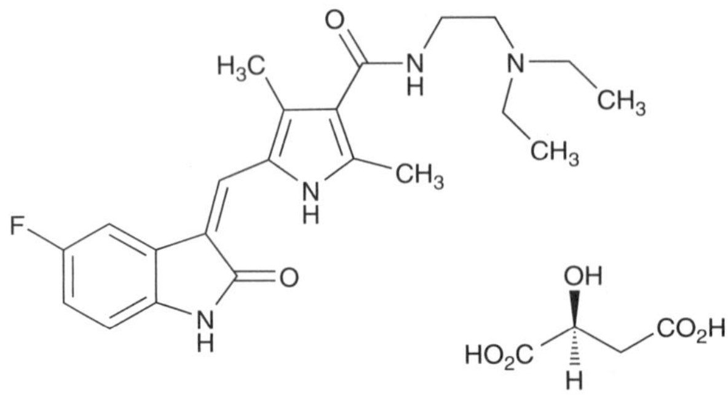

FIGURE 1.3 Structure of sunitinib malate.

single doses of sunitinib from 50 to 350 mg. Plasma sunitinib concentrations and plasma concentrations of SU-12662, an active metabolite, were determined serially after dosing. Likewise, FLT3 phosphorylation was measured at various times after dosing. Subjects were genoptyped for major FLT3 kinase mutations, with FLT3-ITD being associated with a negative prognosis in AML (36). Figure 1.4 shows FLT3 phosphorylation as a function of time after sunitinib dosing. Figure 1.5 shows the correlation between plasma C_{max} of active species (sunitinib and Su-12662) and FLT3 phosphorylation, as well as the correlation of time above 100 ng/ml and FLT phosphorylation. In patients with wild-type FLT3, strong inhibition (> 50%) of FLT3 was associated with C_{max} > 100 ng/ml (consistent with experiments in animal xenograft .experiments noted above)

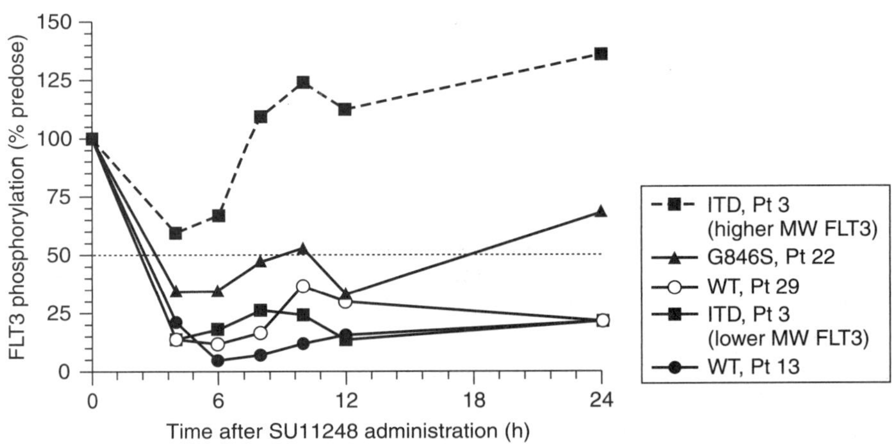

FIGURE 1.4 FLT3 phosphorylation as a percentage of predose values following administration of SU 11248 to AML patients. Points below the dotted line represent strong inhibition of FLT3 phosphorylation. Data from representative subjects are shown. *Source:* Reprinted with permission from the American Association for Cancer Research, *Clinical Cancer Research*, Volume 9, O'Farrell A-M, Foran JM, Fiedler W, et al., pp. 5465–5476, Copyright 2003.

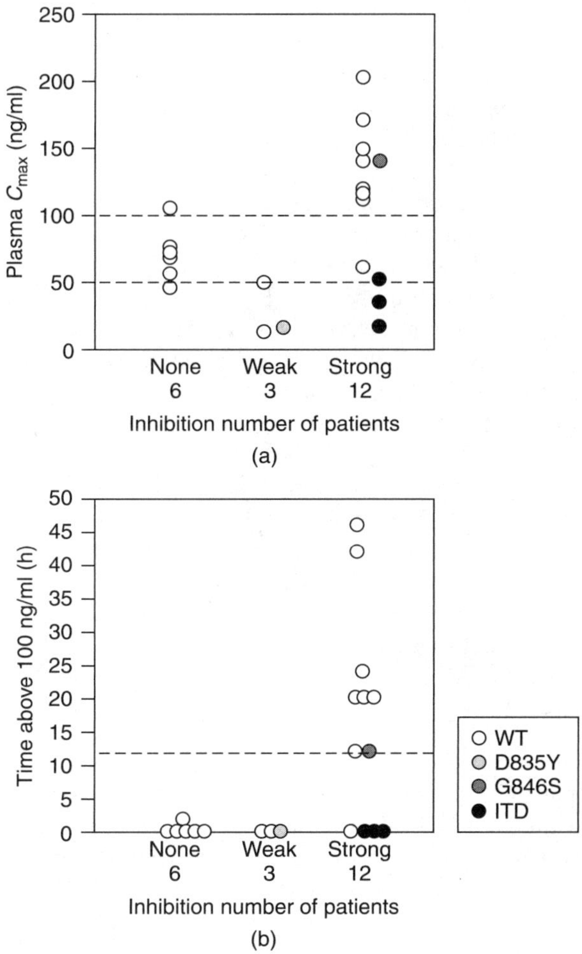

FIGURE 1.5 PK/PD analysis of FLT3 phosphorylation. Plasma C_{max} (combined SU11248 and SU12662; (a) and time exceeding the target plasma concentration of 100 ng/ml (b) are shown for each patient, grouped according to degree of FLT3 inhibition, and color coded based on FLT3 genotype. *Source:* Reprinted with permission from the American Association for Cancer Research, *Clinical Cancer Research*, Volume 9, O'Farrell A-M, Foran JM, Fiedler W, et al., pp 5465–5476, Copyright 2003.

and >10 h above 100 ng/ml of the combined active species in plasma. Interestingly, strong inhibition of FLT3 phosphorylation was observed in patients with ITD mutation. This initial study showed clear modulation of the target biomarker (FLT3 phosphorylation) in humans, and this biomarker was used to establish an effective concentration in AML patients, which was similar to that shown in animal models for AML and other tumor types. The results of this innovative experiment and the use of biomarkers helped to set the stage for future development of sunitinib.

An initial phase I/II trial in GIST patients provides another illustration of the utility of biomarkers in development. In this study (37), 97 patients with GIST were treated with sunitinib using one of three on/off treatment cycles (2 weeks on/1 week off, 2 weeks on/2 weeks off, or 4 weeks on/2 weeks off). Seventy-five of the 96 subjects underwent PET scanning with FDG-PET at baseline, on day 7 of the first cycle, at the end of the first cycle off drug, and during a subsequent cycle while on drug. FDG-PET is a measure of glucose uptake and indicative of metabolic activity in the tumor; decreased tumor activity by this measure has been shown to reflect clinical benefit (37). As such, FDG-PET is a mechanism and outcome biomarker. Figure 1.6 shows the response in one patient, with reduction in tumor activity within 7 days after starting dosing, return of tumor activity in the first off cycle, and continuing reduction in tumor activity during cycle 2. Using a measure of activity, the maximal standardized uptake value (SUV_{max}) for statistical analysis, similar behavior was seen across the cohort that completed all four scans (Table 1.2). Utilizing PET scanning, rapid objective assessment of response was obtained in this study, which set the stage for continued development of the compound.

What then was the implication of the use of biomarkers to drive the development program for sunitinib? The first dose of this drug was administered to a human in 2000, and the product was approved for marketing in the United States in 2006 for the treatment of GIST and renal cell carcinoma.

1.5.2 Maraviroc

In addition to the CD4 receptor being necessary for HIV-1 entry into T cells, more recently, the CCR5 and CXCR4 have been found to be coreceptors needed for HIV-1 entry into cells. The observation that homozygotes for a 32-bp deletion in CCR5 showed natural resistance to HIV-1 and that heterozygotes had a longer disease progression time sparked the development of CCR5 inhibitors for the treatment of HIV infection (38). Maraviroc (UK-427,857, CELSENTRI® Tablets, SELZENTRY® Tablets) (Fig. 1.7) is the first CCR5 receptor antagonist to be approved for HIV infection. Like sunitib, the clinical development and approval of maraviroc was rapid, with initial human dosing commencing in 2001 and approval gained in 2007. Also, like sunitinib, biomarkers played a key role in accelerating development.

An initial phase IIa study with maraviroc was conducted in HIV-infected volunteers who received placebo, 25 mg QD, 50 mg BID, 100 mg BID, and 300 mg BID maraviroc for 10 days. (39, 40) CCR5 receptor occupancy (target biomarker) and viral load (outcome biomarker) were the key measures in this study. Maraviroc reduced viral load as a function of time, with mean reductions $>1.0 \log_{10}$ observed at the 100 and 300 mg BID doses (Table 1.3). Doses at or above 100 mg resulted in >80% CCR5 receptor occupancy. The results from this study, in addition to previously developed HIV-1 disease model (Fig. 1.8) (41), were used to construct a PK/PD model for viral load that was used to predict the efficacy of three additional dosing regimens of maraviroc, which were subsequently studied (39, 40). The model predictions for these dosing regimens, as

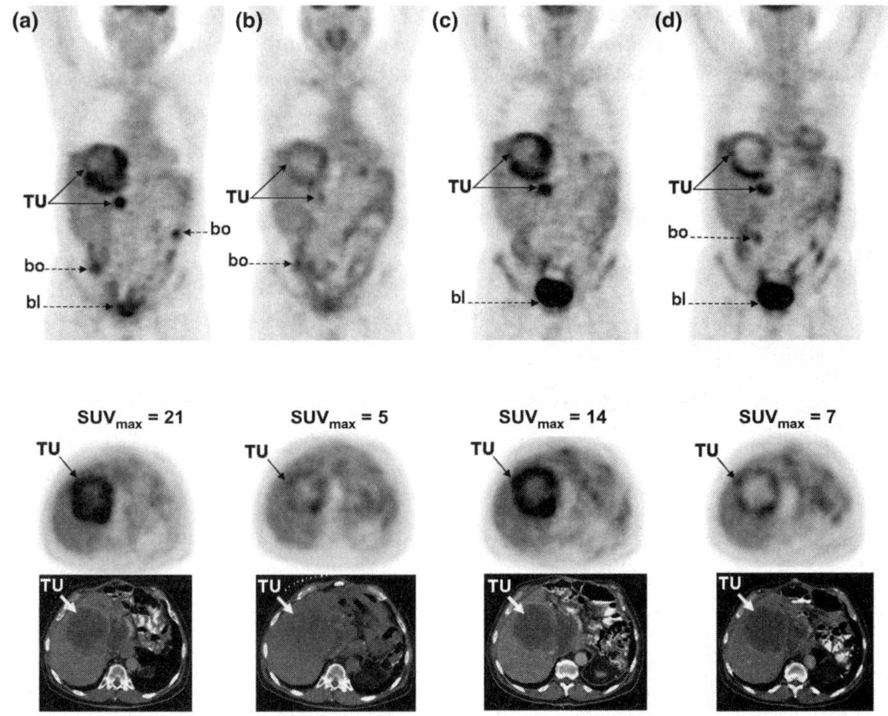

FIGURE 1.6 Coronal (top), axial (middle) FDG-PET slices, and corresponding axial CT slices (bottom) in a patient with GIST metastatic to the liver and anterior abdomen (solid arrows, TU) before sunitinib therapy (a, baseline), during cycle 1 (b), at the end of the resting period before cycle 2 (c, off treatment), and during cycle 4 (d). Physiologic uptake of FDG is seen in the bowel (dotted arrows, bo), and in the urinary bladder (dotted arrows, bl). The baseline FDG-PET (a) shows a large FDG-avid mass with a necrotic center in the liver and a SUV_{max} of 21, and a smaller mass in the anterior abdomen reflecting intense tumor glycolytic activity (solid arrows, TU). A marked decrease in glycolytic activity is noted in both tumor masses (solid arrows, TU) as early as 1 week following treatment with sunitinib during cycle 1 (b). The SUV_{max} of the liver lesion has decreased to 5. Note that the rebound in Glycolytic tumor activity in both masses (solid arrows, TU), as reflected by intense FDG uptake and an increase in the SUV_{max} of the liver lesion to 14, at the end of the resting period before the next cycle of sunitinib (c). During subsequent cycles of sunitinib therapy, a decrease in tumor metabolic activity is again observed (d, cycle 4). The SUV_{max} of the liver lesion has decreased to 7 during cycle 4. Note that the size of the hepatic lesion does not significantly change on the corresponding CTs obtained at the same time points (bottom, white arrows, TU). *Source:* Courtesy of Annick D. Van den Abbeele, MD and Iryna Rastarhuyeva, MD, Dana-Farber Cancer Institute, Boston, MA.

TABLE 1.2 Maximum Standardized Uptake Values (SUV_{max}) for FDG-PET Following the Administration of Several Cycles of Sunitinib in Patients ($n = 74$) with Imatinib-Resistant Gastrointestinal Stromal Tumors[a]

FDG-PET Scan	Baseline	After 7 days treatment in Cycle 1	At the end of first period off drug	During Cycle 4
SUVmax (Standard error)	9.7 (0.6)	5.6 (0.6)	8.0 (0.6)	5.1 (0.5)
Absolute difference in mean log SUVmax from prior scan (95% Confidence interval)		0.32 (0.25–0.39)	0.21 (0.14–0.28)	0.22 (0.15–0.30)

[a]Ref. 37.

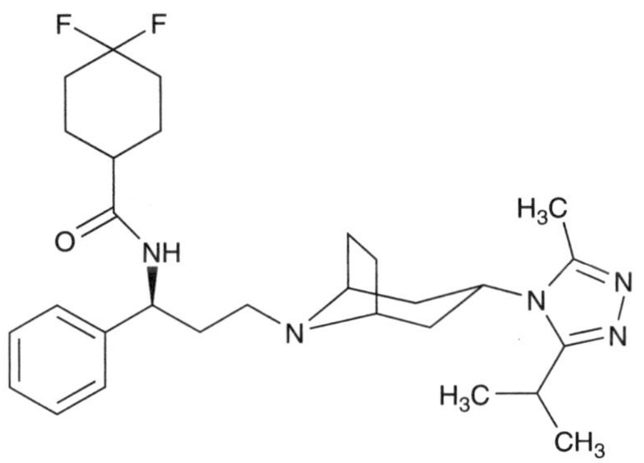

FIGURE 1.7 Structure of maraviroc.

TABLE 1.3 Mean (Range) of HIV-1 RNA $\log_{10}$ Declines after 11 Days of Maraviroc Therapy in Patients[a]

Maraviroc Dose	Mean (Range) HIV-1 RNA $\log_{10}$ Decline
Placebo	0.02 (−0.45 to 0.56)
25 mg QD	− 0.43 (−1.08 to 0.02)
50 mg BID	− 0.66 (−1.37 to 0.40)
100 mg BID	−1.42 (−1.84 to −1.04)
300 mg BID	−1.60 (−2.42 to −0.78)

[a]Ref. 40.

well as the observed values, are shown in Table 1.4. The model predicted the behavior of these new dosing regimens administered over a 10-day period very well. On the basis of clinical trial simulations using this drug and disease model (based on the viral load biomarker), the two phase IIb/III trials that were conducted with maraviroc utilized doses of 300 mg/day or 300 mg twice daily, with

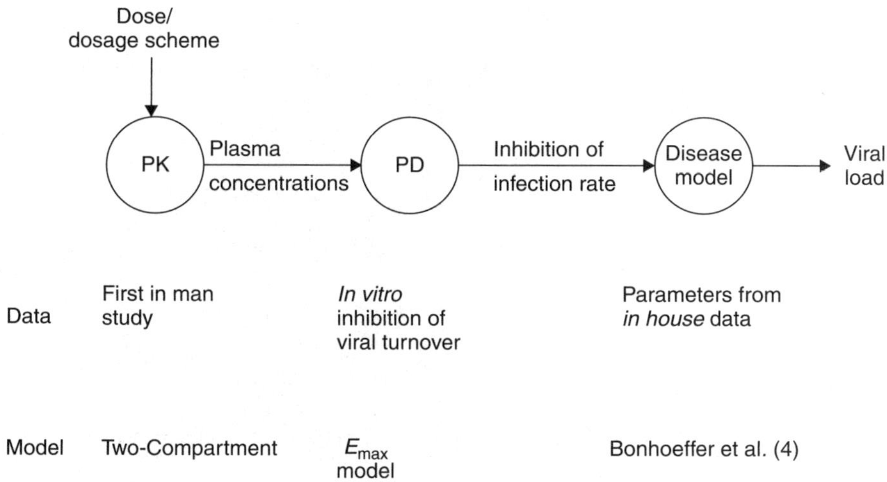

FIGURE 1.8 Schematic representation of pharmacokinetic (PK)-pharmacodynamic (PD)-disease model for an antiretroviral drug. E_{max}, Maximum effect. *Source:* Reprinted by permission from Macmillan Publishers Ltd: *Clinical Pharmacology and Therapeutics*, Rosario MC, Jacqmin P, Dorr P, van der Ryst E, Hitchcock C. A pharmacokinetic-pharmacodynamic disease model to predict in vivo antiviral activity of maraviroc, 78:508–519, Copyright 2005.

TABLE 1.4 Performance of Model in Predicting HIV-1 RNA $\log_{10}$ Declines after 11 Days of Therapy for New Maraviroc Dosing Regimens[a]

Maraviroc Dosing Regimen	Observed Mean (Range)	Predicted Median (90% Confidence Interval)
150 mg bid fasted	−1.45 (−1.71 to 0.90)	−1.30 (−1.67 to −0.82)
150 mg BID fed	−1.34 (−1.79 to −0.51)	−1.12 (−1.52 to −0.58)
100 mg QD	−1.13 (−1.70 to −0.43)	−0.81 (−1.32 to −0.32)
300 mg QD	−1.35 (−1.62 to −0.95)	−1.30 (−1.76 to −0.83)

[a]Ref. 40.

subjects receiving a CYP3A4 inhibitor receiving a dose of 150 mg twice daily (42). This resulted in a streamlined program that supported the rapid development and approval of this new medicine.

1.6 THE PATH FORWARD

In this chapter, examples of the perils of drug development without biomarkers and the use of biomarkers to speed development have been presented. What must happen so that the successes seen in some therapeutic areas may be expanded to others?

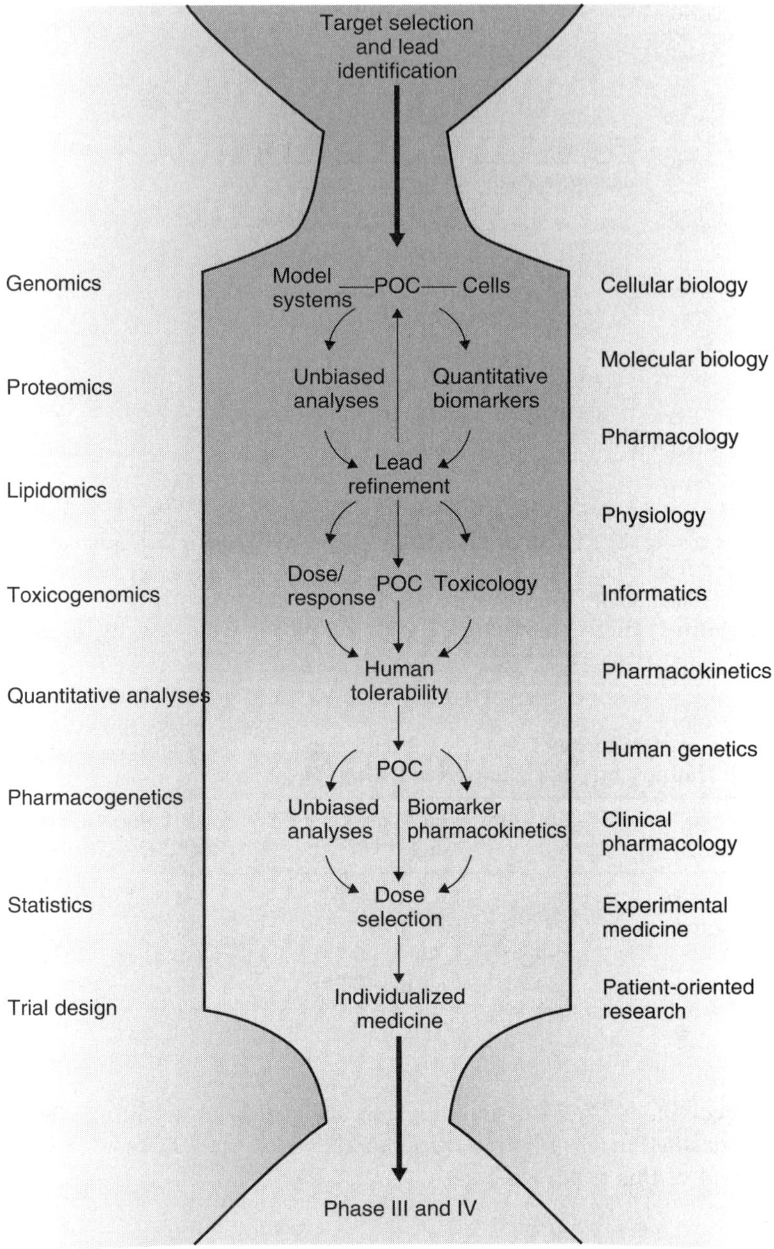

FIGURE 1.9 The spectrum of translational medicine and therapeutics. The translational space imposed on the process of drug development is defined as stretching from proof of concept (POC) in cells and model systems to completion of studies of drug mechanism and variability of response, which afford a basis for individualized dose selection. The conventional disciplines that are drawn on as one progress through the translational channel are indicated. *Source:* Reprinted by permission from Macmillan Publishers Ltd: *Nature Reviews in Drug Discovery*, Fitzgerald GA. Anticipating change in drug development: the emerging era of translational medicine and therapeutics, 4(10): pp. 815–818, Copyright 2005.

To speed decision making in drug development, biomarkers should focus on the aspects of a disease model that can most readily be translated between animals and man—disease pathways and drug pharmacology. Efficacy outcomes are difficult to translate between animals and humans, which is reflected in the current lack of confidence in animal models. Development of translational biomarkers relevant to disease pathways and drug pharmacology must begin when a promising new therapeutic target is identified. These biomarkers should also be developed with a fit-for-purpose mind set. Initially, we want them for decision making in drug development. The biomarker in question may some day be a diagnostic tool or a surrogate biomarker, but its initial development should reflect the limited use for which it is intended.

Since biomarkers are intended to be decision making, all of the stake holders in the decision should be involved in their development. Thus, in addition to the biologists, pharmacologists, and analytical experts needed to identify and quantify biomarkers, a host of others are involved in the analyses and optimal use of these data. Fitzgerald (43) provides an excellent summary of the cross-discipline nature of translational medicine and biomarker development (Fig. 1.9).

Finally, we have to actually use biomarkers, or lack thereof, for decision making. Those who guide drug development decisions must have the fortitude to forego the development of drugs for which there are no biomarkers available and no way to determine whether the drug will actually test the hypothesis regarding a molecular target. They must also be willing to abandon programs early for drugs that do not show the degree of biomarker modulation necessary to justify continued development. Likewise, they must be willing to use the data from fit-for-purpose biomarkers to inform dose selection, patient selection, and other protocol and program design decisions to speed drug development. These further examples of success will further increase the confidence in biomarkers and allow us to move toward the future vision of drug development and patient care conjured by the mapping of the human genome.

REFERENCES

1. Lemonick MD. The genome is mapped. Now what? Time 2000;156:1.
2. Keun HC, Athersuch TJ. Application of metabonomics in drug development. Pharmacogenomics 2007;8:731–741.
3. Kohn EC, Azad N, Annunziata C, Dhamoon AS, Whiteley G. Proteomics as a tool for biomarker discovery. Dis Markers 2007;23:411–417.
4. The Pink Sheet. Wyeth shifting R&D funds to early-stage compound research and licensing. The Pink Sheet 2005;67:19.
5. Mullin R. Drug development costs about $1.7 billion. Chem Eng News 2003;81:8.
6. Wolfe F, Strand V. Radiography of rheumatoid arthritis in the time of increasing drug effectiveness. Curr Rheumatol Rep 2001;3:46–52.
7. Pangelos MN, Schechter LF, Hurko O. Drug development for CNS disorders: strategies for balancing risk and reducing attrition. Nat Rev Drug Discov 2007;6:521–532.
8. Littman BH, DiMario L, Plebani M, Marincola FM. What's next in translational medicine? Clin Sci 2007;112:217–227.

9. Stagnation-Innovation. Critical Path Opportunities Report.US Department of Health and Human Services. Food and Drug Administration; 2004.
10. Mankoff SP, Brander C, Ferrone S, Marincola FM. Lost in translation: obstacles to translational medicine. J Transl Med 2004;2:14.
11. Biomarker Definitions Working Group. Biomarkers and surrogate endpoints: preferred definition and conceptual framework. Clin Pharmacol Ther 2001;69:89–95.
12. Navigating the bench to bedside journey. Refining and adapting established approaches to drug development. Genet Eng Biotech News 2006;26:9. Available at http://www.genengnews.com/articles/chitem.aspx?aid=1665&chid=4.
13. Littman BH, Williams SA. The ultimate model organism: progress in experimental medicine. Nat Rev 2005;4:631–638.
14. Phillips MA, Bitsios P, Szabadi E, Bradshaw CM. Comparison of the antidepressants reboxetine, fluvoxamine and amitriptyline upon spontaneous pupillary fluctuations in healthy human volunteers. Psychopharmacology 2000;149:72–76.
15. Pani L, Pira L, Marchese G. Antipsychotic efficacy: relationship to optimal D_2-receptor occupancy. Eur Psychiatry 2007;22:267–275.
16. Wagner JA. Strategic approach to fit-for purpose biomarkers in drug development. Annu Rev Pharmacol Toxicol 2008;48:631–651.
17. Perel P, Roberts I, Sena E, Wheble P, Briscoe C, Sandercock P, Macleod M, Mignini LE, Jayaram P, Khan KS. Comparison of treatment effects between animal experiments and clinical trials: systematic review. BMJ 2007;334:197–200. Doi:10.1136/bmj39048.407928.BE.
18. DeGraba TJ, Pettigrew LC. Why do neuroprotective drugs work in animals but not in humans? Neurol Clin 2000;18:475–493.
19. Nestler EJ, Gould E, MAnji H, Bucan M, Duman RS, Gershenfeld HK, Hen R, Koester S, Lederhendler I, Meaney MJ, Robbins T, Winsky L, Zalcman S. Preclinical models: status of basic research in depression. Biol Psychiatry 2002;52:503–528.
20. Rockenstein E, Crews L, Masliah E. Transgenic animal models of neurodegenerative diseases and their application to treatment development. Adv Drug Deliv Rev 2007;59:1093–1102.
21. Sultana SR, Roblin D, O'Connell D. Translational research in the pharmaceutical industry: from theory to reality. Drug Discov Today 2007;12:419–425.
22. O'Brien PJ. Cardiac troponin is the most effective translational safety biomarker for myocardial injury in cardiotoxicity. Toxicology 2008;245:206–218.
23. Kavanagh RJ, Kam PCA. Lazaroids: efficacy and mechanism of action of the 21-aminosteroids in neuroprotection. Br J Anaesth 2001;86:110–119.
24. Sato PH, Hall Ed. Tirilazad mesylate protects vitamins C and E in brain ischemia-reperfusion injury. J Neurochem. 1992;58:2263–2268.
25. Marshall LF, Maas AIR, Marshall SB, Bricolo A, Fearnside M, Iannotti F, Klauber MR, Lagarrigue J, Lobato R, Persson L, Pickard JD, Piek J, Servadei F, Wellis GN, Morris GF, Means ED, Musch B. A multicenter trial of the efficacy of tirilazad in cases of head injury. J Neurosurg 1998;89:519–525.
26. The RANTTAS investigators. A randomized trial of tirilazad mesylate in patients with acute stroke (RANTTAS). Stroke 1996;27:1453–1458.
27. The RANTTAS II investigators. High dose tirilazad for acute stroke (RANTTAS II). Stroke 1998;29:1256–1257.

28. Bracken MB, Shepard MJ, Holford TR, Leo-Summers L, Aldrich EF, Fazl M, Fehlings M, Herr DL, Hitchon PW, Marshall LF, Nockels RP, Pascale V, Perot PL Jr, Piepmeier J, Sonntag VK, Wagner F, Wilberger JE, Winn HR, Young W. Administration of methylprednisolone for 24 or 48 hours or tirilazad mesylate for 48 hours in the treatment of acute spinal cord injury. Results of the Third National Acute Spinal Cord Injury Randomized Controlled Trial. JAMA 1997;277:1597–1604.

29. Kassell NF, Haley EC, Apperson-Hanson V, Alves WM. Randomized, double-blind, vehicle-controlled trial of tirilazad mesylate in patients with aneurismal subarachnoid hemorrhage: a cooperative study in Europe, Australia, and New Zealand. J Neurosurg 1996;84:221–228.

30. Haley EC, Kassell NF, Apperson-Hanson C, Maile MH, Alves WM. A randomized double-blind, vehicle-controlled trial of tirilazad mesylate in patients with aneurismal subarachnoid hemorrhage: a cooperative study in North America. J Neurosurg 1997;86:467–474.

31. Hulst LK, Fleishaker JC, Peters GR, Harry JD, Wright DM, Ward P, Fenton CM. Effect of age and gender on tirilazad pharmacokinetics in humans. Clin Pharmacol Ther 1994;55:378–384.

32. Lanzino G, Kassekk NF, Dorsch NW, Pasqualin AL, Brandty L, Schmiedek P, Truskowski LL, Alves WM, and the participants. Double-blind, randomized, vehicle controlled study of high-dose tirilazad mesylate in women with aneurismal subarachnoid hemorrhage. Part I. A cooperative study in Europe, Australia, New Zealand and South Africa. J Neurosurg 1999;90:1011–1017.

33. Lanzino G, Kassell NF. Double-blind, randomized, vehicle-controlled study of high-dose tirilazad mesylate in women with aneurysmal subarachnoid hemorrhage. Part II. A cooperative study in North America. J Neurosurg 1999;90:1018–1024.

34. Mendel DB, Laird AD, Xin X, Louie SG, Christensen JG, Li G, Schreck RE, Abrams TJ, Ngia TJ, Lee LB, Murray LJ, Carver J, Chan E, Moss KG, Haznedar JO, Sukbentherng J, Blake RA, Sun L, Tang C, Miller T, Shirazian S, McMahon G, Cherrington JM. In vivo antitumor activity of SU11248, a novel tyrosine kinase inhibitor targeting vascular endothelial growth factor and platelet-derived growth factor receptors: determination of a pharmacokinetic/pharmacodynamic relationship. Clin Cancer Res 2003;9:327–337.

35. O'Farrell A-M, Foran JM, Fiedler W, Serve H, Paquette RL, Cooper MA, Yuen HA, Louie SG, Kim H, Nicholas S, Heinrich MC, Berdel WE, Bello C, Jacobs M, Scigalla P, Manning WC, Kelsey S, Cherrington JM. An innovative phase I clinical study demonstrates inhibition of FLT3 phosphorylation by SU11248 in acute myeloid leukemia patients. Clin Cancer Res 2003;9:5465–5476.

36. Meshinchi S, Woods WG, Stirewalt DL, Sweetser DA, Buckley JD, Tjoa TK, Bernstein ID, Radich JP. Prevalence and prognostic significance of Flt3 internal tandem duplication in pediatric myeloid leukemia. Blood 2001;97:89–94.

37. Van den Abbeele A, Melenevsky Y, de Vries D, Manola J, Dileo P, Tetrault R, Baum C, Badawi R, Demetri G. Imaging kinase target inhibition with SU11248 by FDG-PET in patients (pts) with imatinib-resistant gastrointestinal stromal tumors (I-R GIST). J Clin Oncol (ASCO Annual Meeting Proceedings). 2005;23(16S), Part I of II (June 1 Supplement), Abstract nr 9006.

38. Carter NJ, Keating GM. Maraviroc. Drugs 2007;15:2277–2288.

39. Fätkenhauer G, Pozniak AL, Johnson MA, Plettenberg A, Staszewski S, Hoepelman AIM, Saag MS, Goebel FD, Rockstroh JK, Dezube BJ, Jenkins TM, Medhurst C, Sullivan JF, Ridgway C, Abel S, James IT, Youle M, van der Ryst E. Efficacy of short-term monotherapy with maraviroc, a new CCR5 antagonist, in patients infected with HIV-1. Nat Med 2005;11:1170–1172.
40. Rosario MC, Poland B, Sullivan J, Westby M, van der Ryst E. A pharmacokinetic-pharmacodynamic model to optimize the phase II development program of maraviroc. J Acquir Immune Defic Syndr 2006;42:183–191.
41. Rosario MC, Jacqmin P, Dorr P, van der Ryst E, Hitchcock C. A pharmacokinetic-pharmacodynamic disease model to predict in vivo antiviral activity of maraviroc. Clin Pharmacol Ther 2005;78:508–519.
42. Meanwell NA, Kadow JF. Drug evaluation: maraviroc, a chemokine CCR5 receptor antagonist for the treatment of HIV infection and AIDS. Drugs 2007;8:669–681.
43. Fitzgerald GA. Anticipating change in drug development: the emerging era of translational medicine and therapeutics. Nat Rev Drug Discov 2005;4:815–818.

2

VALIDATION OF BIOCHEMICAL BIOMARKER ASSAYS USED IN DRUG DISCOVERY AND DEVELOPMENT: A REVIEW OF CHALLENGES AND SOLUTIONS

GABRIELLA SZEKELY-KLEPSER AND SCOTT FOUNTAIN

2.1 INTRODUCTION

Biomarkers have been used for over a hundred years in medical practice and have been playing a key role in drug discovery and development for over half a century. The National Institute of Health Biomarker Definitions Working Group recently defined biomarkers in various biochemical, physiological, imaging, and behavioral characteristics that are objectively measured as indicators of normal or pathologic processes or in response to therapeutic intervention (1). As the examples of two of the most well-known biomarkers, blood glucose for insulin-dependent diabetes and LDL-cholesterol (LDL-C) for hypercholesteremia and cardiovascular disease illustrate the discovery, validation, and application of biomarkers in medical therapy and drug discovery and require interdisciplinary research of biology, medicine, and drug development and can take several years to decades. While the earliest records of diabetes can be found in Egyptian papyrus records in 1552 B.C., mentioning frequent urination as the symptom; and from the eleventh century, diabetes was diagnosed by tasting the urine of subjects as its sweet taste was connected to the disease, it was not until the nineteenth century that the first chemical tests to measure sugar in the urine were developed. It took over 100 years of medical and biology researches to discover and link insulin

Predictive Approaches in Drug Discovery and Development: Biomarkers and In Vitro/In Vivo Correlations, First Edition. Edited by J. Andrew Williams, Jeffrey R. Koup, Richard Lalonde, and David D. Christ.

with type 1 diabetes and 1921 was the year for the first successful treatment of a depancreatized dog with insulin. The first home tests for urinary glucose that became available in the 1960s and A. H. Clemens' patented blood glucose meter in 1971 enabled the easier monitoring and medication management of diabetics (2). Closer monitoring of the glucose biomarker levels combined with adaptive therapy applying more frequent doses and self-adjustments according to individual activity and eating patterns have significantly delayed the onset and progression of long-term complications in diabetic patients, thus illustrating the importance and utility of the biomarker in managing clinical outcomes for this disease (3). As for the history of LDL-C, the concept of using biomarkers in the prevention and treatment of cardiovascular diseases can be traced back to the Framingham study (4). The investigators initiated this study in 1949 to "seek a single essential cause to produce cardiovascular disease." It was soon realized that complex and multifactorial interactions led to the pathogenesis of atherosclerotic cardiovascular disease. However, it is through this study that the quantitatively measured clinical parameters as traditional risk factors for coronary heart disease were identified. Total cholesterol and LDL-C were among these risk factors identified in the Framingham study. Subsequent pharmaceutical research to identify drugs that inhibited cholesterol synthesis led to the discovery of 3-hydroxy-3-methylglutaryl coenzyme A (HMG-CoA) reductase inhibitors that we now know as "statins." Statins were approved in 1987 to lower total cholesterol and LDL-C levels; and in 1994, statins were shown to reduce cardiovascular events (5).

As illustrated by the above examples, there are a number of important roles that biomarkers play in clinical applications including diagnosis, monitoring disease progression or reversal, and patient selection for clinical trial stratification. In recent years, modern drug discovery has recognized the importance of biomarkers in the earlier stages of drug discovery and development. Their utilization is often required in compound selection strategies for preclinical development using the efficacy and/or safety biomarker profile of drug candidates and for the proof of medical hypothesis by linking drug effect to the biological target using relevant and validated target and mechanism biomarkers (6).

Biomarkers are also a means to assess pharmacodynamic (PD) response. Their use in understanding effect and exposure relationships is essential for pharmacokinetic and pharmacodynamic (PK/PD) model-based drug development enabling better predictions of efficacious dose and regimen (7–10). Target and mechanism biomarkers that are directly linked to a target enzyme activity or its mechanism of action (MOA) through a given biochemical pathway are also important. Their full utilization requires translatability between preclinical models and human for the validation of the animal models for preclinical *in vivo* efficacy screening as well as for confirming the MOA in proof of concept (POC) clinical studies.

Translational research is a recent interdisciplinary approach that focuses on successfully advancing fundamental discoveries from the discovery to clinical setting, and is often interpreted to include a bidirectional component where clinical findings are integrated back into the preclinical space (11–13). More broadly,

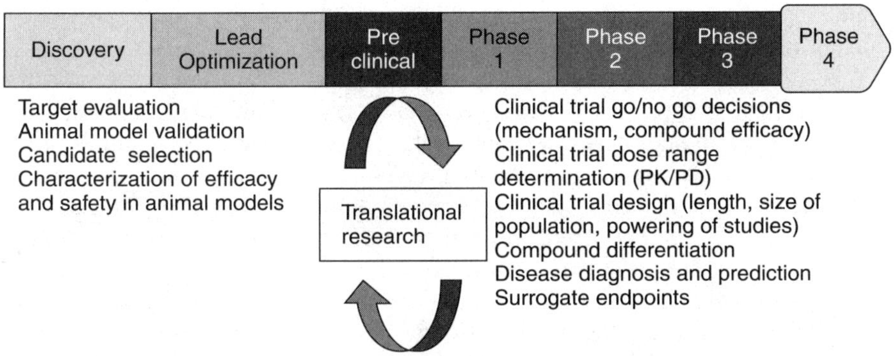

FIGURE 2.1 The multifaceted role of biomarkers in pharmaceutical decision making. Drug discovery and lead optimization utilizes biomarkers for validation of novel targets, validation of new animal models, and selecting lead drug candidates based on their efficacy and safety profile. In preclinical and clinical development, biomarkers can be used to test the medical hypothesis (mechanism of action) and patient and dose selection. Translational research is the bidirectional exchange and integration of preclinical and clinical information linking drug discovery and development.

this bidirectional component can be interpreted to include a "learning and confirming" model of iteratively increasing and applying knowledge (14). Figure 2.1 illustrates the multiple roles the biomarkers can play in the preclinical and clinical stages of pharmaceutical research.

Regardless of which purpose the biomarker is used for, its successful utilization in scientific decision making requires its validation for the intended purpose. This fit-for-purpose validation consists of two critical steps, (i) the technical validation of the analytical method used to quantitatively measure the biomarker and (ii) the biological validation of the biomarker, confirming its linkage to the relevant biological and pharmacological hypothesis that is being tested. These two validation steps often occur concurrently, since a reliable analytical assay needs to be developed first to be able to test the biomarker's linkage to pharmacology in an animal or human model. Once the biomarker linkage to biology is tested and modulation of the biomarker between normal versus disease state or response to a therapeutic intervention has been shown, the analytical method requirements can be finalized and the method validated. Depending on the drug discovery program, multiple potential biomarkers in multiple animal models and species may be under investigation until a decision-making biomarker is selected. The development of multiple biomarker assays in multiple species and matrices and their testing in relevant disease models and clinical populations is a resource intensive process with respect to both cost and time. Therefore, carefully designed and robust strategies for the development and fit-for-purpose validation of translatable biomarkers between preclinical and clinical applications need to be formulated in the form of a biomarker research and operating plan (11, 12). The existence of this research plan containing input from preclinical and clinical biology, pharmacology, pharmacokinetics, analytical, and PK/PD scientists can assure timely availability of validated biomarkers and assays to support preclinical and clinical decision making.

While biomarkers are increasingly used in internal decision making within pharmaceutical companies, there is a focused interest in their utilization for regulatory evaluation of new drug candidates. The Food and Drug Administration (FDA) Critical Path Initiative outlines the framework and evidence needed to qualify biomarkers for regulatory drug evaluation purposes and defines some of the critical biomarker needs in various disease areas (13). This need for high confidence in biomarker data used in scientific or regulatory decision making emphasizes the need for high quality, reproducible, and reliable assay measurements that are translatable between preclinical and clinical applications and validated for their intended purpose. In a recent conference, the American Association of Pharmaceutical Sciences (AAPS) Clinical Ligand Assay Society Biomarkers Workshop addressed the key challenges in biomarker research and summarized validation recommendations for immunoanalytical ligand-binding biomarker assays in a summary report (15). Follow-up publications continued to discuss the need for iterative, fit-for-purpose approach to biomarker method development and validation keeping in mind the intended use of the data as well as the regulatory requirements associated with that use (16, 17).

The two most often used technology platforms for the quantification of biochemical biomarkers accessible in body fluids and tissue extracts are immunoanalytical or ligand-binding assays and more recently, liquid chromatography tandem mass spectrometry or LC-MS/MS. This chapter provides information on the factors that may determine which platform to utilize for biomarker quantification, and discusses the key challenges of biomarker assay validation using these technologies.

2.2 GENERAL CONSIDERATIONS FOR BIOMARKER MEASUREMENTS AND SELECTION OF ASSAY PLATFORMS FOR THE QUANTIFICATION OF BIOCHEMICAL BIOMARKERS

2.2.1 Biochemical Markers

This discussion focuses on a particular subgroup of biomarkers, called *biochemical biomarkers*, or *circulating biomarkers* that are (i) characterized by a known molecular formula and structure or (ii) heterogeneous proteins with or without posttranslational modifications. The discussion does not consider other types of biomarkers such as animal behavioral models, cell type, count and activity, or imaging measurements.

2.2.2 Source Matrix for Biochemical Markers and Sample Collection Considerations

During the identification and selection of a particular biomarker, consideration should be given to the selection of the biological matrix in which the biomarker levels will be monitored. Key considerations for selection of the matrix are (i) relevance to the pharmacology/biology, (ii) assay sensitivity requirements based on the anticipated biomarker levels, (iii) analyte stability in the matrix, and (iv) feasibility of sample collection.

Biochemical markers that are translatable between preclinical and clinical applications need to be easily accessible for sample collection with minimally invasive procedures. Body fluids such as urine and plasma/serum are the most desired because of their easy collection procedures. In some cases, more invasively obtained fluids such as tissue aspirates, synovial and cerebrospinal fluid, or microdialysate can also serve as a source matrix. Tissue extracts may be used in the preclinical setting or from biopsies in case of clinical applications. While collection, handling, and storage of plasma samples are very straightforward in clinical applications, urine is the most easily accessible body fluid to analyze since its collection is noninvasive. However, for urine samples, an additional challenge for robust assay development is the potential for large variation in the concentration of biomarkers in this body fluid because of individual variations in the urinary volume output. Therefore, urinary levels of biomarkers should always be normalized, such as by monitoring creatinine output. Biomarker translation plans need to make sure that the biomarker is measurable in a biological matrix that will be feasible to obtain in clinical studies in repeated collections when necessary, and that stabilization and storage of the samples in the biological matrix is feasible.

Besides ease and noninvasiveness of sample collection, another important factor is the linkage to the originating tissue source for biomarkers measured in the periphery such as urine or plasma. Therefore, during the initial studies of the biomarker and assay development, correlations between biomarker levels measured in target tissues such as brain, liver, and synovial fluid and peripheral biological fluids such as plasma and/or urine should be established. For instance, in the case of a biomarker of central nervous system (CNS) events, the relationship between the biomarker concentrations in the brain (at the site of drug action) and in the plasma systemic circulation needs to be characterized.

Sample collection time is also an important consideration and requires a thorough understanding of the half-life of the biomarker as well as its daily variations and the dynamics of its response to a therapeutic. Biomarker plans should include studies to understand the time course and dose response of the biomarker in the matrix of choice. The time course of the changes in the biomarker levels in normal and diseased and/or treated subjects needs to be characterized to determine optimal sampling time and the impact of diurnal variations. Biomarkers with large diurnal variations may not be useful as a single time-point measurement. Instead, integrated measurements over a given period determining the total area under the curve (AUC) for the biomarker may be more useful (18).

The range of biomarker levels between normal controls and diseased specimens also needs to be understood, to define the technical requirements for acceptable assay variability and to maximize the differentiating power of the assay. Response of the biomarker for a given treatment also needs to be characterized in order to determine the magnitude of change relative to dose and the time course of the change. In some instances, response in biomarker levels can be seen after single-dose treatment, while for other biomarkers, chronic, repeated administration of drug may be necessary to see a response. For instance, mevalonic acid (MVA) is the direct product of the reduction of HMG-CoA by the enzyme HMG-CoA reductase, a rate limiting step in the cholesterol synthesis in the liver. MVA is characterized by a significant diurnal variation since the rate of the cholesterol synthesis is the highest at night. Administration of an HMG-CoA reductase inhibitor, such as a statin, can significantly decrease MVA AUC within 24 h after a single dose of the drug, while changes in LDL-C can only be registered after 4–6 weeks of treatment because of the rate of lipoprotein synthesis and turnover (18).

2.2.3 Choice of Measurement Technology Platform

A variety of factors can influence what technology platform can and will be used to develop the biomarker assay. The major considerations are as follows: technical feasibility and fitness-for-purpose, translatability between preclinical and clinical setting, cost effectiveness, and availability at contract research organizations (CRO) to support large scale studies. For circulating biochemical markers, two intensively used platforms are ligand-binding- and LC-MS/MS-based assays. Table 2.1 summarizes these key characteristics and the criteria that have to be evaluated when selecting the assay platform to support biochemical biomarker measurements.

The assessment of technical feasibility should address the selectivity and sensitivity of the assay platform to be able to accurately and precisely quantify the biomarker in the biological matrix relative to the biological variability and the expected modulation of the biomarker. Enzyme-linked immunosorbent assays (ELISAs) and LC-MS/MS assays have different selectivity and specificity based on the differences in the analytical principles applied. The differences in selectivity are discussed in the next section. Their sensitivity is similar in the low nanogram per milliliter-picogram per milliliter range, although in some cases immunoassays may achieve slightly higher sensitivity. In terms of precision and accuracy, LC-MS/MS assays are usually validated to a 10% greater precision and accuracy than immunoassays. The main reason for this is the application of a stable isotope-labeled internal standard (IS) that enables correction for variability in sample preparation and analysis.

Comparison of the throughput of the two platforms needs to take into account that immunoassays usually use several hours of incubation, sometimes requiring overnight treatment and multiple-wash steps that are followed by a very fast readout time, often less than 5 min for a 96-well plate. In contrast, LC-MS/MS assays typically use a sample extraction methodology followed by on-line sample cleanup and chromatographic separation steps that can be automated but take 2–15 min/sample or longer.

TABLE 2.1 Comparison of Assay Characteristics Between Ligand-Binding Immunoassays and LC-MS/MS for Consideration of Assay Platform Selection

Assay Characteristics	Ligand Binding/Immunoassay	LC/MS/MS
Selectivity	Can be applied for a wide range of analytes and favored method for large MW proteins	Can be applied for <10 kDa MW polar/ionizable analytes
	Based on antibody–antigen interaction	Based on HPLC separation, molecular weight, and fragment ion in SRM detection
	Can measure multiple analytes that cross-react with capture/detection antibody	Typically measures one specific analyte/SRM channel at a given retention time
	Can be multiplexed	Can be multiplexed
Sensitivity	pg/ml	pg/ml-ng/ml
Precision (% CV)	<20%	<15%
Accuracy (% RE)	<20%	<15%
Sample volume	<1 ml	<1 ml
Throughput	Sample preparation in automated parallel processing, incubations few hours to overnight/96-well plate	Sample preparation in automated parallel processing in <30 min/96-well plate
	Reading time in <5 min/plate	LC-MS/MS analysis time 2–15 min/sample
Replicates/sample	Duplicate	Singlet
Sample preparation	Usually none, direct analysis of serum and plasma	Extraction, addition of stable isotope-labeled internal standard
Reagent needs	Antibody reagents are needed	Usually no specific reagents are needed
Reagent preparation	Labeling of reagents are required (source of variability)	Stable isotope-labeled internal standards may need to be synthesized
Equipment availability and cost at CROs	Equipment cost is >$100,000–$60/sample	Equipment cost is >$400,000–$100/sample
Data reduction	No audit trail or customized solutions	21CFR Part 11 compliant
Translatability	Yes, depends on reagent availability	Yes

Abbreviation: MW, molecular weight.

Regarding sample volume requirements, the two platforms are very comparable; however, it is accepted industry practice to analyze samples in duplicates by immunoassay and in singlet by LC-MS/MS, due to the differences in the precision and accuracy of the platforms.

Another aspect of technical feasibility is the availability of the assay method in time for its use. In general, method development efforts for biomarkers on LC-MS/MS platforms can be done in a matter of days to a few weeks, if all standards and reagents are available. LC-MS/MS assays, as mentioned above, usually utilize stable isotope-labeled analyte as IS or a structural analog that may need to be synthesized. The development of immunoassays usually takes longer, a few weeks to several months, because of the need for the identification of suitable antigens, the generation, purification, and screening of analyte-specific antibodies, if not commercially available. While there is a significant difference in the price of immunoanalytical and LC-MS/MS instrumentation, both platforms are readily available both in the industry and at contract research organizations. Owing to the more expensive capital investments need and the complexity of LC-MS/MS separation and detection technology, the analysis cost per sample is usually slightly higher for LC-MS/MS but still comparable to immunoassays. After the above generic considerations for the choice of a platform, next an overview of assay validation requirements and challenges related to the quantification of endogenous analytes are discussed on the LC-MS/MS and ligand-binding assay platforms.

2.3 OVERVIEW OF ASSAY VALIDATION REQUIREMENTS AND CHALLENGES

Assay validation refers to the procedures taking place after the bioanalytical method has been developed and optimized. The validation experiments are aimed at demonstrating that the method is suitable for its intended use ("fit for purpose"). The bioanalytical method validation should include all of the procedures that demonstrate that a particular method used for quantitative measurement of analytes in a given biological matrix, such as blood, plasma, serum, or urine, is reliable and reproducible for the intended use. The fundamental parameters for this validation, in general, include accuracy, precision, selectivity, sensitivity, reproducibility, and stability. In the next section, the particular challenges and solutions related to biomarker assay validations are reviewed.

2.3.1 Selection of the Standard Curve Matrix

Bioanalytical assay usually means analyzing exogenous drug (xenobiotic) and drug-related analytes, such as metabolites, in various biologic matrices (i.e., serum, plasma, urine, tissues). In the broader context, bioanalytical assays are used not only for the analysis of exogenous analytes but also to determine endogenous biomarkers in biologic matrices. There is a key difference between matrix selection for the preparation of the standard curve for exogenous and endogenous analytes. In most bioanalytical assays for xenobiotics, the standard curve is usually prepared in the same biological matrix as the study samples. However, in the case of endogenous biomarker assays, analyte free matrix is not available. Therefore, assay accuracy measurements must take into account the endogenous basal analyte concentration for absolute quantification. Several approaches have been published to accommodate this including the use of stripped or substitute matrices or the use of stable isotope-labeled standards (19, 20). When a substitute matrix is used, it is paramount that parallelism between the sample matrix and the substitute matrix is established. Parallelism in this definition literally means that the standard curve prepared in substitute matrix is parallel with a standard curve that is prepared in the biological matrix of interest, thus the responses of the analyte are the same in the two matrices. During validation, it is demonstrated that the standard curve prepared in a substitute matrix devoid of the analyte can provide accurate and precise quantification of unspiked and spiked quality control (QC) samples prepared in the biological matrix across the full range of the calibration curve. These matrix-based QCs should contain the same amount of biological matrix as the study samples. If the study samples are analyzed without dilution, it is recommended that the biological matrix content of the QCs is at least 95% or greater.

Selection of an appropriate substitute matrix is primarily driven by finding a matrix that enables accurate and precise quantification of matrix-based samples. Intuitively, a substitute matrix that is most similar to the sample matrix is desired, such as immunodepleted or stripped matrix, artificial urine, and plasma from another species that has negligible concentration of the biomarker. However, it should be mentioned that while these matrix options can work well, often their preparation requires complex procedures potentially introducing variability into the assays or they may not be readily available in sufficient amounts. Often, simple buffer systems or buffers fortified with low concentration of protein (i.e., serum albumin) to avoid nonspecific binding of analyte to glass or plastic surfaces work very well, if the sample preparation and recovery from the biological matrix is robust enough while offering a simplified assay procedure.

2.3.2 Selection of Reference Standard

For most small molecule biomarkers, high purity standards that are structurally identical to the endogenous analytes are readily available, and the confirmation of the structure and purity of the reference standard is typically straightforward. The common exceptions are lipids and phospholipids where biological source and mode of preparation can contribute to standard variability. An example of this is sphingomyelin, a major lipid component of myelin sheets in the brain that accumulates in certain forms of Neumann–Pick disease. When extracted from bovine brain, the primary lipid components of sphingomyelin are stearic and nervonic acids, while sphingomyelin obtained from chicken egg yolk primarily contains palmitic acid, underlying the need for species-specific characterization and purification. Reference standards for macromolecular biomarkers pose an even greater challenge. A reference standard identical in structure to the endogenous biomarker with known purity is desired. In reality, the standard and even the endogenous biomarker could be a heterogeneous mixture. This is especially true for proteins that can have a variety of possible posttranslational modifications leading to not only differences in structure but also in biological activity and immunoreactivity. When developing ligand-binding assays for macromolecules, care should be taken to understand the structure, purity, and stability of the reference standards and appropriate selectivity experiments should be designed to understand what the assay is measuring. Recombinant proteins are more frequently used as reference standards in immunoassays, and care should be taken to assure that the recombinant protein is adequate for the quantification of the endogenous, native protein in the assay.

2.3.3 Evaluation of Matrix Effects

Matrix effects can impact both LC-MS/MS and immunoanalytical assays in different ways. In the case of LC-MS/MS-based assays, matrix-ion suppression effects related to variations in ionization response due to matrix components coeluting with the analyte can be empirically determined (21). This is especially important to characterize in cases when a substitute matrix is used to prepare the standard curve since there is an inherent difference in the matrix suppression between the standards and incurred samples. The use of stable isotope-labeled analog as an IS is presumed to compensate for this difference. Normally, when there is ion suppression or enhancement present because of matrix effect, it is assumed that the absolute responses from the analyte and the coeluting IS are impacted the same way, thus leaving the analyte/IS response ratio the same (for a given analyte/IS concentration ratio) and enable accurate and precise quantification. However, the stable isotope-labeled IS does not always compensate for the matrix effect as was illustrated in an example for the determination of MVA in human urine (20). The authors discovered that the matrix suppression was different for MVA and its stable labeled IS in certain urine samples, thus changing the analyte/IS response ratio at a given analyte/IS concentration ratio between different urine samples. To resolve this problem, a method with nearly zero-matrix effects needed to be developed using more elaborate sample preparation and separation procedures. To monitor the matrix effect and its potential impact on the accuracy of the method, Li et al. (22) introduced the use of matrix suppression coefficient to estimate the strength of the matrix effect.

The suppression coefficient is calculated by taking the ratio of the average peak area response of the IS (stable labeled is ideal) spiked in the substitute matrix to its average peak area response in QC samples prepared in actual sample matrix and/or study samples. A suppression coefficient close to the value 1 is desirable since that demonstrates that the ionization response in the substitute matrix and in the biological matrix is identical. If the suppression coefficient is significantly different from the value 1, further optimization of the procedures for sample purification and/or the chromatographic separation should be considered to ensure the accuracy of the method, linear range of dilution, and a successful assay validation.

In the case of immunoanalytical assays, interfering components in the biological matrix can compete with the target analyte for binding with the capture antibody and their immunoreactivity for the secondary antibodies may be different from the target analyte's resulting in varying recovery. Since immunoanalytical assays typically do not use sample extraction and purification procedures, one way they eliminate the matrix effect is by dilution of the samples before analysis to minimize the concentration of interfering components in the sample matrix by applying a minimum required dilution (MRD) (Section 2.3.7).

Evaluation of matrix effects should occur during assay method development and lack of matrix effects should be confirmed for the final validated method for both LC-MS/MS and immunoanalytical methods.

2.3.4 Lower Limit of Quantification (LLOQ) and Dynamic Range

The lower limit of quantification (LLOQ) is defined as the lowest concentration that can be determined with suitable accuracy (%RE) and precision (%CV). This is typically lesser than 20% RE and CV (coefficient of variation) for LC-MS/MS assays and lesser than 25% for ligand-binding assays. Method validation practices require that a QC sample in the biological matrix be placed at the LLOQ level. Matching the exact concentration of the lowest calibration standard with an appropriate lower limit quality control (LLQC) in the biological matrix may not be possible for endogenous biomarkers, since the matrix is not blank and the endogenous baseline level can increase or decrease during disease or treatment. It is recommended to set an LLQC prepared in the biological matrix as close to the lowest standard as possible. If the endogenous basal level in the biological matrix samples used to prepare pooled QC matrix is higher than the proposed LLOQ and downmodulation of the biomarker is anticipated, the LLQC may be prepared by diluting the unspiked pooled matrix. The application of a dilution requires the assessment of dilution linearity. If the dilution linearity is established (see later), application of this diluted QC to cover the lower end of the curve is acceptable. The calibration curve range is chosen to bracket the anticipated sample concentrations. For LC-MS/MS-based assays, linear response is expected as a function of the concentration for a broad dynamic range (up to three orders of magnitude) and the calibration curve is fitted with a linear regression model (Fig. 2.2a). The use of quadratic calibration curve to fit data encompassing a broader dynamic range where the instrument response deviates from linear toward saturation of the detector is discouraged. Instead, sample dilution is recommended if the biomarker concentration range is very broad and concentrations above the upper limit of quantification (ULOQ) are found.

Owing to the nature of the physicochemical interactions between the antibodies and antigens, calibration curves in the case of ligand-binding assays are characterized by a nonlinear, sigmoidal curve is applied to describe the response versus concentration function. Other calibration models may be applicable and the appropriate regression model should be selected during the method development and optimization experiments. For example, Figure 2.2b depicts typical

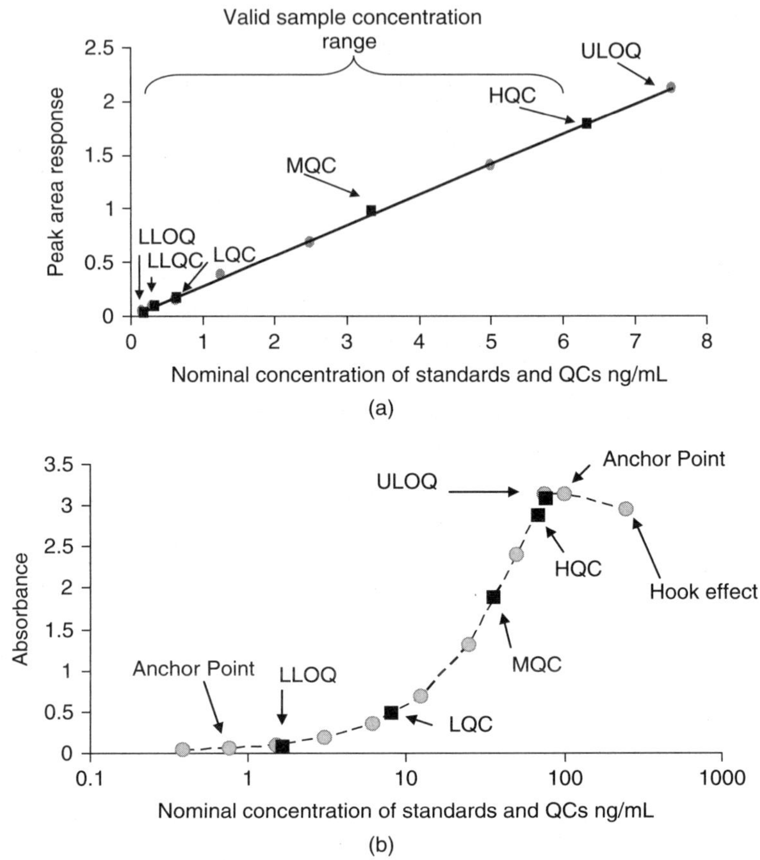

FIGURE 2.2 Illustration of a linear (a) and polynomial (b) calibration curve indicating the placement of the low, medium, and high QC samples (LQC, MQC, and HQC) relative to the lower and upper limit of quantification (LLOQ and ULOQ), respectively. The linear curve is commonly used in LC-MS/MS-based assays, where the peak area response is derived by taking the ratio of the analyte area response and the internal standard (IS) area response at different concentration of the analyte and a constant concentration of the IS. The polynomial curve is usually used for immunoassays where the analytical endpoint is usually colorimetric, so the absorbance produced by different concentrations of the analytical standard is measured at a given wavelength to construct the standard curve. The anchor points are used to aid the parametric curve fitting. The hook (prozone) effect indicating saturation by the antigen at high concentrations and resulting in underestimation of the concentration is also shown.

calibration curve and the placement of standards and QCs on the calibration curve for linear and four-parameter fit models.

2.3.5 Precision and Accuracy

Understanding the intra- and interassay precision and accuracy of a biomarker assay is very important for the full understanding of its differentiating capability between normal and disease states or different treatment groups. The lower the variability of the biomarker data relative to its observed change, the greater its differentiating power to separate disease versus healthy populations or to give an indication of unambiguous response to a therapeutic intervention. There are two sources of variability in any biomarker measurement. One is the inherent variability due to imprecision and bias in the measurement itself. The second source is the biological variability, such as intra- and interindividual variability due to age, gender differences, disease versus normal states, diet, diurnal variations, and medications. Any measured biomarker value will always be impacted by these two sources of variability. The differentiation capability of the assay is dependent on how large the variability is relative to the modulation window of the biomarker, as illustrated in Figure 2.3.

Fit-for-purpose validation experiments should be designed to fully understand the overall variability of biomarker assays, including the biological and technical components. Having this information can then enable the scientists to set appropriate assay performance criteria that can be different from typically used limits and provide confidence in the decision-making power of the biomarker.

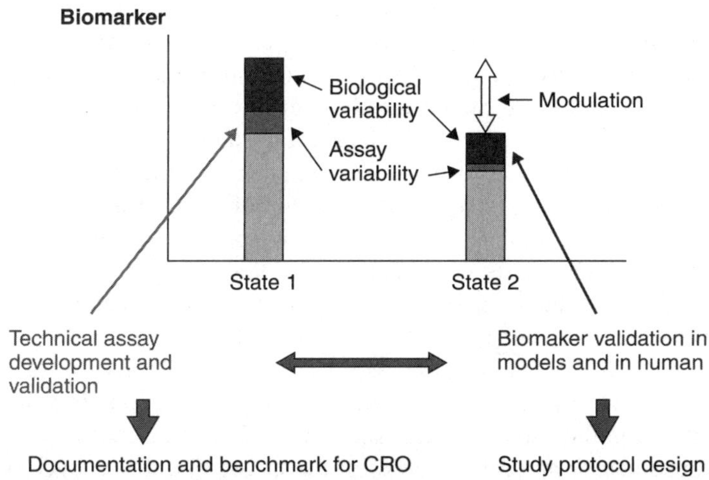

FIGURE 2.3 Schematic illustration of the contribution of the two sources of variability, including the biological and the assay variability, to the overall biomarker measurement and their impact on the technical validation and biomarker validation study designs. States 1 and 2 indicate two instances where the biomarker levels expected to be different (i.e., normal vs disease or treated vs placebo). The biomarker modulation window is the difference between the biomarker signals between the two states. When measuring the biomarker levels, the overall signal variability in each state has a contribution from the biological variability and the assay/measurement variability. The biological variability is inherent in the populations and impacts the study protocol designs. For instance, a biomarker with high degree of biological variability and small differences between different disease and treatment states will require a larger number of subjects in order to differentiate between the two states. The assay variability is related to measurement procedures, and the goal of method development and optimization is to minimize this variability and characterize it during method validation.

One of the goals of method development and optimization is to minimize the variability in the assay. While the biological variability cannot be eliminated, it also needs to be characterized before the application of the biomarker to be considered in subject/patient selection and study design. During technical method validation, the intra- and interassay variability is fully characterized and the fitness of the assay for the intended application is verified based on the biological variability. In general, validation guidelines require LC-MS/MS-based assays to be validated for 15% accuracy and precision with 20% acceptable at the LLOQ, while guidelines for ligand-binding assays require 20% accuracy and precision with 25% at the LLOQ level and better than 30% total error, where the total error is defined as the sum of the precision and accuracy.

Fit-for-purpose validation may require assay acceptance criteria that are more or less stringent than the recommended guidelines. Study designs should take into account the variability of the assay and the requirements for its application in scientific decision making to determine the adequate power of the study using statistical methods. Regarding the technical variability, assay method optimization, and in the case of LC-MS/MS-based assays, the utilization of a stable isotope-labeled standard, can significantly lower the variability resulting in highly accurate and precise measurements. The endogenous nature of the analyte also requires additional considerations for the evaluation of the absolute accuracy of biomarker assays. Since typically a substitute matrix is used for the preparation of the standard curve it is imperative that the accuracy and the precision of the assays is evaluated using biological matrix-based QCs. As illustrated in Figure 2.2, the QC samples are typically prepared with concentrations at or near the LLOQ and through the quantitative range to the ULOQ by spiking appropriate amounts of analyte into the matrix. It is important to note that because of the endogenous nature of the analyte, the preparation of matrix-based QCs inherently requires an estimation of the endogenous basal levels in the biological matrix that is used to prepare the QCs. The accuracy of the determination of spiked QCs ultimately hinges on the accuracy of the determination of the endogenous baseline.

It is not uncommon to observe a shift in the inter-run endogenous baseline concentrations across multiple batch runs, indicating that the recovery of spiked analyte is less variable than the endogenous analyte. This phenomenon can occur when the endogenous analyte exists as a complex within the biological matrix and recovery from this complex is not consistent in the sample preparation step. When shifts in the endogenous baseline concentrations occur, the sample preparation and recovery needs to be further optimized, despite accurate recovery of spiked QCs, since ultimately the study samples are better represented by the unspiked QCs. Another factor that needs to be considered is that the endogenous baseline concentration of the biomarker can be up or down modulated by the treatment or the disease relative to controls and normals. This can pose a problem if biological matrix from a representative treated or disease population is not available at the time of assay development and validation to prepare QCs at or near the LLOQ that is required to measure the down modulated levels. One possible way to overcome this problem is by using diluted unspiked QCs, in which case dilution linearity needs to be validated.

Absolute interassay accuracy is important to demonstrate when the biomarker data is compared across multiple batch runs, potentially from multiple studies, between laboratories. For some biomarker assays, the lack of absolute accuracy may be acceptable, since the ability to precisely measure *changes* between treated and untreated or healthy and diseased populations may be more important than absolute concentration levels. This situation should be an exception rather than recommended practice, since even in early preclinical investigations it is very common to compare biomarker data across multiple studies, so not having a robust assay with proven accuracy and precision can lead to false conclusions or require repeating studies. The unspiked QCs should also be used in the freeze/thaw and storage stability assessments and in the qualification of any new QC preparations using new, independent weighing of the standard, to assure that there is no shift in the endogenous baseline levels and to continue to verify the accuracy of the concentration scale between studies and for new QC preparations.

2.3.6 Selectivity, Specificity, and Recovery

The selectivity of an assay is a measure of the extent to which the method can determine a particular compound in the analyzed matrices without interference from matrix components. A method that is perfectly selective for an analyte or group of analytes is *specific* (23, 24). The validation procedure should confirm the ability of the method to unequivocally assess the analyte in the presence of other components that may be present (e.g., impurities, degradation products, and matrix components). The necessary validation studies depend on the method's use. Lack of specificity of an individual analytical procedure may be compensated by other supporting analytical procedures (23, 24).

The selectivity of LC-MS/MS assays is because of the following: (i) sample preparation, often including liquid/liquid- or solid-phase extractions to remove the majority of interfering components, (ii) reverse phase HPLC separation, (iii) selected reaction monitoring (SRM) detection based on the precursor ion mass-to-charge ratio value (*m/z*) and the *m/z* value of a selected product ion. SRM assays typically measure a single analyte in a given SRM channel with the potential of measuring additional analytes in multiplexed assay using several SRM channels simultaneously or in a time programmed manner. The inherent selectivity of immunoanalytical or ligand-binding assays is derived from the selectivity of the antibody–antigen interaction. Antibodies are heterogeneous with respect to antigen specificity, a known feature that is attributed to the polyclonal B-cell response that leads to their formation (25). This known heterogeneity, combined with variation of antibody avidity can account for considerable variability in specificity and sensitivity for immunoassays. Because of the cross-reactivity of antibodies to structurally similar analytes, immunoanalytical assays have the potential to measure the sum of the response from multiple analytes that interact with a given antibody, providing a total response measurement. This can be advantageous when the expression of a certain protein needs to be determined regardless of minor modifications or truncation of the protein. On the other hand, ambiguity around assay specificity can lead to unexpected results, such as the example of the peptide endothelin 1 (ET-1) (26). Clinical data showed that people with heart failure had higher levels of ET-1 than healthy controls based on immunoassays. However, the assay used to measure ET-1 cross-reacted with big-ET, the precursor of ET-1, the levels of which are actually higher in patients with heart failure. Later on, when a new ET-1-specific assay, which did not cross-react with big-ET, was applied in the clinical study, the data did not show a difference between the normal and diseased populations, thus leading to the termination of the drug candidate (27).

Because of the endogenous nature of the analyte, blank matrix is not available to directly test for the selectivity of the assay as it is done in the case of assays measuring xenobiotics. Therefore, selectivity of biomarker assays is usually demonstrated in multiple ways, where selectivity if inferred through additional assay characterization approaches. These additional approaches can include instrument-based techniques evaluating "cross talk" for LC-MS/MS detection or cross-reactivity for ligand-binding assays. Absolute recovery of analyte can be assessed by comparing the signal detected for the analyte in neat solution and in the matrix after subjecting both sets of samples to the sample preparation and analysis procedures. To establish selectivity for an assay requires optimization of sample preparation and separation procedures ultimately eliminating the interference of matrix components while maintaining robust recovery of the analyte. Because assay selectivity and recovery are ultimately intertwined, assay validation tests can be designed to evaluate the recovery and selectivity in the same set of experiments. Special considerations apply to evaluate recovery from tissue samples, where spiking can only be applied to already homogenized tissue. In these cases, it is not possible to demonstrate the accuracy of the absolute recovery and only the reproducibility and the precision of the analyte recovery can and should be demonstrated (28, 29).

2.3.7 Dilution Linearity and Parallelism

Dilution linearity and parallelism of biological samples must be assessed during validation if study samples are expected to require dilution before assay, the matrix content of the calibration curve or QCs is different than study samples, or the biological matrix content of expected study samples is not consistent (e.g., tissue mass). Dilution linearity can be assessed during validation by preparing spiked dilution QC samples at known concentration exceeding the calibration range. Spiked samples are diluted using the dilution factors anticipated for study samples and analyzed. Linearity of dilution is demonstrated, when the accuracy (%RE) and precision of the mean value of the analyte concentrations corrected for dilution is better than the predetermined validation criteria.

The parallelism assessment in incurred samples is intended to evaluate the linearity of dilution of the endogenous analyte in the biological matrix, diluting from the endogenous baseline level without spiking. This may be required in cases when the endogenous biomarker levels are high and/or expected to decrease in treated samples relative to endogenous baseline and the lower limit QC sample needs to be diluted in order to cover the lower range of the calibration curve. Typically, multiple different lots of biological matrix containing the endogenous analyte should be serially diluted (in substitute matrix used for calibrating standards and QCs or dilution buffer used to dilute study samples) from high concentration to the lowest anticipated concentration and analyzed to show the dilution linearity in incurred samples (parallelism).

In some immunoassay methods, there may be a significant matrix effect such that biological samples cannot be evaluated with acceptable precision and accuracy. This may be because of the presence of interfering components in the matrix, which results in a nonparallel signal response (relative to the calibration curve) and/or a nonspecific background signal. To overcome this matrix effect, all samples must be diluted before analysis; the MRD is the minimum magnitude of dilution that a sample must be subjected to (with a specified diluent). Illustration of the need to evaluate dilution linearity, MRD in biological matrix representative of study samples and parallelism in incurred study samples is shown in Figure 2.4 for an IL-18 human plasma assay. For this assay, QC samples were prepared in normal human plasma, which showed dilution linearity in spiked samples as well as in unspiked matrix and an MRD of 1:4 was validated. However, when incurred study samples were tested for parallelism in plasma from rheumatoid arthritis patients, parallelism was not observed below an MRD of 1:100, indicating the presence of a significant matrix effect in the disease patient samples.

Minimum required dilution may also be necessary in case of biomarkers that have high concentrations in the biological sample, so each sample needs to be diluted before analysis to avoid saturation of the detector signal. This was the

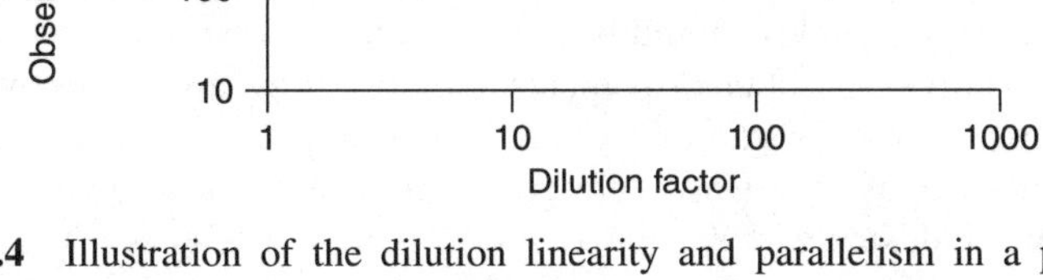

FIGURE 2.4 Illustration of the dilution linearity and parallelism in a plasma IL-18 ELISA assay. The dilution linearity was validated at a minimum required dilution (MRD) of 1:4 or greater, as shown by the symbols (♦); however, the dilution of the incurred plasma samples showed lack of parallelism below an MRD of 1:100, as shown by the data represented by the symbols (▲). Therefore, dilution linearity was revalidated at higher spiked concentrations and MRDs of 1:10 and 1:100, indicated by the symbols (■).

case for myo-inositol in rat brain tissue, a mechanistic biomarker in the PI3 kinase pathway, where the biomarker concentrations were in the range of microgram per milliliter in the undiluted brain homogenate, while the linear range of the assay was from 5 to 1000 ng/ml. Therefore, a minimum of fivefold dilution was required before analysis (28).

2.3.8 Stability

Stability of standard reagents, stock solutions, and the analyte in the biological matrix in the presence of any added anticoagulant and/or stabilizing reagent needs to be evaluated for the duration of use and analysis and under the representative application and storage temperatures. Usually, standard reagent stock and analyte stability in matrix is evaluated at room temperature at least for the duration of a typical sample preparation procedure. Short- and long-term storage stability, as well as freeze/thaw stability of the analyte is evaluated at typical storage temperature of -20 or -70°C. In addition, process stability (or final extract/autosampler stability) is also evaluated in case sample reanalysis is required.

It is recommended that analyte sample storage and freeze/thaw stability are evaluated early in method development, as soon as a reliable measurement is established. This assures that appropriate sample collection and storage conditions can be recommended to the clinicians even before a final validated method is available. Also, while lack of analyte stability does not invalidate the analytical method itself, serious considerations should be given to a biomarker that has poor or limited stability, since its utility and robustness can be questionable. For longitudinal studies, long-term stability QC samples need to be prepared to be analyzed at interval relevant to study sample storage durations to confirm analyte stability (30).

2.3.9 Assay Reproducibility in Incurred Samples

Validated analytical methods regardless of whether they quantify xenobiotics or endogenous substances are expected to be robust and reproducible in order to produce reliable data. Therefore, during method development and validation, factors such as sample stability, metabolite formation, degradation, sample homogeneity, matrix suppression or enhancement, and effect of hemolysis, should be studied and characterized. In addition to characterizing the assay using spiked QC samples, incurred study samples should be studied since QC samples spiked in pooled matrix from a limited number of subjects sometimes do not represent the whole range of biological matrix variability that can be expected in a study.

The importance of this was highlighted in recent discussions between the pharmaceutical industry and the FDA. As an outcome of these discussions, the conference report of the Third AAPS/FDA Bioanalytical Workshop (Crystal City III) endorsed the concept that validated bioanalytical drug assays must demonstrate assay reproducibility and accuracy in incurred study samples. A follow-up AAPS workshop on the same topic issued a report on clarification and recommendations regarding incurred sample reassay (ISR) (31).

Understanding assay reproducibility in incurred study samples is especially important for endogenous biomarker assays since the characterization and validation of these assays is more difficult as analyte free matrix is not available. During the method development phase, efforts should be taken to understand the range of biomarker concentration in the studied subject populations and understand the stability of the analyte and matrix effects in incurred (unspiked samples) as well as spiked QCs. For instance, it is not uncommon that while analyte recovery and stability are demonstrated in spiked QC samples, the actual baseline of the endogenous biomarker levels is changing in the unspiked matrix. The reasons for this are analyte and matrix dependent, and can be due to multiple factors such as enzymatic interactions between a matrix component and the analyte, complex formation between the analyte and matrix components, perturbation of the biological sample when preparing spiked QCs, and influence of sample preparation on analyte recovery. Method development and characterization should minimize matrix effects and elucidate the sample stability under various conditions. In-study method validation should demonstrate the robustness and reproducibility of the method, and the reanalysis of incurred study samples is a good practice to demonstrate this. Therefore, the practice to establish incurred sample reproducibility is recommended for biomarker assays as well.

2.4 FUTURE TRENDS, EMERGING TECHNOLOGIES

2.4.1 Protein Quantification Using LC-MS/MS Platforms

LC-MS/MS using SRM has become the primary analytical tool for the quantification of xenobiotics in biological matrices and is gaining acceptance as a tool for the quantification of small molecule (typically <5 kDa) biomarkers. In addition, it has been used for more than a decade for the relative quantification of proteins using stable isotope labeling approaches such as the ICAT (isotope-coded affinity tag) method and its variations (32–34). These original labeling methods have suffered from the disadvantage of limited throughput and the need for numerous control samples to enable pair-wise comparisons. More recently, other approaches have been developed for absolute quantification of proteins using proteolytic digestion followed by SRM quantification of signature peptides specific to the protein of interest and multiplexed quantitative mass spectrometric multiple reaction monitoring assays were developed for 137 major plasma proteins (35). The extension of this method to quantify lower abundance proteins utilizes the enrichment of specific target peptides by antipeptide antibodies is called *SISCAPA* (stable isotope standards and capture antipeptide antibodies). Coupling immunoaffinity techniques with mass spectrometry greatly enhances the sensitivity for low abundance proteins/peptides (36). The method depends on the ability to generate peptides that are specific to the protein of interest quantitatively using a digestion method. These methods are highly specific and since they detect surrogate peptides generated from denatured proteins, they are insensitive to folding and intersubunit associations, which can be detrimental for immunoassays. On the other hand, to detect posttranslationally modified peptides/proteins or genetic variants, single amino acid modifications the assays have to be specifically designed up-front. Since these issues can be taken into account and appropriate assays can be designed, it is clear that because of the existence of robust quantitative LC-MS/MS platforms and expertise, the relatively short time required for method development, mass-spectrometry-based assays for large molecule protein biomarkers can become a viable alternative to immunoanalytical platforms.

2.4.2 Multiplexing Biomarker Assays

Multiplexing biomarker assays is often useful in investigations where the measurement of multiple analytes is required from the same sample in order to fully understand the correlation with the underlying biological pathway and/or to investigate multiple biomarker candidates simultaneously before selecting a decision-making biomarker. Multiplexing can minimize variability between different separations and also can enable measurement of biomarkers from studies where limited amount of sample is available (i.e., mouse studies). Multiplexing assays on an LC-MS/MS platform is fairly straightforward and has been demonstrated for over 100 protein analytes (35). Immunoanalytical platforms amenable to multiplexing are also available and use flow-cytometry-based technology or planar array technologies (37). The Luminex platform uses color-coded microspheres, up to 100 different colors, that can each be coated with antibody reagents to selectively bind analytes of interests from biological samples. The captured analytes are detected by secondary antibodies coupled with reporter dye molecules. The individual analyte-specific beads or sets of beads are added to the sample to be analyzed and the mixture is injected into the sheath flow of a flow cytometer and excited by laser light, which enables the detection of multicolored beads and the reporter dyes, thus enabling the quantification of up to 100 protein analytes in a sample (38, 39). Using the principle of flow cytometry, now other companies (BD Biosciences, BenderMedsystems) also offer cytometric bead-based assays that enable multiplexed analysis of biomarkers using regular flow cytometers.

Mesoscale Discovery Inc.'s platform uses a planar microarray format, enabling the detection of up to 10 analytes in a single well of a 96-well plate using antibody capture and secondary antibody detection with an electrochemiluminescent readout. The platform offers greater sensitivity and broader dynamic range than regular ELISAs or flow-based assays while reducing the number of required wash steps.

The sensitivity of immunoassays for biomarkers is greatly improved by technologies developed by Singulex, Inc. The technology underlying the assay is the Erenna system, a highly sensitive capillary design, incorporating laser-induced fluorescence detection capable of detecting signal from a single fluorophore molecule (40).

Avantra Biosciences' proprietary protein microarray technology uses an ultra-thin film of nitrocellulose on a glass substrate for binding proteins and antibodies. Their new instrument uses microfluidic MAX BIOCHIP(TM) immunoassay cartridges that enable fully automated and multiplexed immunoassays for the analysis of serum and plasma samples right at the site of collection without the need for freezing and storage of clinical samples (41).

Validation of these new multiplex assay platforms and novel detection technologies will be required before their clinical applications.

2.5 CONCLUSIONS

The current renaissance of biomarkers in the pharmaceutical industry because of efforts to gain leverage on late stage clinical trial failures in order to increase the efficiency of new drug development is resulting in the identification of a new set of biomarkers. Introducing these biomarkers in the earlier phases of drug discovery and development enable the validation of new drug targets and test the target and medical hypothesis early on using a variety of new chemical entities enabling the researchers to "fail earlier and fail cheap." If a new target cannot be reliably linked to the medical hypothesis or if the newly discovered chemical entity is not inhibiting the validated target, decisions to terminate the prosecution of the target or the chemical entity can be made earlier in drug development, thereby reducing the risk of later stage failures that can be more costly. These novel biomarkers require extensive validation before their applications and the assays that measure them should be well characterized and their fitness for intended purpose validated. Only when both the biomarker and its measurement are validated for their intended application can they fulfill their promise of facilitating earlier and reliable decision making.

2.6 GLOSSARY

Accuracy: The closeness of agreement between the measured result and its nominal or known true value; expressed as percentage relative error (%RE).

$$\%\text{RE} = \left[\left(\frac{\text{Observed}}{\text{Nominal}}\right) - 1\right] \times 100$$

ALQ: Acronym referring to calculated sample concentrations that are *a*bove the *l*imit of *q*uantification.

Analyte: A specific chemical moiety being measured, which can be intact drug, biomolecule or its derivative, metabolite, endogenous molecule, and/or degradation product in a biologic matrix or nonbiologic matrix.

Anchor Point: A calibration standard that is outside the range of quantification and can be used to facilitate or optimize standard curve fitting. Mainly used for immunoassays that use nonlinear parametric standard curve fitting models.

Biologic Matrix: A discrete material or tissue of biologic origin (e.g., serum, plasma, urine, feces, saliva, sputum, and various other discrete tissues).
Biomarker: An endogenous molecule that can be used as an indicator of normal biologic processes, pathological states, or a physiological response to a therapeutic intervention.
BLQ: Acronym referring to calculated sample concentrations that are *b*elow the *l*imit of *q*uantification.
Calibrator: A sample to which a known amount of the target analyte has been added. Calibrators are used to construct calibration curves from which the concentrations of analytes in QCs and in unknown study samples are interpolated.
Coefficient of Variation (CV): The measure of precision expressed as a percentage of the standard deviation relative to the mean (see also "Precision").
CRO: Acronym for *C*ontract *R*esource *O*rganization, a domestic or international service organization providing contracted research services to the pharmaceutical or biotechnology industries on a fee-for-service basis.
Fit-for-Purpose Validation: Assay validation should be tailored to meet the intended purpose of the method (i.e., "fit-for-purpose") with the necessary rigor commensurate with the intended use of the data.
GLP: *G*ood *l*aboratory *p*ractices, as regulated by Code 21 of the United States Federal Regulations, Part 58.
Hook Effect: Is a phenomenon that is inherent with some "sandwich" immunoassay designs. The antigen at very high concentrations in the sample binds to all available sites—saturating them—on both the antibody-coated solid phase and the antibody-labeled conjugate used for detection of the antigen, thereby preventing the formation of the "sandwich." Under these conditions, the measured levels of the analyte maybe significantly lower than the actual concentration in the sample. The high dose hook effect refers to the hook that is observed in the dose response (standard curve) when the data is plotted as a signal versus analyte concentration. This high dose hook effect is also called the *prozone effect*.
Linearity of Dilution: The ability of the analytical method to obtain test results that are directly proportional to the concentration of the analyte within the test sample on dilution of high concentration samples (typically those spiked with the reference standard) through the quantitative range of the assay.
Lower Limit of Quantification (LLOQ): The lowest amount of an analyte in a sample that can be measured with suitable precision and accuracy.
Matrix Effect: The alteration or interference in response because of the presence of interfering substances in the sample.
Nominal Concentration: Theoretical or expected concentration.
Parallelism: The assessment of the dilution linearity of incurred study samples without spiking. The term may also be used to reference the parallel nature of standard curves prepared in biological matrix and substitute biological matrix, when analyte free biological matrix is not available.
PK: Acronym for pharmacokinetics, a scientific discipline focused on quantification of the time course of drug and its metabolites in the body and the development of appropriate models and parameters to summarize observations

and predict kinetic outcomes in situations other than in which the data were obtained.

PD: Acronym for pharmacodynamics, the relationship between the effects produced and the systemic exposure to the drug with time following drug administration.

Precision: A measure of the random variation between repeated measurements from multiple sampling of the same sample under known conditions, expressed as coefficient of variation (CV).

$$\%CV = \frac{SD}{Mean} \times 100$$

Upper Limit of Quantification (ULOQ): The highest analyte concentration in an assay method that is validated to be measured with acceptable precision and accuracy.

Relative Error (RE): A measure of the closeness of an observed result to its theoretical true value, expressed as a percent relative difference (see also "Accuracy").

Selectivity: The ability of the bioanalytical method to accurately measure the analyte in the presence of other matrix components (e.g., metabolites, impurities, and binding proteins), which could potentially interfere with the analyte detection. A method that is perfectly selective for an analyte or group of analytes is *specific*.

Specificity: The ability of the method to unequivocally assess the analyte in the presence of other components. Specificity should be validated during method validation.

Total Error: A concept that expresses the closeness of agreement between a measured test result and its theoretical true value. The term *total error* describes the summation of the systematic (mean, %RE) and random (precision, %CV) error components.

Upper Limit of Quantification (ULOQ): The highest amount of an analyte in a sample that can be quantitatively determined with acceptable precision and accuracy.

Validation Run: A complete set of analytical samples with appropriate number of standards and QCs for their validation. Several runs (or batches) may be completed in 1 day, or 1 run (or batch) may take several days to complete.

REFERENCES

1. Biomarkers Definitions Working Group. Biomarkers and surrogate endpoints: preferred definitions and conceptual framework. Clin Pharmacol Ther 2001;69:89–95.
2. Mendosa D. History of blood glucose meters. http://www.mendosa.com/history.htm. Accessed 2009 Sep 17.
3. Diabetes Control and Complications (DCCT) report: the effect of intensive treatment of diabetes on the development and progression of long-term complications in insulin-dependent diabetes mellitus. N Engl J Med 1993;329:977–986.

4. Nambi V, Ballantyne CM. Role of biomarkers in developing new therapies for vascular disease. World J Surg 2007;31:676–681.
5. Scandinavian Simvastatin Survival Study Group. Randomised trial of cholesterol lowering in 4444 patients with coronary heart disease: the Scandinavian Simvastatin Survival Study (4S). Lancet 1994;344:1383–1389.
6. Mackay M, Street S, McCall J. Risk reduction in drug discovery and development. Curr Top Med Chem 2005;5:1087–1090.
7. Miller R, Ewy W, Corrigan BW, Ouellet D, Herman D, Kowalski KG, Lockwood P, Koup JR, Donevan S, El-Kattan A, Li CSW, Werth JH, Feltner DE, Lalonde R. How modeling and simulations have enhanced decision making in new drug development. J Pharmacokinet Pharmacodyn 2005;32(2):185–197.
8. Pillai GC, Mentre F, Steimer J-L. Non-linear mixed effect modeling- From methodology and software development to driving implementation in drug development science. J Pharmacokinet Pharmacodyn 2005;32(2):161–183.
9. Barrett JF, Gupta M, Mondick JT. Model-based drug development applied to oncology. Expert Opin Drug Discov 2007;2(2):185–209.
10. Sarker D, Workman P. Pharmacodynamic biomarker for molecular cancer therapeutics. Adv Cancer Res 2007;96:213–268.
11. Sultana SR, O'Connel D, Roblin D. Translational research in the pharmaceutical industry: from theory to reality. Drug Discov Today 2007;12(9,10):419–425.
12. Wagner JA, Williams SA, Webster CJ. Biomarkers and surrogate end points for fit-for-purpose development and regulatory evaluation of new drugs. Clin Pharmacol Ther 2007;81(1):104–107.
13. U.S. Department of Health and Human Services, Food and Drug Administration. 2006. Innovation-Stagnation, Critical Path Opportunities List. Available at http://www.fda.gov/ScienceResearch/SpecialTopics/CriticalPathInitiative/default.htm Accessed 2009 Sep 17.
14. Sheiner LB. Learning versus confirming in clinical drug development. Clin Pharmacol Ther 1997;61(3):275–291.
15. Lee JW, Weiner RS, Sailstad JM, Bowsher RR, Knuth DW, O'Brien PJ, Fourcroy JL, Dixit R, Pandite L, Pietrusko RG, Soares HD, Quarmby V, Vesterqvist OL, Potter DM, Witliff JL, Frichte HA, O'Leary T, Perlee L, Kadam S, Wagner JA. Method validation and measurement of biomarkers in nonclinical and clinical samples in drug development: a conference report. Pharm Res 2005;22(4):499–511.
16. Lee JW, Devanarayan V, Barrett YC, Weiner R, Allinson J, Fountain S, Keller S, Weinry I, Green M, Duan L, Rogers JA, Millham R, O'Brien PJ, Sailstad J, Khan M, Ray C, Wagner JA. Fit-for-purpose method development and validation for successful biomarker measurement. Pharm Res 2006;23(2):312–328.
17. Lee JW, Figeys D, Vasilescu J. Biomarker assay translation from discovery to clinical studies in cancer drug development: quantification of emerging protein biomarkers. Adv Cancer Res 2007;96:269–298.
18. Parker TS, McNamara DJ, Brown CD, Kolb R, Ahrens EJ Jr, Alberts AW, Tobert J, Chen J, De Schepper PJ. Plasma mevalonate as a measure of cholesterol synthesis in man. J Clin Invest 1984;74:795–804.
19. Li W, Cohen L. Quantitation of endogenous analytes in biofluid without a true blank matrix. Anal Chem 2003;75:5854–5859.

20. Jemal M, Schuster A, Whigan DB. Liquid chromatography/tandem mass spectrometry methods for quantitation of mevalonic acid in human plasma and urine: Method validation, demonstration of using a surrogate analyte, and demonstration of unacceptable matrix effect in spite of use of a stable isotope analog internal standard. Rapid Commun Mass Spectrom 2003;17:1723–1734.
21. King R, Bonfiglio R, Fernandez-Metzler C, Miller-Stein C, Olah T. Mechanistic investigation of ionization suppression in electrospray ionization. J Am Soc Mass Spectrom 2000;11(11):942–950.
22. Li W, Nemirovskiy O, Mathews R, Fountain S, Szekely-Klepser G. Clinical validation of an immunoaffinity LC-MS/MS assay for the quantification of collagen type II neoepitope peptide: a biomarker of MMP activity and osteoarthritis in human urine. Anal Biochem 2007;369(1):41–53.
23. EURACHEM/WELEC Guidance Document No. WGD2 Accreditation for Chemical Laboratories: Guidance on the interpretation of the EN 45000 Series of standards and ISO Guide 25. Laboratory of Government Chemist. Teddington, UK; 1993.
24. International Conference on Harmonization. Validation of analytical methods: definitions and terminology. European Agency for the Evaluation of Medicinal Products; London, UK; 1995.
25. Plotz PH. The autoantibody repertoir: searching for order. Nat Rev Immunol 2003;3:73–78.
26. Baker M. In biomarkers we trust? Nat Biotechnol 2005;23(3):297–304.
27. Wei CM, Lerman A, Rodeheffer RJ, McGregor CJ, Brandt RR, Wright S, Heublein DM, Kao PC, Edwards WD, Burnett JC Jr. Endothelin in human congestive heart failure. Circulation 1994;89:1580–1586.
28. Kindt E, Shum Y, Badura L, Snyder P, Fountain S, Szekely-Klepser G. Development and validation of an LC/MS/MS procedure for the quantification of endogenous *myo*-inositol concentrations in rat brain tissue homogenates. Anal Chem 2004;76(16):4901–4908.
29. Szekely-Klepser G, Wade K, Woolson D, Brown R, Fountain S, Kindt E. A validated LC/MS/MS method for the quantification of pyrrole-2,3,5-tricarboxylic acid (PTCA), a eumelanin specific biomarker, in human skin biopsies. J Chromatogr B 2005;826:31–40.
30. Tworoger SS, Hankinson SE. Collection, processing and storage of biological samples in epidemiological studies: sex hormomones, carotenoids, inflammatory markers and proteomics as examples. Canc Epidemiol Biomarkers Prev 2006;15:1578–1581.
31. Fast DM, Kelley M, Viswanathan CT, O'Shaughnessy J, King SP, Chaudhary A, Weiner R, DeStefano AJ, Tang D. Workshop report and follow-up—AAPS Workshop on current topics in GLP bioanalysis: Assay reproducibility for incurred samples—Implications of Crystal City recommendations. AAPS J 2009;11(2):238–241.
32. Pan S, Aebersold R. Quantitative proteomics by stable isotope labeling and mass spectrometry. Method Mol Biol 2006;367:209–218.
33. Gygi SP, Rist B, Gerber SA, Turecek, F., Gelb, MH Aebersold, R. Quantitative analysis of complex protein mixtures using isotope-coded affinity tags. Nat Biotechnol 1999;17:994–999.
34. Gygi SP, Rist B, Griffin TJ, Eng, J., Aebersold R. Proteome analysis of low-abundance proteins using multi-dimensional chromatography and isotope-coded affinity tags. J Proteome Res 2002;1:47–54.

35. Anderson L, Hunter CL. Quantitative mass spectrometric multiple reaction monitoring assays for major plasma proteins. Mol Cell Proteomics 2006;5:573–588.

36. Ackermann BL, Berna MJ. Coupling immunoaffinity techniques with MS for quantitative analysis of low-abundance protein biomarkers. Expert Rev Proteomics 2007;492:175–186.

37. Pavlickova P, Schneider EM, Hug H. Advances in recombinant antibody microarrays. Clin Chim Acta 2004;343(1,2):17–35.

38. duPont N, Wang K, Wadhwa P, Culhane J, Nelson E. Validation and comparison of Luminex multiplex cytokine analysis kits with ELISA: Determinations of a panel of nine cytokines in clinical sample culture supernatants. J Reprod Immunol 2005;66(2):175–191.

39. Liu MY, Xydakis AM, Hoogeveen RC, Jones PH, O'Brian Smith E, Kathleen W. Nelson KW, Ballantyne CM. Multiplexed analysis of biomarkers related to obesity and the metabolic syndrome in human plasma, using the Luminex-100 system. Clin Chem 2005;51:1102–1109.

40. St. Ledger K, Agee SJ, Kasaian MT, Forlow SB, Durn BL, Minyard J, Lu QA, Todd J, Vesterqvist O, Burczynski ME, Analytical validation of a highly sensitive microparticle-based immunoassay for the quantitation of IL-13 in human serum using the Erenna® immunoassay system, Journal of Immunological Methods 2009;350:161–170.

41. Heymach JV, Tran HT, Fritsche HA, et al. Lower baseline levels of plasma hepatocyte growth factor (HGF), IL-6, and IL-8 are correlated with greater tumor shrinkage in renal cell carcinoma patients treated with pazopanib using the Q400 Biomarker Workstation. Presented at: AACR-NCI-EORTC International Conference on Molecular Targets and Cancer Therapeutics; November Boston, MA. Abstract A11. 2009;15–17.

3

PROTEOMIC METHODS TO DEVELOP PROTEIN BIOMARKERS

RUTH A. VANBOGELEN AND DIANE ALESSI

Protein biomarkers are often the macromolecules of choice for biofluid biomarkers because proteins are a "surrogate" for the dynamic biology occurring in the organism. While there are only a *few* protein markers currently in use for diagnostic purposes (e.g., Prostate Specific Antigen (PSA)), *many* protein biomarkers are being used in clinical trials and *hundreds* of proteomics studies have been published, which report the discovery of new biomarkers (1). Thus, the pipeline for protein biomarkers is rich with opportunities. Technologies and reagents are improving every year, offering opportunity to increase the success rate and decrease the time line for developing protein biomarker assays. The pace of improvements in proteomics technologies may seem slower than the pace of improvements in genomics technologies, but in fact, maybe the pace is similar if one considers both the amount of protein ($\sim$10% of the mass of cells) and the number of different protein species in each sample type (cells, tissues, biofluids) compared to the significantly smaller quantity and diversity of genes.

There are likely a trillion different protein species in human biology. The number of potential protein biomarkers encoded by the human genome is calculated as follows. The human genome contains approximately 20,000 protein-encoding genes. On average, each of these genes has five splice variants; that is, each gene can direct the synthesis of five different mRNA molecules, which, in turn, can each encode a unique protein molecule. Thus, roughly 100,000 (20,000 $\times$ 5) human proteins are possible (even without considering genetic variants). However, we must also consider posttranslational modification of proteins. If, on

Predictive Approaches in Drug Discovery and Development: Biomarkers and In Vitro/In Vivo Correlations, First Edition. Edited by J. Andrew Williams, Jeffrey R. Koup, Richard Lalonde, and David D. Christ.

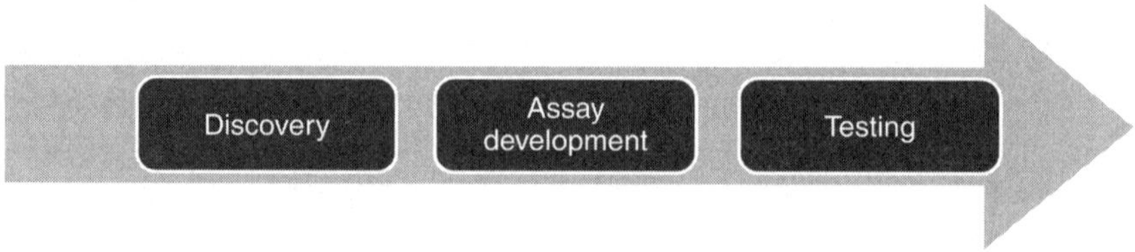

FIGURE 3.1 The three major steps in biomarker development.

average, each of the 100,000 proteins might contain any combination of 10 post-translational modifications, then the number of possible human protein isoforms equals 100,000 times 10 factorial. This calculation estimates that 3.6×10^{11} different proteins exist in human biology.

Proteins are difficult molecules to monitor. Methods to study cellular proteins have been around for well over a century, but even today, proteomic methods are considered "immature," especially compared to methods for studying genomes (DNA analysis) and cellular transcripts (mRNA profiling). The first technical issue is the "depth" of the current methods is not sufficient considering the broad range (12 orders of magnitude) of protein concentrations in biological samples. The dynamic range of most methods is three to four orders of magnitude. A second technical issue is the development of assays to monitor all the candidate biomarkers. The number of protein biomarkers that can be routinely and accurately monitored in preclinical and clinical settings is very small ($\sim$500) relative to the possible 0.4 trillion proteins in human biology.

The objective of this chapter is to describe *to biologists* the different proteomic methods that are being used to develop protein biomarkers. Many biologists, not familiar with the different proteomic methods, find it difficult to determine which method is optimal for their project. We hope this chapter will serve as a consultant. This chapter focuses on methods for quantifying the *level* of proteins rather than methods for monitoring protein *activity*. The chapter is divided into three parts, corresponding to the three major steps in biomarker development (Fig. 3.1).

Introductory level information for most of the methods described in this chapter is available on the internet. In particular, most are described in Wikipedia, which also provides original publications and often other references. The theme throughout this chapter is to describe the application of these methods for biomarker development and to help the biologist know what questions to ask when discussing how to execute a proteomic study.

The first step in any proteomics experiment, whether for discovery, assay development, or testing, is sample collection and sample preparation.

- Sample collection is the most critical step in any proteomics study. High quality samples are often difficult to obtain, and to date, there are no standard tests to evaluate the quality of proteins in samples. Below are considerations for sample collection.

- *Amount of time between when the sample is removed from the organism and when it is frozen (typically ultracold, −80°*C). Removal of a sample from an organism (37°C for mammals) and exposing it to room temperature (typically 22°C) is a cold shock, and changes in proteins are known to occur very quickly (seconds). Fewer changes will occur if the sample is taken to 4°C immediately.
- *Steps between when the sample is removed and when it is frozen*. For example, many tissues will contain a significant amount of blood in the vascular of the tissue. Removal of the blood with a cold perfusion step is often recommended. Separation of serum or plasma from total blood sample is another common step. Standard operating procedures (SOPs) for plasma collection should be established and followed. The Early Detection Research Network (EDRN) has an SOP on the internet (http://edrn.nci.nih.gov/resources/standard-operating-procedures/biological-specimens/plasma-sop.pdf). Human tissue samples are often prepared for pathology purposes first (fixed and embedded), and these protocols can affect the proteins in the sample.
- *Freeze–thaw cycles*. The stability of proteins and peptides to freeze–thaw cycle is unique to each protein, but in general, freeze–thaw cycles are considered detrimental to samples used for proteomics. Aliquoting the samples is highly recommended. Tissue samples can be frozen and while still frozen "powdered" (mortar and pestle or a hammer) so that small portions can be removed.
- *Storage of samples*. The effects of long-term storage of samples even under optimal conditions are also likely different for each protein and peptide. Tests are often done during the validation of an assay to evaluate for the target proteins/peptides and the effects of storage of samples.

- Sample preparation is the second critical step in the analysis of samples. Below are key considerations for sample preparation.
  - *Disruption of cellular material or matrix*. Tissue samples are needed to be homogenized before the next step. Beware of heating samples and damaging proteins during this step.
  - *Solubilization of proteins*. The array of characteristics of proteins is broad, and thus, no one method will solubilize all proteins in a sample. Detergents work well for many proteins, but for some of the proteomics methods (e.g., mass spectrometry (MS)) described in this chapter, detergents cannot be used, as they interfere with the chromatography steps. For embedded samples (e.g., FFPE (formalin-fixed paraffin-embedded) tissue), there are commercial products for solubilizing proteins for proteomic analysis (see Expression Pathology, Qiagen and others).
  - *Fractionation of the complex protein mixture*. Most biological samples are a very complex mixture of proteins, including thousands (potentially millions) of different protein species that vary in abundance over 8–12 orders

of magnitude (12 orders of magnitude spans the range of a satellite image of the earth to viewing the mosaic features of the *Drosophila* eye). Fractionation of the complex mixture is a common step in proteomic platforms. However, if protein quantification is the goal of the study (which is almost always the situation for biomarkers), then a method for reassembling the data is critical. Fractionation methods include depletion of abundant proteins (e.g., albumin, IgG) from biofluids, separation of the proteins on a 1D SDS-PAGE gel and then segmenting the gel lane, chromatography (strong cation exchange (SCX), reverse phase (RP), etc.).

- *Protein digestion*. For protein identification and for protein quantification using MS, the enzymatic digestion of the proteins into peptides is required. Trypsin is the enzyme of choice for this step because the cleavages are most reproducible. For identification of proteins, high reproducibility of the digestion is not required. However, for protein quantification, highly reproducible digestion is critical.
- *Removal of nonprotein/peptide molecules*. Some sample types (e.g., urine, synovial fluid) contain a wide array on nonprotein molecules that interfere with the analysis of proteins by some of the proteomic methods. Two common methods for removal of these interfering molecules are electrophoresis through polyacrylamide and solid phase extraction (SPE). However, any method that removes interfering molecules has the potential to also remove some of the molecules of interest.

3.1 PROTEIN BIOMARKER DISCOVERY

The overall objective of a protein biomarker discovery study is to find proteins whose levels change in response to some state/condition (disease state, mutation, treatment, etc.). A typical experiment compares a control condition to a test condition. Rarely, however, does a biomarker discovery study include processing merely two samples.

The variables to consider include the following.

- *Technical Variability.* How does technical variability affect the design of the experiment? The higher the technical variability, the higher the number of technical replicates that need to be done (an average of the replicates is a better representative of the sample). For some proteomic methods, the technical variability is high. Technical variability can be evaluated by comparing data from replicates of the same sample. When data from the replicates is compared, a correlation coefficient (R^2) can be calculated. A good "rule of thumb" is that an $R^2 > 0.9$ indicates low technical variability. An R^2 between 0.8 and 0.9 is acceptable, while an $R^2 < 0.8$ indicates very high technical variability, and thus, it will be difficult to detect a biological signal.

- *Biological Variability.* Biological variability can be due to many things, including genetic diversity among individual biological specimens (even gender) and differences between the environment of each biological specimen (e.g., diet).

No one proteomic method has emerged as the method of choice for all protein biomarker discovery studies, as each method has its own set of advantages and disadvantages. The key factors affecting the success of a protein biomarker study are

- the quality of the samples
- the depth of analysis of proteins (how many proteins are detected)
- the technical variability of the method.

When considering which platform is best for a study, the key questions to ask are

- How many samples (and what quantity of each) are available for the study?
- What is the time line for completion of the study?
- What funds are available for the study?

The methods that provide the most in-depth detection and quantification of proteins will also be the most expensive and will require the longest amount of time for analysis.

Table 3.1 provides an overview of the various proteomic methods that are used for biomarker discovery. Further information about each method is then detailed in the Boxes 3.1–3.4.

TABLE 3.1 Comparison of Five Common Proteomics Methods Used for Protein Biomarker Discovery

	2D Gels	SILAC	iTRAQ	Label-Free (dMS)	GeLC/MS
Quantitation method	Protein stain image analysis	Isotope label	Isotope label	Peak area	Spectral counts, peak areas
No. of samples	~50	Pair wise	4-plex, 8-plex	100s	<20
Throughput	Medium	Low	Medium	High	Low
No. of proteins	Low to Medium	High	Low to Medium	Medium	Very high
Sample requirement	High	Special low	Medium	Medium	Low

Abbreviations: SILAC, stable isotope labeling with amino acids in cell culture; iTRAQ, isotope tags for relative and absolute quantitation; dMS, differential mass spectrometry.

Box 3.1 2D GELS; TWO-DIMENSIONAL POLYACRYLAMIDE GEL ELECTROPHORESIS

Description

With this method, proteins are denatured into their component polypeptides. Proteins in a mixture are first separated in a tube gel or on a strip of gel by their isoelectric point (pI). The tube gel or strip is then placed on a slab gel (SDS-PAGE) where the proteins are further separated, this time by their molecular weight. In order to identify which protein(s) is present in the various gel "spots," individual spots are excised from the gel and then subjected to proteolytic digestion. The resulting peptides are then analyzed by mass spectrometry, yielding, in successful cases, the identity of the protein(s) present in each analyzed gel spot (http://en.wikipedia.org/wiki/Two-dimensional_gel_electrophoresis).

Opinion of authors

This is not the method of choice for biomarker discovery when studying complex biofluids (e.g., plasma, CSF, or urine) because many of the gel spots will contain multiple proteins, making individual protein identification difficult and the process of drawing conclusions nearly impossible. However, this method may be acceptable when studying cells or tissues. 2D gel analysis is very useful when studying less complex organisms (e.g., bacteria).

Typical study

Any type of biological sample can be used. A typical experiment entails a comparison of two conditions, typically a control condition and a test condition (e.g., treatment, mutation).

Sample requirements

Protein at 40–400 μg concentration is typically loaded onto each gel.

Typical time lines

Sample preparation and the running and staining the 2D gels takes 2–3 days. Analysis of the gel images can be performed manually ("gel gazing") or with the use of image analysis software. Typically in 1 day, a dozen images can be examined manually, and the protein spots that change twofold or more can be determined. Image analysis software provides numerical values indicating the volume of each protein spot, thus enabling statistical analysis. Software-assisted image analysis of a dozen gel images typically takes 1 week for an experienced user.

Typical costs to outsource

Small format gels cost approximately $300 per gel; large format gels cost approximately $600 per gel. Protein identifications cost approximately $300 per excised spot.

Example data

Data may be presented as a 2D gel image. If image analysis has been performed, data is typically presented in a spreadsheet with each row representing a gel spot and each column representing an individual sample. Rarely are all the proteins on a gel identified. Instead, the investigator typically selects a subset of spots that are of interest and from which proteins are to be identified.

Statistical analysis

If image analysis software is used, the data is amenable to most statistical analysis methods.

Advantages

A major advantage of 2D gels is that sequence of the genome is not required to perform quantitative analysis. Thus, biomarker identification studies in many agricultural and animal science species are possible. The second advantage is that the synthesis rates of proteins can be analyzed. In order to achieve this, a biological sample is metabolically radiolabeled (e.g., with ^{35}S-met) before preparing a gel sample. The resulting gel image is captured on X-ray film or with a phosphoimager screen. The third advantage is 2D gels are the least expensive proteomic method.

Disadvantages

Gel spots often contain more than one protein.

Terms and acronyms

IEF, isoelectric focusing; MW, molecular weight; SDS, a detergent used to denature polypeptides; PAGE, polyacrylamide gel electrophoresis; SDS-PAGE, a method that separates proteins by their molecular weight; DIGE, differential gel electrophoresis, a method involving the labeling of proteins with a dye before running 2D gels so that different samples can be mixed together and run on the same gel; spots, when visualizing stained or radiolabeled proteins on a 2D gel, round or oval shapes (i.e., spots) are typically observed; image analysis, the process of aligning multiple 2D gel images so that they may be compared to each other.

Box 3.2 PROTEIN LABELING METHODS (e.g., SILAC, iTRAQ, ICAT)

Description

SILAC (stable isotope labeling with amino acids in cell culture) studies use proteins that are metabolically labeled (2). ICAT (isotope coded affinity tag) studies use proteins that are labeled after they have been extracted from a biological sample. iTRAQ (isotope tags for relative and absolute quantitation) and TMT (tandem mass tag) studies use peptides that are derived from proteins that have been extracted from a biological sample and proteolytically digested. These resulting peptides are subsequently labeled with an isobaric tag. Peptides from different samples are labeled with different tags (or no tag at all). This allows the peptides from different samples to be mixed together for mass spectrometry analysis and quantification. Typically, the mixed peptide sample is fractionated to reduce the complexity of the peptide mixture before mass spectrometry analysis. Each fraction is analyzed (often in duplicate). Specialized software is needed for analysis of the data (e.g., (3)) (http://en.wikipedia.org/wiki/SILAC; http://en.wikipedia.org/wiki/Isotope-Coded_Affinity_Tags; http://en.wikipedia.org/wiki/ITRAQ).

Opinion of authors

This proteomic method is appropriate for a study involving four to eight samples, as this number of samples can easily be mixed together for analysis. Larger sample sets require multiple mixtures, where the peptides that are detected in one mixture may not be detected in another mixture. This leads to a final spreadsheet with many missing data points.

Typical study

Control + experimental each with $N = 4$.

Sample requirements

Protein at 20–100 μg per sample.

Typical time lines

One mixture of samples can be analyzed in a week even with 10 fractions collected.

Typical costs

Analysis of 10 fractions from one mixture costs approximately $12,000.

Example data

Data is presented as the ratio of an experimental sample compared to a control sample.

Statistical analysis

The ratio data is amenable to most statistic analysis methods.

Advantages

Costs are low because multiple samples can be combined before mass spectrometry analysis.

Disadvantages

When used for large sample sets (e.g., 50 or more samples), there can be a significant amount of data missing. With some instruments, the data is compressed (e.g., a twofold change may be reported as a ratio of merely 1.2) because the ratio for a single protein is typically derived by "rolling up" the data from the protein's multiple peptides into a single ratio for the protein. Rarely do all the peptides from one protein have the same ratio, and in fact, the ratios for the different peptides from a single protein are often quite different from each other.

Terms and acronyms

ICAT, iTRAQ, and TMT, commercial or patented names of specific products; isobaric tags, the various reagents used to label proteins or peptides; SCX, strong cation exchange, a method for fractionating protein or peptide samples.

Box 3.3 LABEL-FREE MASS SPECTROMETRY (e.g., DIFFERENTIAL MASS SPECTROMETRY (DMS))

Description

In this method, proteins are extracted/solubilized and proteolytically digested, the resulting peptides are fractionated, and then each fraction is analyzed by MS. The LC-MS data (peptide m/z) are aligned and quantified. Statistical analysis across samples is performed at this stage. Once differentially expressed peptides/proteins are found, the peptides are further analyzed by MS to identify the differentially expressed proteins (4).

Opinion of authors

This method works well for large sample sets of biofluids. As the quality and quantity of fractionation methods used is increased, so does the potential number of proteins that might be detected in the study's samples.

Typical study

A total of 50–100 samples from two or more groups that are to be compared.

Sample requirements

Protein at 0.02–1 mg.

Typical time lines

A study of 100 samples wherein each sample yields 20 fractions, and each fraction is subsequently analyzed in duplicate 2-h MS runs, requiring 1 year to complete.

Typical costs

For a 100-sample study, 20 fractions per sample study costs approximately $250,000.

Example data

A list of proteins each with a ratio value or a composite peak value is provided. Not all proteins in the sample are identified. All unidentified proteins are listed as "features" observed and are not linked to a gene name.

Statistical analysis

Specialized statistical tools integral to the data analysis and processing software are used.

Advantages

This method focuses on the proteins that are differentially expressed across groups.

Disadvantages

High cost, long time lines.

Box 3.4 GeLC/MS

Description

In this method, a sample is prepared and run on a lane of 1D SDS-PAGE gel. The lane is then cut into segments, typically 20–40. The proteins in each gel segment are proteolytically digested and then analyzed by MS. Proteins are identified, and the data from all the segments from one gel lane are combined to provide a nonredundant list of proteins along with the quantitation data for each protein. The data from samples run in different gel lanes are then compared (5).

Opinion of authors

This is a powerful method for biomarker discovery for small- or medium-sized studies. Additionally, this method provides the information that is required for developing protein assays using the peptide MRM method.

Typical study

Control and experimental samples are compared. Often the biological replicates within a group are pooled, so that the gel lane represents the average of the replicates. Data for individual proteins are monitored with a peptide MRM assay. Data from a GeLC/MS discovery study generally correlates very well with the data generated during protein assay development.

Sample requirements

Optimal is 20 μg, although good results can be obtained with even 100,000 cells (e.g., from laser capture microdissection).

Typical time lines

A GeLC/MS comparison of three samples per pool in which each gel lane is cut into 40 segments can be completed in 1–2 weeks.

Typical cost

Approximately $5000–$11,000 per sample or pool.

Example data

Quantitative data (either spectral counts or average peak areas) from a GeLC/MS study is typically presented in a spreadsheet, with each row representing an individual protein and each column representing an individual sample. An example of spectral count data for five proteins and four samples is shown below.

Statistical analysis

Data from a GeLC/MS study is amenable to standard statistical analysis methods.

Advantages

Separation of proteins by molecular weight using SDS-PAGE is superior to fractionation methods such as SCX. With GeLC/MS, a specific protein isoform will typically appear in only one or two gel segments, whereas the same protein isoform may appear in multiple SCX fractions. Additionally, GeLC/MS data can reveal different molecular weight isoforms of a single protein.

Disadvantages

Cost and time line for a study with a large number of samples.

3.1.1 Statistical Analysis

In order to identify proteins differentially expressed among the different groups of samples in a discovery experiment, statistical analysis of the data should be done. The amount of statistical analysis will depend on the number of technical replicates and the number of biological replicates. Standard analysis includes calculating technical averages and assessing technical variance, and then using the technical averages to calculate biological averages and variance. Fold change calculations across groups are typically done, although the cutoff for identifying a protein differentially expressed is typically arbitrary (1.5- or 2-fold is commonly used). p values (t test) are calculated if three or more biological replicates were analyzed. False discovery rate (FDR) calculations (6) are also done when sufficient biological replicates are analyzed. p value and FDR criteria are also arbitrary, although typically p values less than 0.05 are considered significant and FDR less than 20% are significant. Multivariate analysis is also useful. Principal component and hierarchical clustering methods are most common, but many types of multivariate analysis can be performed.

There are core laboratories within many universities and large pharmaceutical companies that perform these proteomic methods for their institution's investigators. There are also companies that perform this work on a fee-for-service basis. When utilizing either a core laboratory or a fee-for-service company for conducting proteomic studies, an investigator should also request the laboratory to analyze the data, select the best biomarkers (based on the data), and provide a plan for how the subsequent protein assay will be developed. This is important, because the proteomics groups understand the data and the caveats in the data and have typically developed, with the help of statisticians, the best analysis method. The investigator should carefully review the analysis and should make the final selection of any candidate biomarkers. One consideration the biologists will add to the analysis is whether the candidate biomarkers can be related back to the biology or not. Often the candidate biomarkers will not be "rationale"; that is, current knowledge does not explain why the protein would be a biomarker for the system of interest. However, these "unknown" biomarkers are often statistically significant and thus worthy of further consideration.

3.2 ASSAY DEVELOPMENT

Once the candidate biomarkers have been chosen, the next step is to confirm the biomarkers. First, an assay for the protein(s) of interest needs to be developed. There are two common types of protein assays that can be developed, namely, antibody-based assays and MS-based assays. Table 3.2 provides an overview of the features of both types of assays. Antibody-based assays have been used for years, and most biologists are familiar with these assays. However, most biologists are less familiar with the MS-based assays, and thus, detailed information about this assay method is provided.

These two protein assay approaches are complementary and are typically validated to perform well. New commercially available assays are introduced continually. The hPDQ (human proteome detection and quantitation) project has been proposed, which calls for the rapid development of a protein assay for each of the 20,500 human genes (7). Currently, if an assay is not available, the researcher needs to have assay(s) developed for the biomarkers discovered.

TABLE 3.2 Comparison of Antibody-Based and Mass-Spectrometry-Based Assay

Feature	Antibody-Based Assay	Mass-Spectrometry-Based Assay
Sensitivity (ability to detect low abundance proteins)	High	Medium
Specificity (ability to detect only the protein and protein isoform of interest)	Medium to high	High
Multiplexing capability (ability to monitor more than one protein)	1–10 proteins is the practical limit	25 proteins (maybe more)
Typical time line to develop assay	4 months to 5 years	6 weeks for relative quantitation, 3 months for absolute quantitation
Typical cost	\$25,000 per protein	\$1800–\$12,000 per protein
Typical success rate	Low—it is difficult to find a pair of antibodies with high specificity	High
Cost for testing	Low	Low for high multiplexed assay (25 proteins per assay); high for single protein assay

3.2.1 Antibody-Based Assays

The five most commonly used methods are presented below.

3.2.1.1 Western Immunoblot Assays. These are the simplest assays if antibodies for the proteins of interest are available. In this type of assay, samples are run on a 1D SDS-PAGE gel, the proteins are transferred from the gel to a membrane, a primary antibody (or antibody mixture) is allowed to interact with the bound proteins, and then the resulting bound protein–antibody complexes are detected with secondary antibodies that are conjugated to a reporter enzyme, allowing visualization after supplying the appropriate enzyme substrate. Data from this type of assay is qualitative, not quantitative. Many investigators use this method for preliminary confirmation of candidate biomarkers (http://en.wikipedia.org/wiki/Western_blot).

3.2.1.2 ELISA (Enzyme-Linked Immunosorbent Assay). ELISA is used in research, development, and clinical applications. Proteins (antigens) are immobilized on a surface and allowed to interact with a specific antibody. As with Western immunoblots, a conjugated secondary antibody reacts with the immobilized primary antibody-protein complex and allows for detection of the complex. ELISA is typically quantitative, and standard curves can be used to convert the "signal" to a protein concentration. A standard curve is developed using purified protein as the antigen. Typically, five to eight different concentrations of the purified protein are used in the assay to create a standard curve (known concentration vs signal) so that the concentration of this same protein in an experimental sample can be determined from the signal that is generated with the sample. Alternatively, ELISA data can be used for relative quantitation (RQ) (ratio of a test sample to a reference sample). Several assay validations are typically performed in order to evaluate selectivity, specificity, accuracy, precision, and equivalence. There are many commercially available ELISA kits (http://en.wikipedia.org/wiki/ELISA).

3.2.1.3 Platform-Specific Assays. There are commercial platforms for antibody-based assays (e.g., Luminex and Meso Scale Discovery) that are designed to allow multiplexing of proteins into a single assay, thus increasing the throughput of testing. The quantitation derived from these assay platforms is antibody dependent.

3.2.1.4 Reverse Arrays. Reverse arrays allow many antibodies to be tested against a set of arrayed samples. Commercially available sample arrays can be purchased for testing. Alternatively, client-specific samples can be spotted onto arrays by companies (e.g., Theranostics Health and Baypoint Biosystems) that specialize in this technology (http://en.wikipedia.org/wiki/Reverse_phase_protein_lysate_microarray).

3.2.1.5 Mass-Spectrometry-Based Assays. One standard name for MS-based assays has not emerged. Instead, these assays have been alternatively referred to as *peptide MRM (multiple reaction monitoring)* assays, pMRM assays, MRM protein assays, SID (stable isotope dilution) MRM assays, MRM peptide assays, SRM (selected reaction monitoring) assays, or Absolute Quantitation (AQUA) assays. This chapter uses the term *MRM protein assay* to refer to a MS-based assay. MRM protein assays are similar in many ways to small-molecule MRM assays that have been used for more than 20 years to evaluate drug concentrations in preclinical and clinical samples. With MRM protein assays, the small molecule being assayed is a small peptide (a portion of a protein) instead of a compound (drug). While some MRM protein assays were developed before 2004, the number of publications reporting the use of MRM protein assays has escalated since 2004. It is predicted that the use of MRM protein assays will significantly increase the number of validated medically relevant protein biomarkers (7). In May 2009, the authors of this chapter published a tutorial in Genetic Engineering News, describing how MRM protein assays are part of a biomarker workflow designed to significantly reduce the time line for biomarker development (8). Like ELISA, these assays can provide RQ data or, if a standard curve is used, absolute quantitation (AQ) (i.e., protein concentration) data. There are three major advantages of MRM protein assays over antibody-based assays: (i) high specificity for the protein or protein isoform of interest, (ii) high multiplexing of the assay to include 25 or more proteins in a single assay, and (iii) short time lines for assay development.

MRM protein assays are a "targeted" proteomics method. Peptides specific for the proteins of interest (biomarkers) are selected for use in an assay and then the mass spectrometer is directed to monitor only those peptides. Figure 3.2 provides some basic information about the vocabulary associated with MRM protein assays.

Steps in the development of an MRM protein assay:

1. Start with a protein list and sample type for the assay.
2. Select peptides and fragment ions that can be detected by MS.
    - GeLC/MS (biomarker discovery method described above) directly provides this information (empirical selection of peptides), while many other discovery platforms do not. *In silico* selection of peptides can be performed, but typically the success rate is low.
    - Peptides should be unique to the protein of interest.
3. Tune the mass spectrometer to look for the peptides and fragment ions in the sample type to be used.
    - First, individual analysis of each peptide.
    - Second, multiplexed for multiple peptides and fragment ions per protein.
4. Conduct iterative testing in order to select the best peptides for the assay.
5. Test the assay with 3–10 samples in order to assess variability (analytical, technical, biological).

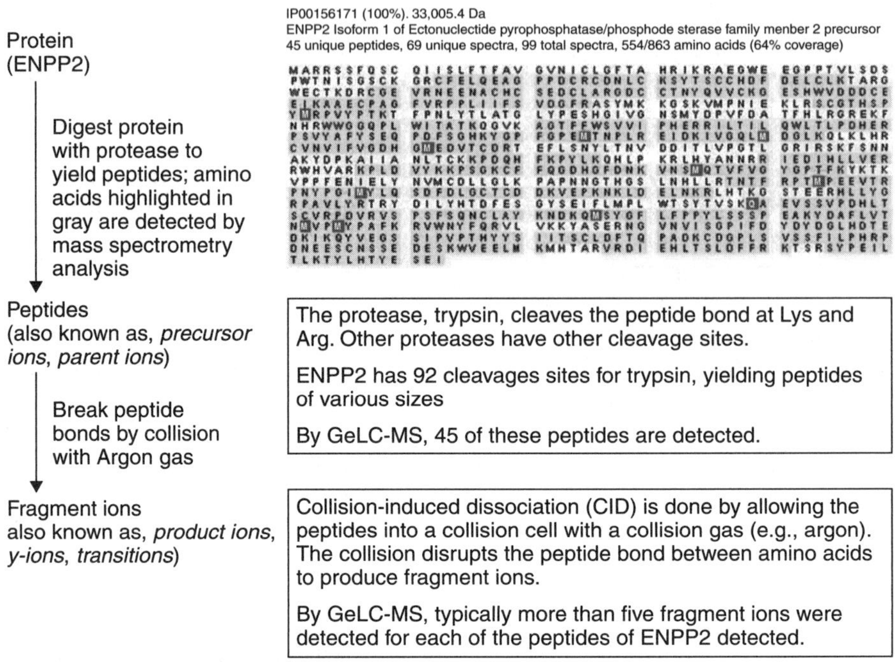

FIGURE 3.2 This figure provides an example of the analytes used for MRM protein assays. Along the left portion of the figure are the different levels of analytes, namely, the protein, the peptides, and the fragment ions. The top right portion of the figure is the complete amino acid sequence (single letter abbreviation for amino acids) of the human protein, Ectonucleotide pyrophosphatase/phosphodiesterase 2, whose gene name is ENPP2. The amino acids highlighted in gray are amino acids in tryptic peptides that were detected by mass spectrometry. In MRM assays, the quantitation value is derived from the peak areas of the fragment ions for a peptide.

6. Refine the assay and develop a final list of peptides and product ions.

At this point, the assay is ready for testing samples to provide RQ. The data provided in a RQ assay is a unitless value. These values can be compared across samples. Typically, one sample is chosen as the reference sample to which all other samples are compared.

If AQ (protein concentration) is desired, continue with additional assay development steps.

7. Order labeled and unlabeled standards for each peptide.
8. Develop calibration curves with the peptide standards.
9. Test the assay with calibration curves and with labeled peptide standards in each sample.
10. Perform fit-for-purpose assay validation.

In order to develop a calibration curve (or standard curve) for an MRM protein assay, labeled and unlabeled peptide standards need to be synthesized for each peptide in the assay. Peptide synthesis services are offered by several companies.

- *Unlabeled Peptides.* These are also known as *light peptides*. They are identical to the endogenous (or native) peptides that are derived from the samples after the proteins in the sample are proteolytically digested. These unlabeled peptide standards are used in the calibration curves.
- *Labeled Peptides.* These are also known as *heavy peptides* or *AQUA peptides*. When synthesized, one stable-isotope-labeled amino acid is incorporated into the peptide. As a result, labeled peptide has a slightly higher molecular weight (an additional 6–10 Da) than the unlabeled peptide. When the native peptide and labeled peptide are mixed together, both the peptides will elute together during LC (liquid chromatography) and will ionize with the same intensity; however, the mass spectrometer will distinguish the two peptides based on the difference between their molecular weights.

Note: if the MRM protein assay is being designed to monitor a posttranslational modification (e.g., a phosphorylation), then the peptide standards must be synthesized with the posttranslational modification.

A calibration curve allows MS data to be converted to protein concentration data. Figure 3.3 illustrates an example of a six point calibration curve.

1. Make six solutions. Each contains the same amount of the labeled peptide and different concentrations (fmol/μl) of the light peptide

	1	2	3	4	5	6
Light (fmol/μl)	10	5	2.5	1.25	0.625	0.3125
Heavy (fmol/μl)	2.5	2.5	2.5	2.5	2.5	2.5

2. Analyze samples by LC-MRM/MS.

Solution No.	Unlabeled peptide area	Labeled peptide area	Ratio of unlabeled to labeled area	Concentration
1	2,839,576	720,597	3.9406	10
2	1,434,129	692,338	2.0714	5
3	706,468	741,791	0.9524	2.5
4	342,945	671,144	0.5110	1.25
5	175,869	727,662	0.2417	0.625
6	87,497	699,403	0.1251	0.3125

3. Plot data to produce calibration curve

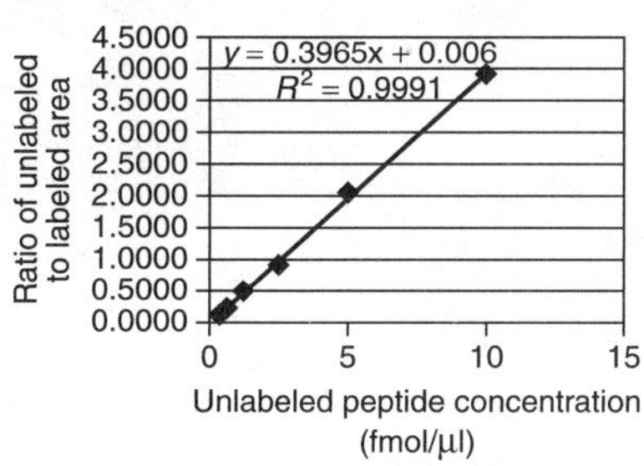

FIGURE 3.3 This figure provides an overview of the three steps in generating a calibration curve.

3.2.2 Detection Level with MRM Protein Assays

Typically, if a protein is detected by a proteomic method (e.g., GeLC/MS) during biomarker discovery, then the protein will be detected with the MRM method. For proteins selected as biomarkers on the basis of literature review or transcriptomics analysis, it is likely that the proteins exist in low abundance in the sample type of interest. In these cases, an enrichment strategy is required. There are three ways to perform enrichment:

1. *Immunocapture with Protein-Specific Antibodies.* In this case, the antibodies are used only for enrichment and not for quantitation.
2. *Protein or Peptide Fractionation Method.* For example, many phosphoproteins can be enriched using methods developed for capture of phosphorylated serines and threonines (e.g., MassPrep™ by Waters).
3. *SISCAPA Method.* This method, described by Leigh Anderson (9), uses antibodies to capture specific peptides of interest from a sample.

There are several important steps in the development of an MRM protein assay.

1. Key to the success of an assay is the detection of the fragment ions for each peptide should be well separated on the X- and Y-axes and sufficient data points should be obtained for each peak (Z-axis). Figure 3.4 is a 3D view of an MRM protein assay that is multiplexed for 57 peptides.

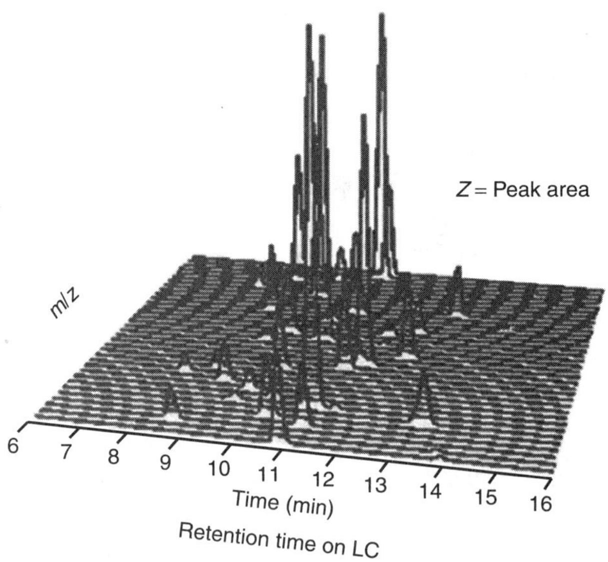

FIGURE 3.4 This figure is a three-dimensional plot of peptides in a multiplex MRM protein assay. The X-axis is the separation of peptides based on their retention time in liquid chromatography (LC), the Y-axis is the separation of peptide based on their mass over charge (m/z) value, and the Z-axis is the sum of the peak areas for the fragment ions for each peptide.

2. Assessment of the analytical and technical variabilities for each peptide is important. Ideally, the combined analytical and technical variability (CV (coefficient of variation)) should be less than 10%. Biological variation may be high and, in fact, is what provides the data range for the assay. Figure 3.5 shows the results of analytical, technical, and biological variation for 10 peptides.

The peptides for APO M and APO C3 display high technical variability and should be replaced with different peptides that perform better (i.e., display acceptable levels of technical variability) in the assay.

3. Because multiple isoforms may exist for a given protein, different peptides from that protein can easily give different quantitative values for that protein. In fact, in some cases, it is desired for an assay to specifically monitor these different isoforms, such as when monitoring the phosphorylated and unphosphorylated states of a protein. In other cases, multiple peptides from a particular protein are deliberately used to gain confidence in the data generated by the assay. Figure 3.6a displays an example data set from an assay using three peptides from one protein. In this case, the three peptides exhibit the same trend line as each other across all 13 different test samples.

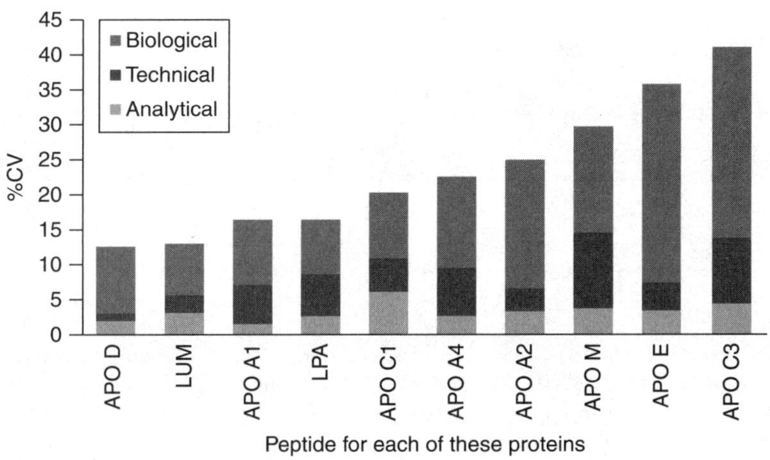

FIGURE 3.5 This figure displays on a stacked column plot the combined analytical, technical, and biological variances (%CV) for 10 different peptides as observed with an MRM protein assay. Analytical variance (%CV) is the result of analyzing the same sample three times (three injections into the mass spectrometer). Technical variance (%CV) is measured by preparing the same sample three times and then analyzing the triplicates three times by mass spectrometry. The technical %CV is calculated using the average of the analytical replicates. The biological variance is measured by preparing 10 different samples, running each sample in analytical triplicate, averaging the analytical triplicates, and then calculating the %CV for the 10 average values. The peptides with high analytical or technical %CV (e.g., >20%) are not suitable for an MRM assay.

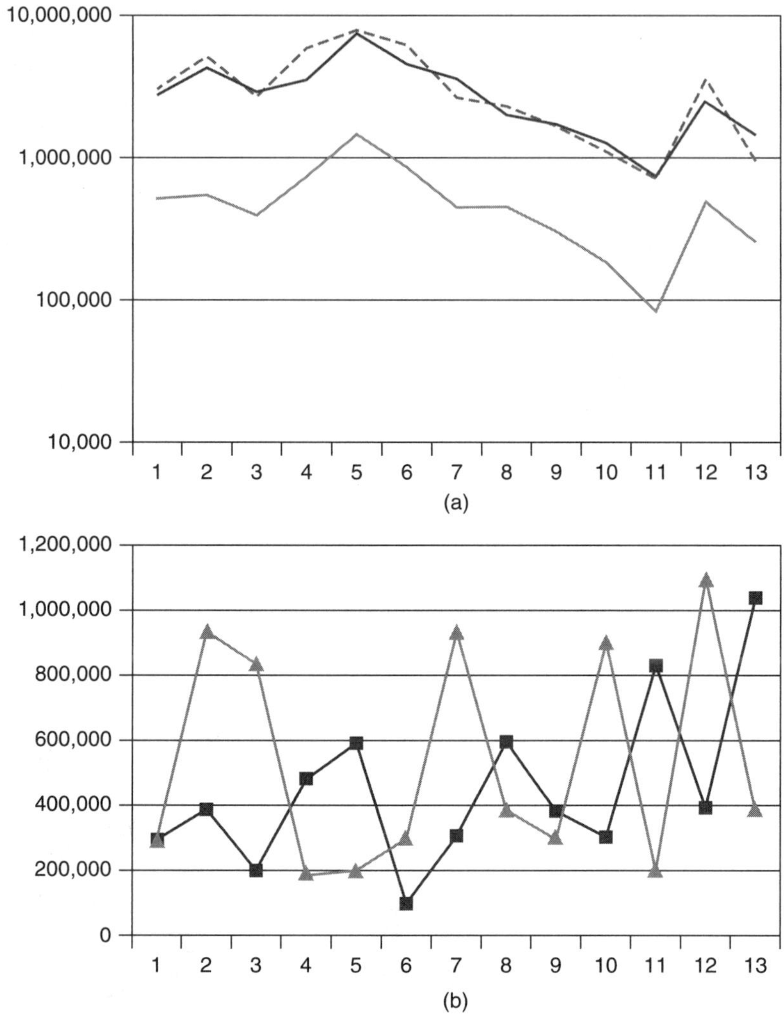

FIGURE 3.6 This figure displays peptide trend lines for two different proteins. Thirteen samples were tested using MRM protein assay. While the peak area for each peptide can be very different because the peak area is the combined results of the abundance of the peptide and the ionization of the peptide in the mass spectrometer, the trending across the 13 samples should be very similar for all peptides of a single protein. (a) Two peptides of this protein have similar peak areas in all 13 samples (hovering between 1,000,000 and 10,000,000), and the third peptide of the protein has a peak area of roughly 10 times less. The line for all three peptides has a very similar shape across the 13 samples. These are peptides well suited to represent the protein in the MRM protein assay. (b) The lines for the two peptides of this protein have a very different shape across the 13 samples. These peptides are not well suited to represent this protein in an MRM protein assay because the variance of the level of this protein in the 13 samples would be very high for most of the 13 samples.

However, as can be observed in Figure 3.6b, the trend line of different peptides from one protein may be quite different from each other in some cases. Both results are real and reveal the biology, but the investigator needs to know whether to expect the same or different results from each peptide from a protein.

4. The analytical system (e.g., the LC and MS components) must perform robustly throughout an entire batch of samples. Part of assay development is the determination of how many samples can be tested in a batch before, say, an instrument needs cleaning and/or recalibration or a column needs changing. This determination is typically performed by spiking an internal standard into a batch of samples and then monitoring the internal standard during the assay. Figure 3.7 presents data generated by one such experiment. In this example, three internal standards (all peptides) were monitored across 90 samples. The %CV for each of the three peptides is less than 10%. Thus, this assay is considered stable (for at least 90 samples per batch).

During assay development, the sample requirements, sample preparation method, cycle time, and number of samples per batch for MS analysis are determined. Below are a few more recommendations to consider during assay development.

- Methods should be "locked down" after assay development = Procedure/SOP.
- Sample preparation should, when feasible, be conducted in a 96-well format in order to increase throughput and decrease cost.
- Internal standard(s) should be added to each sample.
- Reference sample(s) should be included in each sample batch.
- Instrument time should be considered when planning a work schedule (the cycle time for an LC/MRM/MS assay for 25 proteins is typically 20 min).
- Methods for data analysis and reporting should be standardized.

If an assay is designed to provide RQ, then one sample should be chosen as the reference sample for all batches of samples to be tested.

RQ data is typically sufficient for confirming biomarkers. With MRM protein assays, it is relatively easy to exchange peptides in and out of an assay. An example MRM protein assay development workflow is presented below.

- Fifty candidate biomarkers are discovered using GeLC/MS.
- A RQ assay is built for 25 of these proteins.
- Fifty samples are tested using the RQ assay.
- Of the 25 biomarkers, 14 are confirmed.
- The 11 proteins that are not confirmed are removed from the assay and 11 additional candidates are added to the assay.

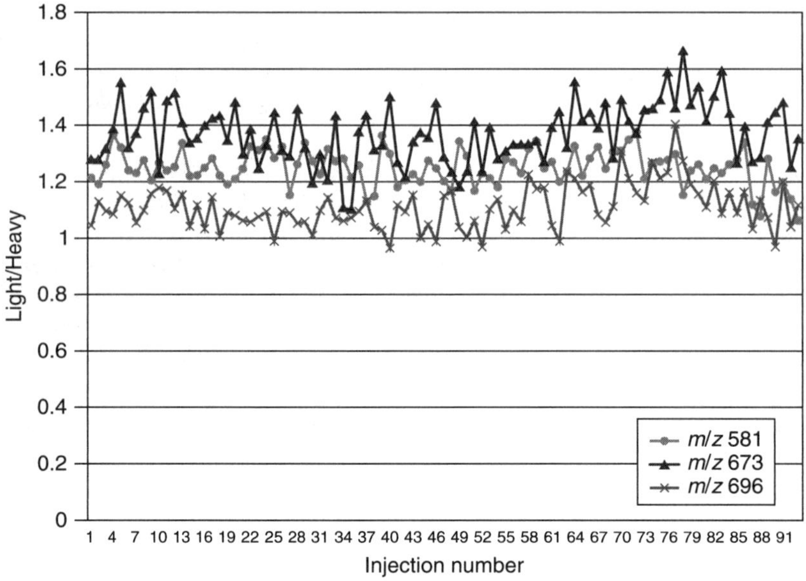

m/z 581		*m/z* 673		*m/z* 696	
Average	1.25	Average	1.38	Average	1.11
Standard deviation	0.06	Standard deviation	0.11	Standard deviation	0.08
%CV	5.03	%CV	7.73	%CV	7.07

%CV below 10 is very good

FIGURE 3.7 This figure shows the data for a test to determine the batch size for testing samples. The X-axis represents 90 different analyses of the same sample. The Y-axis is the response ratio value for each of the three peptides tested. Response ratio is the ratio of the peak area for the light peptide divided by the peak area for the heavy peptide. The table provides %CV for the three peptides. This is an example of a batch size of 90. Typically, this test would be repeated with the same number of injections (analyses) but with numerous samples to ensure that the system is stable with a variety of samples.

- The 50 samples are tested again; this time with the revised assay.
- Of the 25 proteins, 22 are confirmed.
- It is decided that the RQ assay will be converted to an AQ assay for 22 proteins.
- Labeled and unlabeled peptide standards for each peptide are ordered.
- Calibration curves are developed with each peptide standard.
- The AQ assay is tested with the calibration curves and with labeled peptide standards spiked into each sample.
- Fit-for-purpose assay validation is performed.

AQ assays provide protein concentration data. This is accomplished with standard curves (also referred to as *calibration curves*). For example, the use of a

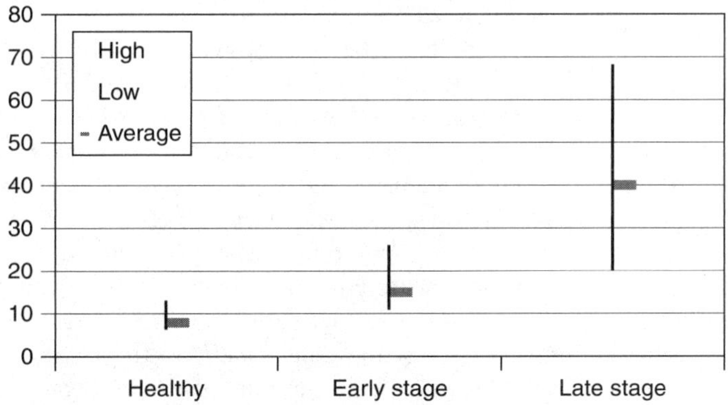

FIGURE 3.8 When an absolute quantitation MRM protein assay is used, the data values are the protein concentration, so that range plots like the one shown in this figure can be produced. The data for numerous samples in each of three groups (healthy, early stage, and late stage) are plotted as a vertical line (no points) to show the concentration range for the sample. The average value for each group is shown on each vertical line. The data for this particular peptide shows little biological variation among the healthy group and early-stage group, although the ranges for these two groups are different. The range for the late-stage group is much broader and the average is significantly higher than that of the other two groups.

standard curve would enable one to convert the relative data presented above to absolute protein concentration data. Most of the protein biomarkers that are monitored in medical practice are analyzed by quantitative assays. The protein concentration range for "normal" and "abnormal" can be defined. For example, Figure 3.8 shows a plot of the protein concentration range for a biomarker that could potentially be used to distinguish healthy individuals from individuals who are in early or late stages of a disease.

3.2.3 Assay Validation

For AQ MRM protein assays, validation of an assay beyond what is described in Section 3.2 is often desirable. The assay validation should be "fit for purpose," as assay validation adds cost to both the assay development and to the use of the assay for testing samples. Listed below are three levels of assay validation that can be considered.

1. *Validation of research projects*. Very little additional validation is required beyond what is described above.
2. *Validation for performing biomarker validation studies*. If the data is intended to be provided to a regulatory agency for determining if a biomarker is validated for a specific purpose, then the regulatory agency should be consulted to determine if they have specific requirements. In addition, these agencies often have a program for presenting preliminary

data on biomarkers, such as biomarker discovery data and data from commercially available antibody-based assays and/or RQ MRM protein assays. In general, the additional validation of the assay will include tests to demonstrate the specificity of the assay for the proteins of interest, the accuracy and precision of the assay, and the linearity of the assay (including both dilutional linearity and equivalence between the sample and the matrix used for the standard curves). Ideally, the matrix used for the standard curves (or calibration curves) should be a sample (or a mixture of multiple samples). However, because the samples contain endogenous levels of the same peptides (light peptides) that are being used to create the calibration curves, the data generated from these calibration curves will represent the sum of the amount of the endogenous peptides in the sample and the amount of light peptide spiked into the sample. Figure 3.9 demonstrates an example of this situation.

3. The most stringent assay validation is performed when the data will be used as part of a filing for an IND (Investigational New Drug) application or an NDA (New Drug Application) during drug development. A draft FDA Guidance for industries was issued in 2005 by the Center for Drug

No endo			
	HP	LP	RR
1	5000	600,000	0.008333
2	10,000	600,000	0.016667
10	50,000	600,000	0.083333
20	100,000	600,000	0.166667
100	500,000	600,000	0.833333

$y = 0.0083x - 2E - 19$

with endo @ 700 fmol/μl			
	HP	LP	RR
1	5000	4,100,000	0.00122
2	10,000	4,100,000	0.002439
10	50,000	4,100,000	0.012195
20	100,000	4,100,000	0.02439
100	500,000	4,100,000	0.121951

$y = 0.0012x - 3E - 18$

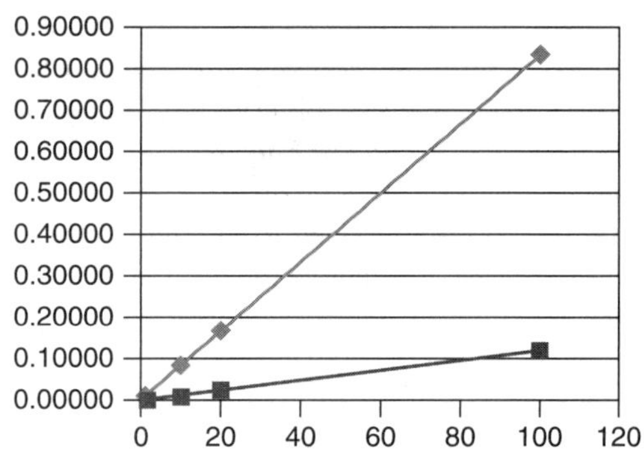

FIGURE 3.9 This figure provides an example of reverse curves used for calibration curves. Reverse curves are sometimes used when the matrix for the calibration curve is the same as the sample matrix. The top table is a chart of the amount of the light and heavy peptides added to the matrix for the calibration solutions. The bottom table is the same chart, but the contribution of the endogenous peptide from the sample; the response value for the light peptide is the combination of the addition of synthesized light peptide plus the endogenous peptide. The figure shows the difference in the calibration curve slope when sample matrix is used (line with small squares) compared to the slope when a proxy matrix is used (line with bigger squares). HP, heavy peptide; LP, light peptide; RR, response rate.

Evaluation and Research (CDER). This document, entitled "Safety Testing of Drug Metabolites," is being used as a guide for validating MRM protein assays with appropriate modifications to accommodate the fact that peptides are being monitored rather than drug metabolites (http://www.fda.gov/downloads/AboutFDA/CentersOffices/CDER/ucm119069.pdf).

Once the MRM protein assay is developed, it can be used repeatedly for testing biological samples.

3.3 TESTING

Figure 3.10 details an example workflow for testing samples with an MRM protein assay, including steps for obtaining AQ. Figure 3.11a is an illustration of the data from a RQ assay, which is reported as protein concentrations. Figure 3.11b shows an example of how the data from such an assay would be presented.

3.3.1 Statistical Analysis

Antibody-based and MRM protein assay data are provided in a format that is amenable to standard statistical analysis and multivariate analysis. The methods typically used for statistical analysis of data from testing are the same as those

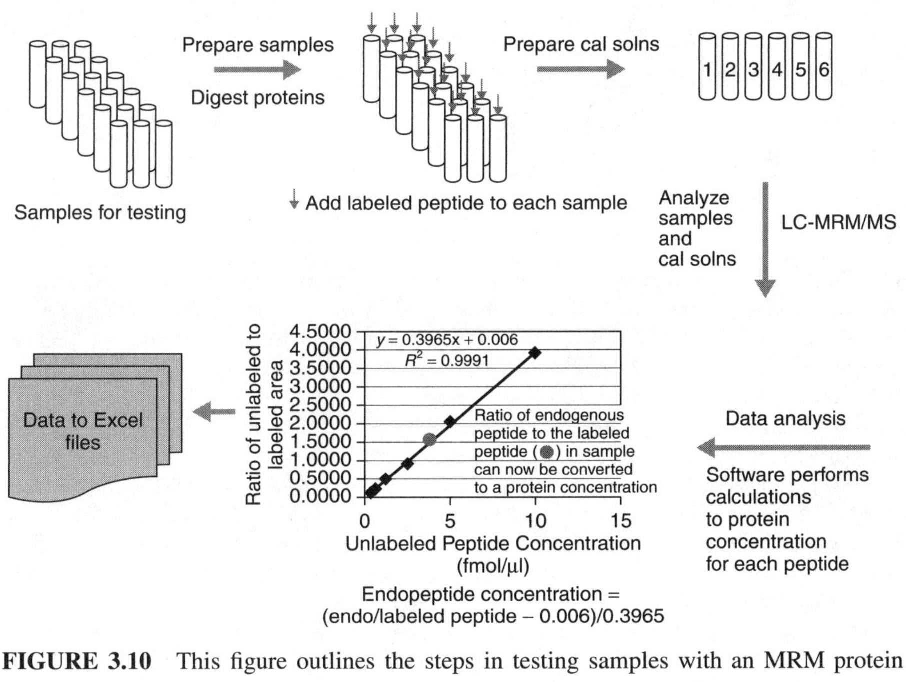

FIGURE 3.10 This figure outlines the steps in testing samples with an MRM protein assay.

used for biomarker discovery (see description of statistical analysis in Section 3.1). Additional methods are often used to assess the discriminatory power of the biomarker(s) and the selectivity and specificity of the biomarker(s).

3.4 SUMMARY

The solution—for a cure or for an effective management plan of many debilitating diseases that cost the society billions of dollars per year—is being able to

			Data from mass spectrometer		Calculation		
	Peptide	Sample	Sample analyte area (SAA)	IS analyte area (IAA)	SAA/IAA	Sample X/Ref	
Control samples	Protein 1_Peptide 1	1*	1,283,475	685,940	1.87	1.00	Different peptides for the same protein will have very different SAA numbers and SAA/IAA numbers (reflects ionization not abundance), but the Sample X/Ref should be similar for all peptides for the same protein
	Protein 1_Peptide 2	1*	3,086,764	706,518	4.37	1.00	
	Protein 2_Peptide 1	1*	98,437	672,221	0.15	1.00	
	Protein 2_Peptide 2	1*	476,883	727,096	0.66	1.00	
	Protein 1_Peptide 1	2	1,411,823	658,502	2.14	1.15	For this sample, the sample X/Ref value is very similar to sample 1 (the ref sample)
	Protein 1_Peptide 2	2	3,395,440	679,081	5.00	1.14	
	Protein 2_Peptide 1	2	93,515	740,815	0.13	0.86	
	Protein 2_Peptide 2	2	534,109	706,518	0.76	1.15	
Treated/disease samples	Protein 1_Peptide 1	3	3,208,688	631,065	5.08	2.72	For these two samples, the level of protein 1 goes up compared to samples 1 and 2 and protein 2, while the level of protein 2 goes down
	Protein 1_Peptide 2	3	9,568,968	720,237	13.29	3.04	
	Protein 2_Peptide 1	3	44,297	692,799	0.06	0.44	
	Protein 2_Peptide 2	3	267,054	665,362	0.40	0.61	
	Protein 1_Peptide 1	4	3,978,773	605,822	6.57	3.51	
	Protein 1_Peptide 2	4	8,025,586	691,428	11.61	2.66	
	Protein 2_Peptide 1	4	55,125	665,087	0.08	0.57	
	Protein 2_Peptide 2	4	214,597	638,747	0.34	0.51	

Sample	Sample analyte area (SAA)	IS analyte area (IAA)
1 is the ref sample	This number will be different for each peptide for the same protein, but should be similar for the same peptide across samples (except for biological variation)	This number should be very similar for each sample and each peptide because it is the IS added to each sample

(a)

FIGURE 3.11 This figure is a display of the data outputs from a set of samples tested with (a) relative quantitation and (b) absolute quantitation assays. (a) The column on the far left indicates which samples are from the control and treated groups. The peptide column indicates which row is for the four different peptides (two proteins) in the assay. The data for four samples is shown (samples 1–4). The two data values (peak areas) obtained from the mass spectrometer are the peak area for the peptide of interest (sample analyte area (SAA)) and the peak area for the internal standard (IS analyte area (IAA)). First the response ratio for each peptide is calculated (SAA/IAA). For relative quantitation assays, one sample is chosen as the reference sample (sample 1 in this case) and the relative quantitation ratio for other samples relative to the reference (sample X/Ref) is calculated. Protein 1 is very similar in the two control samples but has a higher level in both treated/disease samples. Protein 2 is very similar in the two control samples but has a lower level in both treated/disease samples. (b) This figure shows the results only for protein 1. In this case, the SAA/IAA value is used to calculate the protein concentration using a calibration curve (not shown).

	Peptide	Sample	Data from mass spectrometer: Sample analyte area (SAA)	Data from mass spectrometer: IS analyte area (IAA)	Calculation: SAA/IAA	Calculation: Protein Concentration	
Control samples	Protein 1_Peptide 1	1*	1,283,475	685,940	1.87	4.70	Each peptide has its own equation for calculating the protein concentration
Control samples	Protein 1_Peptide 1	2	1,411,823	658,502	2.14	5.38	For this sample, the concentration of protein 1 is very similar to sample 1
Treated/disease samples	Protein 1_Peptide 1	3	3,208,688	631,065	5.08	12.79	For these two samples, the concentration of protein 1 is much higher than that in Samples 1 and 2
Treated/disease samples	Protein 1_Peptide 1	4	3,978,773	605,822	6.57	16.55	
			↑ This number comes from the *endogenous peptide* & will be different for each peptide for the same protein but should be similar for the same peptide across samples (except for biological variation)	↑ This number comes from the *labeled peptide* and should be very similar for each sample and but different for each peptide. Note: only one peptide is shown here.			

(b)

FIGURE 3.11 (*continued*)

diagnose the disease with biomarkers. The current gap (Fig. 3.12) is the availability of biomarker tests that will identify the affected patients in the early stages of the disease.

Alzheimer's disease is one disease affected by this current gap in developing biomarker tests.

For example, Alzheimer's disease affects 10% of the population over the age of 65. This is a debilitating disease for the individual and the family and a costly disease for society (millions of dollars per year for care). A new medicine that would prevent this disease is the next potential blockbuster drug. Much has been learned about what causes Alzheimer's disease, and the pharmaceutical industry has excellent hypotheses about how to design medicines to halt the progression of the disease. In order to get approval to sell new medicines, pharmaceutical companies have to demonstrate the effectiveness of the new medicine.[1] Clearly, it is not difficult to diagnose a patient with late-stage Alzheimer's disease. But at this stage of the disease, irreversible damage has been done. In order to demonstrate the effectiveness, patients in early-stage Alzheimer's disease need to be identified. In order to identify these patients, biomarkers need to be discovered, assays developed, biomarkers confirmed, assays validated, and biomarkers validated and the test needs to be approved for use in patient testing.

[1]At this website, http://alzheimers.about.com/library/blbrain.htm, are images of a brain of a normal individual (left) and the brain of an Alzheimer's patient (right).

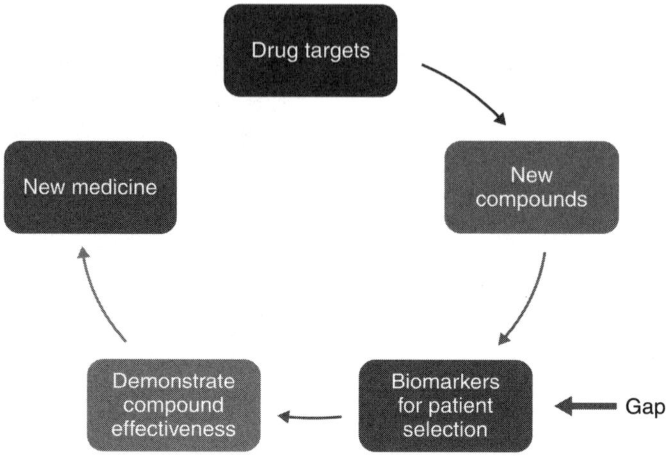

FIGURE 3.12 This figure shows the general steps in drug discovery and development, including a step for selecting the patients to demonstrate the effectiveness of the compound. It is this step that is the current gap in many drug discovery and development programs.

Osteoarthritis affects nearly 50% of the population, and in too many cases, the pain caused by this disease forces people into retirement in the prime of their lives. As with Alzheimer's, the treatment needs to start before severe damage has occurred. Biomarker tests are needed to identify patients with fast progressing forms of this disease.

About 30% of the population over the age of 75 is affected by age-related macular degeneration (AMD). AMD is another example of a debilitating disease that could be prevented with early diagnosis biomarkers. The current treatment is typically started once some vision loss has already occurred. The treatment is shots with hypodermic needles directly into the eye.

Cancer is another disease in desperate need of biomarkers for early diagnosis. In fact, there are many diseases that currently cannot be prevented. In addition, there are many diseases where the diagnosis is a subjective judgment rather than definitive data of a biomarker test. Examples include ADHD (attention-deficit/hyperactivity disorder), depression, and schizophrenia.

All in all, biomarker tests would benefit 90% of the diseases that affect humans. Biomarkers are currently in use for most research and development program and for evaluating patients. However, the number of biomarkers in use is still relatively small. In order to advance medical science to the new level of expectation, personalized medicine, the number of validated biomarkers needs to escalate to hundreds or thousands. In the past, the time line for validating even a single biomarker for a patient assessment was 10–15 years. Clearly, these time lines need to be shortened substantially (Fig. 3.13). A major bottleneck for protein biomarkers is assay development, which is the step in between discovery biomarkers and confirming biomarkers. Figure 3.14 provides a summary of the

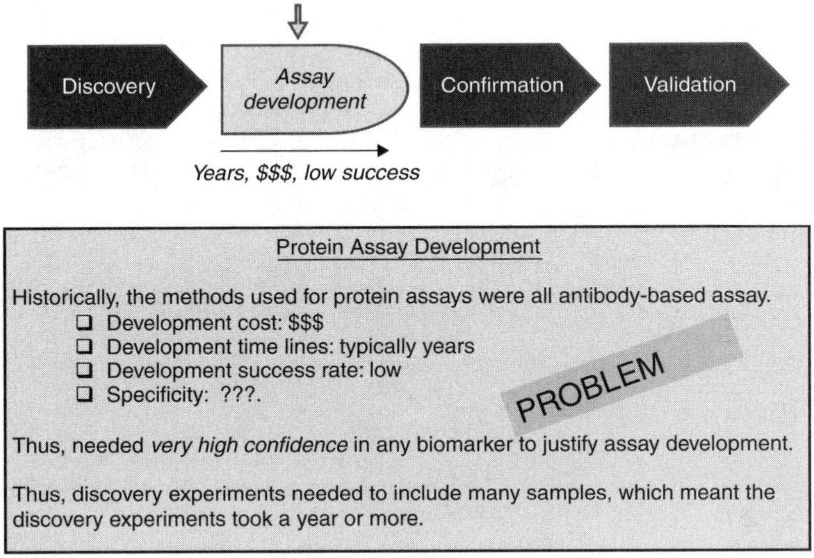

FIGURE 3.13 This figure is similar to Figure 3.1 but emphasized the intermediate step of developing a protein assay, which is the current bottleneck in biomarker development.

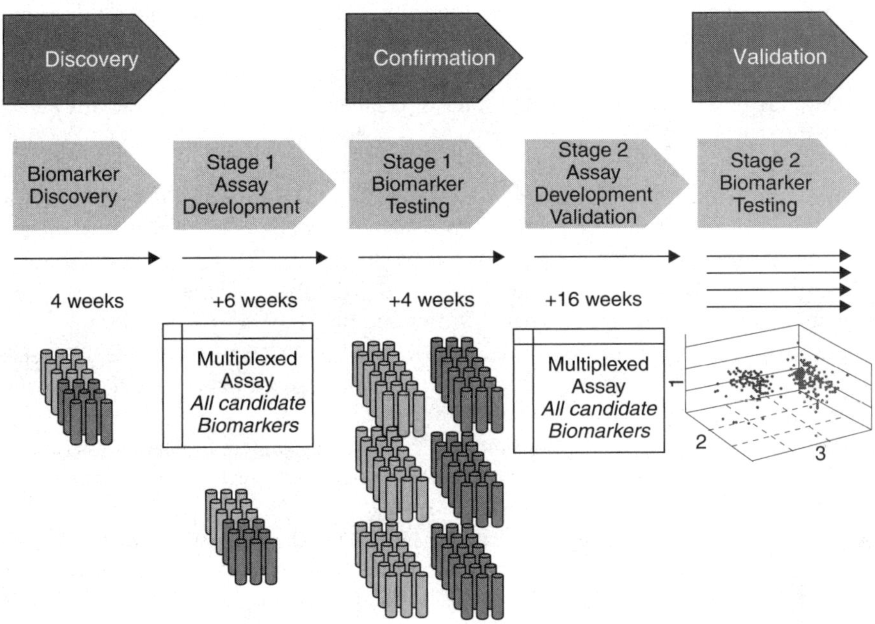

FIGURE 3.14 A workflow that addresses the bottleneck in biomarker development by decreasing the time line for protein assay development.

biomarker development workflow presented in this chapter, which reduces the time line from years to months.

The hPDQ project (7) has begun, and the goal of having a protein assay for every human gene is doable and, most importantly, is essential for medical science to move to personalized medicine programs and to cure the most common diseases of mankind.

ACKNOWLEDGMENTS

Thanks to Richard Jones and Michael J. Ford for providing some of the data and figures used in this chapter.

REFERENCES

1. Anderson NL, Anderson NG. The human plasma proteome: History, character, and diagnostic prospects. Mol Cell Proteomics 2002;1.11:845–867.
2. Ong SE, Blagoev B, Kratchmarova I, Kristensen DB, Steen H, Pandey A, Mann M, Stable Isotope Labeling by Amino Acids in Cell Culture, SILAC, as a Simple and Accurate Approach to Expression Proteomics. Mol. Cell. Proteomics 2002;1:376–386.
3. Cox J, Matic I, Hilger M, Nagaraj N, Selbach M, Olsen JV, Mann M, A practical guide to the MaxQuant computational platform for SILAC-based quantitative proteomics. Nat. Protoc. 2009;4:698–705.
4. Meng F, Wiener MC, Sachs JR, Bruns C, Werma P, Paweletz CP, Mazur MT, Deyanova EG, Yates NA, Hendrickson RC. Quantitative Analysis of Complex Peptide Mixtures using FTMS and Differential Mass Spectrometry. J Am Soc Mass Spectrom 2007;18:226–233.
5. Wilm M, Shevchenko A, Houthaeve T, Breit S, Schweigerer L, Fotsis T, Mann M. Femtomole sequencing of proteins from polyacrylamide gels by nano-electrospray mass spectrometry. Nature 1996;379:466–469.
6. Benjamini Y, Hochberg Y. Controlling the false discovery rate: a practical and powerful approach to multiple testing. J R Stat Soc Ser B 1995;57:289–300.
7. Anderson NL, Anderson NG, Pearson TW, Borcheers CH, Paulovich AG, Patterson SD, Gillette M, Aebersold R, Carr SA. A human proteome detection and quantitation project: hPDQ. Mol Cell Proteomics 2009;8:883–886.
8. VanBogelen RA, Jones RC, Qoronfleh MW. Managing protein biomarker assay workflow, multiple reaction monitoring platform addresses obstacles in multiplex testing. Genet Eng Biotechnol News 2009;29:34–35.
9. Anderson NL, Anderson NG, Haines LR, Hardie DB, Olafson RW, Pearson TW. Mass spectrometric quantitation of peptides and proteins using stable isotope standards and capture by anti-peptide antibodies (SISCAPA). J Proteome Res 2004;3:235–244.

4

OVERVIEW OF METABOLOMICS BASICS

QIUWEI XU AND WILLIAM H. SCHAEFER

4.1 INTRODUCTION

Metabolomics, or metabonomics, is quantitative metabolic profiling of endogenous metabolites that include enzymatic products resulting from gene expression (transcription, translation, and posttranslational modification), nutrients, xenobiotics, and products of gut flora. Metabolites are the functional readout of biochemical or metabolic pathways and are subject to internal perturbation by gene regulation and external stimuli such as diurnal changes, senescence, physiological state, diet, nutrition, stress, and the environment. Therefore, quantitative profiles of metabolite abundance yield molecular phenotypes that can be associated with underlying biochemical changes (such as gene mutation or regulation, enzyme inhibition or activation, changes in nutritional status, disease, or possibly changes in gut flora), and such phenotypes can be analyzed in the framework of systems biology.

The aim of metabolomics is to identify and quantify all metabolites (i.e., metabolome) in biological systems that can include whole organisms, cells, culture broth, tissues, or biofluids (e.g., urine, plasma, cerebrospinal fluid (CSF)) (1, 2). The metabolome includes all small molecules of endogenous metabolites including amino acids, sugars, nucleosides or nucleotides, lipids and fatty acids, steroids, cofactors, and intermediary metabolites in metabolic pathways such as the tricarboxylic acid (TCA) cycle, glycolysis, the pentose phosphate pathways, and fatty acid β-oxidation or synthesis. A multitude of metabolites are present in widely different biofluids (such as urine, blood, CSF, saliva) as well as in

Predictive Approaches in Drug Discovery and Development: Biomarkers and In Vitro/In Vivo Correlations, First Edition. Edited by J. Andrew Williams, Jeffrey R. Koup, Richard Lalonde, and David D. Christ.

tissues (such as liver, kidney, heart, skeletal muscle) but have the same chemical structures no matter where they are located.

The chemical nature of the metabolome is often more diverse than its counterparts of genomics and proteomics. Genes are linearly constructed from only four different bases (although some modified bases can be present) by phosphodiester linkages, and proteins are linearly built from 20 different amino acids in addition to several modified amino acids by amide linkages. Identical linear backbone linkages render genes and proteins decipherable by sequential analyzers. However, the metabolome is complicated by the diverse range of chemical structures and their vastly different physicochemical properties. In addition, quantification of endogenous metabolites is complicated because they can be present at drastically different concentrations spanning many orders of magnitude. On the other hand, the genome is made, for example, in humans, of about 23,000 protein-coding genes that yield an exceedingly large number of proteins with posttranslation modification. The estimated number of small-molecule metabolites is on the order of 2500–5000. From this perspective, metabolic profiling may seem relatively less complex.

The whole range of endogenous metabolites cannot be detected or quantified by a single modern analytical instrument because of the diversity in their physical-chemical properties and concentrations. Combinations of several complementary analytical techniques are often required for different types and concentrations of metabolites. There are generally two distinct approaches to conduct metabolic profiling. Metabolites can be profiled, identified, and quantified based on a chemical reference database that includes chromatographic retention time, mass spectra, and NMR spectra. Alternatively, samples may be profiled by capturing chromatographic retention time and spectroscopic features such as m/z values (in mass spectra), or line shapes and chemical shifts (in NMR spectra) that are used to group and contrast samples without identifying any chemicals initially. After the key changing metabolites (or "features") have been highlighted, they can be identified based on the reference compound databases. Metabolic profiling methodologies rely on the high sensitivity of the spectrometers coupled with their resolving power (in addition to the resolving power of coupled chromatographs), as well as their ability to detect a wide range of chemical species. Because of the extreme complexity of the acquired data, highly sophisticated data processing software and bioinformatics algorithms are critical.

Metabolites are often analyzed by two different approaches: targeted detection of a limited number of analytes of a specific class (e.g., fatty acids) and broad, unbiased profiling of all metabolites. Targeted analysis can be used for a small number of specific analytes that have been proven to be significant for their biological or diagnostic roles (such as high blood glucose indicative of diabetes, and high blood cholesterol signaling the risk of atherosclerosis). In these cases, analysis of a limited number of metabolites is adequate to characterize a particular disease or toxicity. Although particular molecules can be valued as important biomarkers for diagnosis, metabolic profiling enables broad coverage of temporal or spatial modulation of biochemical pathways, for example, due to pharmacodynamics, pharmacokinetics, or toxicity. This opens up the possibility of fully evaluating and understanding affected biochemical pathways or mechanisms, with the integration of transcriptomics and proteomics. Metabolic fingerprinting can be useful for quick grouping or classification based on a profiling analysis (e.g., bacterial classification based on membrane lipid profiles). The development of metabolomics and its application to diverse research areas (such as pharmacology and toxicology (3), disease diagnoses (4), diet and nutrition (5), plants (6), microbes (7, 8), metabolic engineering (9), and cells (10)) owe a large part to the development of modern LC/GC-MS and NMR.

Metabolic profiles in biofluids are favored for accessible biomarkers. Urine is an easily accessible biofluid. Under a normal physiological condition, only limited numbers of endogenous metabolites are excreted in urine. Many nutrients or essential metabolites that are often reabsorbed by healthy kidneys show up in urine only when renal toxicity is present. Overproduction of metabolites in other tissues can find their ways to plasma or urine; for example, dicarboxylic acids produced in peroxisomes because of inhibition of fatty acid β-oxidation in mitochondria are found in urine (11). Toxicity induced by hypoglycin A found in ackee fruit leads to urinary excretion of organic acids (12, 13).

Metabolic profiling of plasma and serum is often complicated by the presence of plasma proteins and lipids, and by the continuous effort of the body to maintain homeostasis of blood metabolites. Many metabolites are present in blood or plasma at low concentrations, but their changes in response to diseases or toxicity can be very useful for diagnoses. For example, small changes in blood glucose and cholesterol may show progressive deterioration in clinical conditions.

Cerebrospinal fluid is relatively difficult to collect and can often be only available in small quantities for small animals. In addition, the concentrations of CSF components tend to be low relative to plasma. However, at least 46 small molecules are identified in human CSF from metabolic profiling (14).

Metabolic profiling in tissues is valuable for investigating ongoing biochemical processes in that tissue (90). This is direct evidence of changes that can be associated with a disease state, therapeutic treatment, or toxicity. In addition, metabolic profiling in tissues can be a critical follow-up to biomarker detection in systemic fluids such as plasma and urine. This ensures that the diagnostic biological chemicals that are detected in plasma or urine actually originate from the tissue where the biochemical changes occur. However, snapshots of metabolic profiles in tissues absolutely require fast inactivation of enzymatic activities during tissue collection, storage at cryogenic temperature, and inactivation of enzymes (e.g., by perchloric acids or methanol) before tissue homogenation and metabolite extraction.

In vitro cell models provide simpler systems that can be amenable to metabolic profiles *in vivo* or *ex vivo* (138). The exact biochemical pathways or processes *in vivo* can sometimes be obscured by rapid exchange of endogenous metabolites from organs into continuous blood circulation and excretion from body. As a self-contained and closed system, metabolic profiles inside *in vitro* cells and in media are extremely useful in constructing biochemical sequences by pathway analyses.

4.2 ANALYTICAL METHODS

Comprehensive profiling of endogenous metabolites requires analytical techniques that are quantitative, and can detect a diverse range of chemical structures. ^{1}H NMR and chromatography-coupled mass spectrometry (GC-MS and LC-MS) meet these criteria, and are especially well suited for metabolomics profiling. GC-MS and LC-MS yield mass data (molecular weight and fragment spectra) and retention time that can provide structural information, or can be used to identify compounds based on a reference database. One dimensional ^{1}H NMR provides chemical shift and peak multiplicity information that allow structural identification of metabolites and quantification (1). After standardized analytical methods are developed and established, known reference chemicals can be analyzed to generate a reference database. This requires a substantial initial investment in chemicals and analysis time, but it is critical to facilitate rapid identification of metabolites for metabolite profiling. The use of standardized methods can also facilitate comparison of results between different laboratories.

The signals for metabolites generated by GC/LC-MS or NMR can provide a fingerprint or pattern that can be characteristic of a particular sample. Although fingerprinting can sometimes be used to differentiate samples, algorithms are often needed for pattern recognition and differentiation. Fingerprinting without chemical identification has inherent risks but can be minimized with positive identification of chemical structures. Pattern recognition based on peaks of unknown chemical structures can succumb to pitfalls of overlapping chromatographic or spectroscopic peaks and to variation of instrument settings.

Structure characterization of unknown chemicals enables results to be compared between different laboratories collected at different times and on different analytical platforms, although peak identification is often a time-consuming process and can require expensive instrumentation. A known structure also allows a link to pathway analysis and biochemical understanding. Most importantly, it contributes to holistic knowledge of underlying biochemical mechanisms (15). Although chemical structures of most endogenous metabolites are readily known or can be inferred from metabolic pathways, a few may still elude our comprehension because of their transient presence or trace amounts in biological systems.

Knowledge of chemical structures is important, but quantification is absolutely required to compare results between treatment and control groups, although relative quantification is useful and helpful for metabolomics. Absolute quantification facilitates comparison of results between different types of samples (e.g., cell, tissue, organ, and species) as well as between experiments conducted years apart. Comparison of a metabolite between two different organs (e.g., kidney and liver), or the same organ (e.g., liver) from different species, can be done based on the amount of metabolite in the sample normalized to the weight of the tissue sample analyzed. Establishing baseline levels of a metabolite in a normal cell or tissue under normal physiological conditions requires absolute quantification, as is the practice in clinical chemistry. Although clinical chemistry focuses on targeted analyses of specific analytes, metabolomics has the potential for broad quantitative metabolite profiling. Absolute quantification of metabolites inside a cell, for example, aids in understanding cellular processes based on their influence on free energy and rates of metabolic conversions (16). In addition, quantitative values provide a means to evaluate the reliability of the whole process (sample collection, preparation, and analytical quantification) of metabolic profiling. For example, in control liver samples, ATP levels that are less than 1 nmol/mg of liver raise concerns about quality of the tissue or sample preparation procedures. Even if a relative trend in the metabolic data is observed, it can be an artifact of poor sample quality. Quantitative analyses should aim at reducing process variability, expanding chemical diversity, and increasing dynamic range of measurements.

4.2.1 NMR

NMR is a nondiscriminative and nondestructive method that is very useful for global metabolite profiling (17–20). NMR measures molecules by exciting particular nuclei (such as ^{1}H, ^{13}C, ^{31}P, and ^{15}N) in a strong magnet using electromagnetic power, and detecting their signals decaying over time (called *free induction decay* (FID)) at radio frequencies. Given the uniformity of nuclei excitation with radio frequency (RF) power, molecules containing excited nuclei are readily detected regardless of chemical structures. Since NMR is nondestructive, the samples in the NMR tubes are not altered or consumed, and they can be used for further analyses by NMR or other analytical methods. A minimum requirement for NMR samples is often the addition of deuterated solvents (such as deuterium oxide and deuterated chloroform) to provide an electromagnetic lock to maintain a constant NMR field frequency during the analysis of a sample. The sensitivity of NMR analyses is often low because the population difference in nuclei aligning parallel and antiparallel to a magnet field is minuscule. The NMR detection levels range from micromolar to millimolar concentrations for chemicals in solution which is adequate for many important biochemicals. In addition, identification of metabolites with overlapping signals or those that are present in trace quantities can be challenging. Spectrum fitting or peak deconvolution has been developed to help overcome the problem of characterizing overlapping signals.

4.2.1.1 Magnets. It is essential to have a stable and homogeneous magnet facilitating batch NMR analyses. In addition, NMR sensitivity and spectral resolution depend on magnet field strength. High magnet field strength offers improved signal sensitivity and peak dispersion. In theory, both sensitivity and peak dispersion are proportional to NMR magnetic field strength. The commonly used NMR magnets for metabolomics include 11.7, 14.1, and 16.4 T. These field strengths are often referred to in terms of proton detecting frequencies, that is, 500, 600, and 700 MHz, respectively. Field strengths lower than these values are sometimes used but lack the sensitivity and resolving power that is advantageous for metabolomics analyses. Large magnets are sometimes demonstrated to be superior, although with much higher investment cost.

Besides field strength, field spatial homogeneity is also important in maintaining good signal-to-noise (S/N) ratios and robustness for analyses of large batches of samples. Magnetic field inhomogeneity experienced by molecules in NMR tubes often causes poor NMR peak line shape with broad line width, leading to severely compromised deterioration of S/N ratios. To compensate for magnet inhomogeneity, electronic coils are built into the upper barrel enclosing an NMR sample tube and generate a correcting and compensating electromagnetic field. This is often called as *shimming* in the NMR terminology. Another factor affecting field homogeneity is field drift. The magnetic field strength of a superconductive magnet field does not stand still, and its field strength gradually weakens ("drift"), although at an extremely slow pace (it takes several decades or more for a superconducting magnet to degrade to unacceptable field strength). An electromagnetic field is generated to compensate for the small decrease in the magnetic field during the analysis of a sample. This is often accomplished using an electronic feedback loop by keeping the deuterium frequency constant during the NMR analysis; therefore, deuterium solvents are generally needed to serve as a reference to "lock" the magnetic field.

4.2.1.2 ***Probes.*** Probes can be categorized based on sample accommodation (such as tube or flow probes) or thermal noise reduction (such as room temperature or cryogenic/cold probes). Probe selection requires consideration of the nuclei to be detected as well as sample size (e.g., 5- or 3-mm NMR tubes). The nuclei that are most often used for metabolites profiling include ^{1}H, ^{31}P, and ^{13}C in order of sensitivity. NMR probes contain a couple of concentric coils that send out electrical pulses and detect the signals after excited nuclei return to the equilibrium state in a magnet. The coil that detects the nucleus of interest (e.g., ^{1}H, or ^{13}C) is usually placed on an inner coil closest to a sample tube. For example, a ^{1}H coil is usually set at an inner position closer to a sample tube, and x-nuclei (e.g., ^{13}C or ^{31}P) are usually detected by the outer coils. If ^{3}C or ^{31}P is the nucleus of interest, the placement of its detecting coil on an inner coil helps improve the sensitivity of detection. Since the proton exhibits the largest intrinsic sensitivity among all NMR nuclei, direct detection of protons is often utilized for metabolomics; therefore, indirect detection is utilized for x-nuclei. Indirect detection of x-nuclei is accomplished using pulse sequences to detect proton signals directly, with x-nuclei detected indirectly after exciting both proton and x-nuclei, such as NMR spectra of HSQC (heteronuclear single quantum coherence spectroscopy) (21) and HMQC (herteronuclear multiple quantum coherence spectroscopy) (22). The advantage of indirect detection of x-nuclei through proton excitation is increased sensitivity that is proportional to the gyromagnetic ratios of proton over x-nuclei. The increased factor is equivalent to 4.0 for ^{13}C and 2.5 for ^{31}P. Additional enhancements can be gained by reducing background signals or noise observed when dominant proton signals that are chemically uncoupled to x-nuclei are filtered out. This leaves the x-nuclei-coupled proton peaks with improved S/N ratios. With full natural abundance of proton and its highest sensitivity for NMR detection, an inverse probe with direct proton detection is often a preferred choice for saving data collection time.

Selection of NMR probes to accommodate different tube sizes is usually determined by available sample volume or chemical concentration. For a given amount of chemical, concentrated samples dissolved in small volumes offer much improved S/N values and reduced data collection time to reach an acceptable S/N ratio when being analyzed inside probes of correspondingly small inner diameters. A sample with twice higher concentration will require four times shorter acquisition time. The improvement of sensitivity is due to an improved probe quality factor (Q), reduction of thermal noise arising from bulk solvent molecules, and good chance of magnetic field homogeneity in a small volume. Available biological sample sizes can vary greatly from abundant urine in rats or large animals to scant CSF or aqueous humor in mouse. Selection of tube or probe diameter is often constrained by practical limitations such as solubility of all possible chemicals in a concentrated solution, available options for handling and pipetting into small NMR tubes manually or with assistance of robots, and the cost of tubes. Both 3- and 5-mm tubes are often used for NMR metabolomics work. A typical 5-mm NMR tube requires about 600–700 μl of solution sample, but a typical 3-mm NMR tube needs no more than 200–240 μl. Therefore, selecting a 5-mm NMR probe is a compromise but with its versatility to accommodate either 5- or 3-mm NMR tube. It is common to exchange room temperature probes to accommodate different diameters of NMR tubes; however, changing cryogenic or cold probes can be delicate and very time consuming.

Cryogenic or cold probes offer additional improvements in S/N ratios relative to standard probes. The gain in sensitivity is achieved by cooling the receiver coils to cryogenic temperatures (e.g., 25 K) where the coils have low resistance and reduced thermal noise but high probe quality factor (Q) (23). Theoretical improvements can be as large as 4 (i.e., equivalent to a reduction of 16× in acquisition time). However, with high salt concentrations (e.g., 100–150 mM), the sensitivity improvement may drop to 2× (24). Varian's Salt Tolerant Cold Probe improves the S/N ratio of the cryogenically cooled probe by reducing the large electric field perturbation with modified probe geometry (e.g., squash-shaped or oblong cross-sectional NMR tubes) (25). Although squash-shaped tubes are preferred, normal tubes (such as 3, 4, and 5 mm) can be used as well.

4.2.1.3 NMR Tubes and Flow Cells. Routine NMR analyses can be carried out with either tubes of different sizes (such as 5, 3, 1.7 mm) or flow cells. Flow-cell-based NMR analyses can save the cost of NMR tubes and the time of sample transfers into NMR tubes. In addition, shimming is very consistent with the same fixated flow cell, as long as sample conditions (such as solvent, concentrations, ionic strength, stability) remain similar. Alternatively, tube-based NMR analyses can avoid cross contamination between samples and allow samples to be reanalyzed readily if needed.

4.2.1.4 Solvents (Buffers and References). Deuterium oxide is the most commonly used solvent for metabolic profiling of polar and hydrophilic metabolites. Buffering deuterium oxide with potassium or sodium phosphate helps keep the solution pH at a fixed value in order to minimize pH-dependent changes in chemical shifts. This is important for identification of chemicals based on closely matched peak positions (i.e., chemical shifts) and peak multiplicities. The buffer strength may depend on sample type. Urine may require strong buffering capacity (e.g., 100 mM phosphate), but CSF may need only a weak buffering solution (e.g., 40 mM phosphate). Chemicals such as DSS-d_6 (2,2-dimethyl-2-silapentane-5-sulfonate-d_6 sodium or 3-(trimethylsilyl)-1-propanesulfonic-2,2,3,3,4,4-d_6 sodium) and TSP-d_4 (3-(trimethylsilyl)propionic-2,2,3,3-d_4 sodium) are often used to define the zero ppm (parts per millionth of chemical shift in hertz). Any other peaks are labeled in positive ppm with respect to DSS-d_6 or TSP-d_4. The chemical shift values increase from right to left (e.g., 10 ppm on the left to 0 ppm on the right) (Fig. 4.1).

Quantitative analyses using NMR have the advantage of simultaneous and uniform response factors for each particular nucleus (e.g., ^{1}H, ^{13}C, or ^{31}P). The detected peak intensity (or area) is proportional to the amount of a particular nucleus (e.g., methyl protons) in a chemical (e.g., acetate) and the amount of the chemical in an analysis solution. In other words, protons have the same response factor no matter what chemicals are incorporated. Therefore, one single

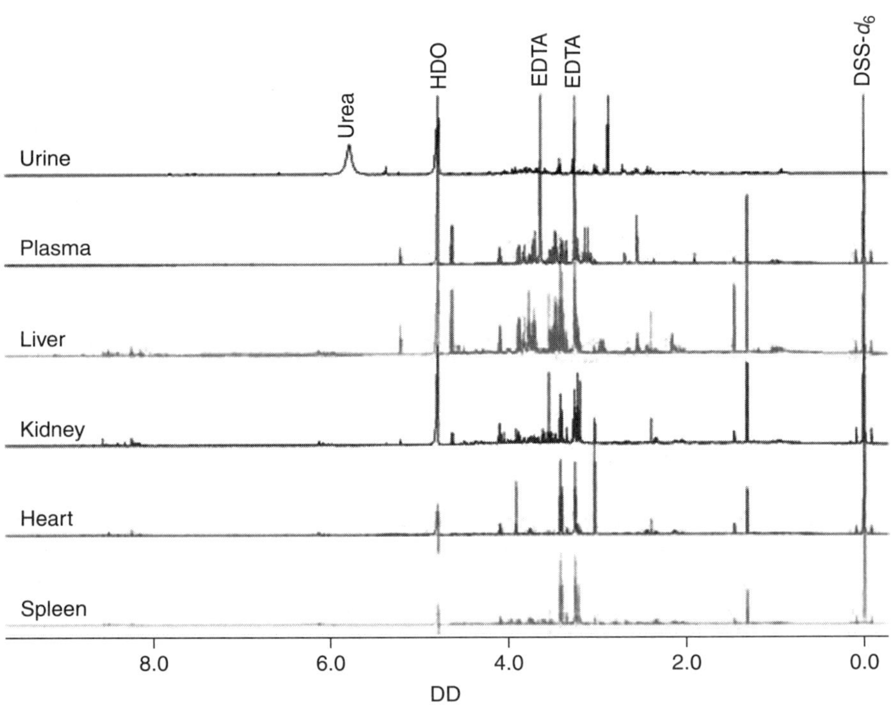

FIGURE 4.1 Typical 1D proton NMR spectra of mouse urine, plasma, liver, kidney, heart, and spleen (top to bottom) in deuterated phosphate buffer at pH 7. Lipids and proteins in plasma, liver, kidney, heart, and spleen were removed by 80% methanol.

internal chemical of known concentration can be used for quantification of all detectable chemicals in a solution. The same chemical (e.g., DSS-d_6) that is used as a chemical shift reference or any other excipient chemical can be used as a concentration (26). For example, deuterated phosphate buffers supplied by SigmaAldrich are quality controlled for consistent DSS-d_6 concentration as a quantification reference.

Excipient chemicals can be added to measure *in situ* solution pH. For example, the trimethylsilyl (TMS) group in difluorotrimethylsilanylphosphonic acid (DFTMP) can accurately measure solution pH between 4.3 and 8.2, with a root mean squared error (RMSE) of 0.02 pH units (27). With respect to the 0 ppm signal of DSS-d_6, TMS of DFTMP shifts between 0.195 and 0.225 ppm at a pH between 4.3 and 8.2. Divalent cations (e.g., magnesium and calcium) can strongly affect chemical shifts of chemicals, especially those that may form chelating complexes. Adding excipient chelating chemicals such as EDTA can reduce this chemical shift variation for chemicals such as citrate because of their interaction with Mg^{2+} or Ca^{2+} (28).

Analyses of lipids and hydrophobic chemicals require deuterated organic solvents such as chloroform, acetone, methanol, and acetonitrile. Absolute quantification based on an internal excipient chemical can be less accurate in volatile organic solvents (such as methanol, acetone, and acetonitrile); however, quantification based on relative peak areas is still reliable. Quantification in less volatile organic solvents such as chloroform can be carried out with an internal excipient chemical as a quantification (29).

4.2.1.5 Solvent Suppression. Concentrations of metabolites are usually an extremely small fraction of the solvent concentrations. For example, water concentration is 55 M (or 100 M protons), but detectable metabolites are usually in the low millimolar or micromolar range in biofluids. When organic solvents are used with NMR, they are often fully deuterated, so they have a minimal effect on proton spectra. Even though deuterated buffers can be added to biological samples, the water present in the original urine or plasma sample can still overwhelm the NMR detection receivers and make the small metabolite signals less detectable within the dynamic range of the spectrometer. Fortunately, the signal for water proton (or other solvent proton) can be suppressed by the spectrometer during excitation of nuclei before data acquisition to reduce these huge solvent signals and enable the detection of the small metabolite signals.

Several pulse sequences available for water suppression include WET (water suppression enhanced through T_1 effects) (30) and noesy1d (i.e., first time increment of 2D nuclear Overhauser effect spectroscopy (NOESY)). For reliable quantification, a pulse sequence should minimize perturbation of signals outside the water peak, and protons of different molecules should not exhibit different relaxation decay rates before detection. Although noesy1d is a highly reproducible and robust method to suppress the water peak and produce a flat baseline, the peak intensities of individual analyte signals are susceptible to alteration due to mixing times by a variety of mechanisms (e.g., different T_1 relaxation time,

NOE, and chemical exchange effect among protons in a molecule or between different molecules). WET avoids the above problems associated with noesy1d, but it requires good shimming and calibration on individual samples for acceptable water suppression. For automation, this requires stable instrumentation and accommodating software that can optimize individual samples to achieve effective water suppression.

A key first step with WET-based water suppression is optimal shimming to ensure consistent peak shape and narrow line width. A narrow line width for the water signal can reduce the excitation width, thus avoiding the suppression of peaks proximate to water (e.g., the anomeric proton of β-glucose). Effective suppression of the water peak, especially with a narrow suppression window (e.g., 50 Hz), requires careful calibration of the WET pulse power and frequency position to hit the water signal squarely and produce a spectrum with a minimum water signal. With such meticulous optimization, spectra with adequately suppressed water signal can be used for reliable metabolite quantification.

4.2.1.6 Automation. Automation is critical to facilitating metabolic profiling of tens or hundreds of samples daily. Equally important to the speed of the analyses are the reliability of operation (i.e., automation control, spectroscopy performance, variable temperature control, autoshimming), consistent spectral quality (i.e., peak line shape, water suppression), and robust spectral collection (i.e., consistency in data acquisition on different days and by different operators). It often requires seamless integration of software and hardware for both spectroscopy and a sample handling robot, a well-planned workflow of individual steps in hardware operation, and precise execution of pulse sequences for spectral calibration and data acquisition.

When large numbers of samples are being analyzed as a batch with automation, refrigeration of the samples on the autosampler (or robot) can be critical to preventing analyte degradation and bacterial growth. For example, Peltier coolers can maintain sample temperatures near refrigeration temperatures (4–8°C). Purging the sample racks with dry air or nitrogen gas can prevent condensation of moisture on the tubes. In addition, bacterial growth can be suppressed by adding deuterium oxide to the samples (e.g., 50% or above) (31), which are full of nutrients and an excellent growth medium.

Both Varian's AS768 and Bruker's SampleJet robots currently offer the ability to handle several hundred samples per batch, with an option of sample refrigeration. Varian's AS768 utilizes 4-inch × 5-mm NMR tubes in 96-well sample holders that are placed over Gilson's Peltier refrigeration units. It can accommodate as many as eight 96-well sample holders, equivalent to 768 sample tubes. NMR tubes of 5–3 mm (i.e., a 3-mm or smaller tube with a 5-mm stem on top to accommodate the AS768 robot for handling 5-mm tubes) can also be used for analyses of small sample volumes (e.g., <200–250 μl). The AS768 robot manages NMR sample loading and retrieval from the magnet as scheduled by the NMR spectrometer software (i.e., VnmrJ), as well as insertion into and extraction of the NMR tubes from the seven sample turbines. Bruker's SampleJet can handle five 96-well NMR tubes racks (i.e., 480 samples) using 4-inch NMR sample tubes of various sizes (from 1- to 5-mm tubes). Customized caps for those tubes bypass the need for sample turbines. Samples can be kept at 4–6°C in the SampleJet.

The ability to associate sample attributes and descriptions, or metadata (e.g., study numbers, animal numbers, treatment conditions, notebook pages), with the sample and the spectral data is very important when dealing with large numbers of samples. This is preferably done at the stage of sample submission. For example, a spreadsheet can be prepared to tabulate the sample positions in a 96-well sample holder and the corresponding information for each sample. The NMR automation interface can accept this information from the spreadsheet (e.g., as a tab-delimited text format). This differs from the open access automation (e.g., routine medicinal chemistry analyses) where samples and sample descriptions are submitted individually.

Spectral consistency and quality are absolutely critical for analyte quantification and metabolite identification. This must be managed through many steps along the analytical process including the collection and preparation of the samples themselves before data acquisition. Instrument setup is also important, including continuously updated gradient and shimming maps. Sample analysis should begin with automated gradient shimming, calibration of the RF pulses, and effective water suppression. These will ensure consistent peak line shape that will facilitate reproducible peak intensities with a calibrated 90° pulse and robust suppression of the water peak, with little perturbation on nearby peaks such as the anomeric proton of β-glucose. Individual spectra should be saved with unique identifiers for individual samples that can be later grouped and analyzed using batch data processing.

4.2.2 Mass Spectrometry

MS detects molecules after ionizing them based on their molecular mass. It offers high sensitivity (e.g., ion trap) in detection, large linear dynamic range (e.g., quadrupole) for quantification, and mass accuracy (e.g., time of flight (TOF), Fourier transform ion cyclotron resonance (FT-ICR), and Orbitrap) to facilitate both quantitative and qualitative analyses (32). For metabolite identification, in general, the approach of accurate mass measurement combined with fragmentation analyses is very powerful. In addition, the use of an isotope ratio filter is an effective technique to limit the number of analytes for interpretation.

Technological improvements through innovation over last decade have led to mass spectrometers with a variety of specific performance and analytical capabilities. The choices of which mass spectrometer to use can depend on predefined goals of studies (as well as availability); therefore, familiarity with the different types of MS instruments helps in the judicious selection of which mass spectrometer to use and in the development of analytical methods. The following sections highlight strengths and limitations of MS configurations that are commonly used for metabolomics.

4.2.2.1 Single-Staged MS

Single Quadrupole. The single quadrupole LC-MS is a simple MS instrument that is commonly used to obtain molecular weights for analytes and for quantitative analyses, thanks to its large linear dynamic range (usually three to five orders of magnitude) and the fact that it is readily coupled with HPLC or GC. Although it lacks the ability to perform selective MS/MS experiments like a triple quadrupole, it does have the capability to perform in-source collisions to obtain dissociation data by adjusting specific voltage settings in the ion source. Selected ion analyses can provide increased sensitivity for quantitative analyses by monitoring a limited number of molecular ions rather than scanning over a range of masses. In this way, a quadrupole MS acts as a mass filter, only allowing ions to pass for a given RF and voltage setting. Some LC-MS instruments are capable of quickly switching between positive and negative ion modes, with alternating scans. This broadens the coverage of analytes that can be detected in a single analysis, and it can be very efficient for metabolic profiling.

Ion Trap. Ion trap MS instruments provide good reproducibility and sensitivity (33) and are valuable for profiling and structure elucidation. They are capable of performing multiple-MS/MS experiments (i.e., MS^n) with MS/MS (or MS^2) on a parent ion with subsequent MS/MS (MS^3, MS^4, etc.) on product ions that are formed with good scan speed (e.g., 6 scans/s) and wide dynamic range (three to four orders). The instruments usually have automated data-dependent routines for automated MS/MS (product ion scan) analyses, where a full-scan MS spectrum can be acquired and the data system can perform MS/MS on the most abundant ions in that spectrum. This can be done by synchronizing with chromatography owing to its fast scan speeds and the lack of necessity of introducing any additional collision gas. MS/MS product ion spectra provide information about molecular structure on the basis of unique fragmentation patterns of a parent ion. MS^n can be used to decipher fragmentation pathways and mechanisms for further structure elucidation. Because ion traps are primarily scanning instruments, there is often little enhancement of signals by performing experiments with selected ion monitoring (SIM) or selected reaction monitoring (SRM) that is often used for quantification on quadrupole instruments.

Time of Flight (TOF). Time-of-flight MS offers higher resolution and higher mass accuracy than the single quadrupole instruments. TOF-MS instruments are often coupled to either GC (e.g., Waters Micromass GCT mass spectrometer or LECO Corp. GC/TOF-MS) or HPLC (e.g., Micromass LCT Premier) (34). TOF instruments have rapid scan times that are amenable for chromatographic interface, and they offer very high sensitivity because of their high duty cycle. Another very valuable feature of TOF-MS instruments is their ability to operate at high resolution and acquire accurate mass data. Accurate molecular mass data can be used to determine the empirical chemical formulas for test analytes, and they can be especially useful for characterizing unknown metabolites. In addition, the high mass resolution can help resolve metabolites that may be coeluting chromatographically and have the same nominal (i.e., integer) mass but have different chemical formulas and different accurate masses. For example, the Waters Micromass LCT oa-TOF (orthogonal acceleration-TOF) is such a stand-alone benchtop TOF-MS coupled with LC-electrospray (35). It has two orthogonal paths, and they are independent for continuous ionization and TOF mass analysis. Ions are passed from one path to the other with a reflector. Its parallel detection of all masses in a spectrum allows fast acquisition (e.g., a full spectrum of large mass range can be recorded every 50 μs), thus enabling collection of a good number of data points across a narrow peak produced by a highly efficient separation chromatography such as UPLC (ultra performance/pressure liquid chromatography).

FT-ICR. Fourier transform-ion cyclotron resonance-mass spectrometry (FT-ICR-MS) instruments are capable of operating at even higher resolution than TOF instruments. They can be especially valuable for rapid profiling analyses with flow injection that does not have the benefit or burden of a chromatographic separation. The high resolution of FT-ICR instruments can also facilitate extremely high mass accuracy within 1 ppm. Molecular formulas with the same nominal mass (i.e., integer mass) but slightly different exact masses can be differentiated, for example, down to four decimal places for a nominal mass of 100. The power of resolving exact masses depends on the strength of a superconducting magnet. For example, a 14.5-T hybrid linear quadrupole ion trap FT-ICR-MS can reach mass accuracy less than 0.3 ppm and 200,000 resolving power (36). However, the benefit of high mass accuracy comes with a high cost of a superconducting magnet, routine maintenance of liquid helium, and liquid nitrogen refills.

Orbitrap. Like FT-ICR-MS, an Orbitrap mass analyzer (Thermo Finnigan, San Jose, CA) can provide very high resolution and mass accuracy (i.e., <2 ppm) (37, 38). It also has the advantages of being free of a superconducting magnet and the associated cost of refilling with liquid nitrogen and helium. Ions are instead restrained by a radial electric field in space around a special spindle-shaped electrode. The ion masses are calculated by Fourier transforming the superimposed frequencies of the harmonic oscillations along the electrode.

4.2.2.2 Tandem MS Tandem arrangements of the single-staged mass spectrometers described above can provide mass spectrometers with great agility for elucidating chemical structure by selectively scanning for precursor and product ions or neutral loss, as well as improved sensitivity and specificity using SRM (also called *MRM* (*multiple reactions monitoring*)) for quantification. Various tandem mass spectrometers are offered by major vendors. Those that are often used for metabolomics include a triple quadrupole, quadrupole-linear ion trap, quadrupole-TOF (Q-TOF), ion trap-FT-ICR (LTQ-FTICR) MS, and ion trap-Orbitrap (LTQ-Orbitrap).

Triple Quadrupole MS. Triple quadrupole MS can perform MS/MS for structural elucidation using a number of different scanning functions (39), including product ion scans, precursor ion scans, and neutral loss scans. For a product ion scan, the instrument selects the molecular ion of interest in the first quadrupole, dissociates (or fragments) it in the second quadrupole, and scans the product ions in the third quadrupole. Thus, a product ion spectrum shows the ions that arise from a given precursor ion. A precursor ion scan determines which precursor (or parent) ions give rise to a particular product ion. In this analysis, the first quadrupole scans through the mass range, the second quadrupole dissociates the scanned ions, and the third quadrupole fixes on the mass of the product ion of interest. This is very useful to identifying all the components in a mixture that have a common chemical moiety. A related experiment is a neutral loss scan, in which both the first and third quadrupoles scan with their mass offset by the value of the neutral loss. Thus, a precursor ion that fragments with the loss of a neutral component of interest will give rise to a corresponding product ion that is detected in the third quadrupole. Both scans of precursor ion and neutral loss provide information about analytes in the sample that have common structural moieties and can be tracked through fragmentation as either a loss of a charged fragment (precursor ion scan) or an uncharged (or neutral) fragment (neutral loss scan). Neutral loss can be quite informative to identify, for example, glycine conjugates of short-chain organic acids whose levels can be elevated in the urine from patients experiencing toxicity due to inhibition to fatty acid β-oxidation (13).

A triple quadrupole MS is also a key instrument for highly sensitive and selective targeted metabolite profiling. The quantification of targeted metabolites with predetermined molecular fragmentation patterns can be facilitated by selected reaction monitoring (SRM) or multiple reaction monitoring (MRM). If chromatography is used, the specific transitions in SRM or MRM must coincide with the correct retention time, adding further specificity to the analysis. Each SRM transition for each analyte takes between 0.02 and 0.1 s. Multiple SRM transitions (MRM) can be set up to run in sequence throughout an analysis, or specific transitions can be analyzed at specific times during a chromatographic separation.

Linear Ion Trap MS. A linear ion trap has the advantage of being able to function as either an ion trap or a quadrupole mass analyzer. A tandem linear quadrupole ion trap mass spectrometer combines quadrupoles for ion separation and detection with a collision cell (in the second quadrupole) with the third quadrupole that can function as either a conventional quadrupole mass analyzer or an ion trap for MS/MS analyses. The increased capacity of ion trapping by linear ion traps is facilitated by its geometry and increased trapping volume in the center of a quadrupole in comparison to a circular ion trap. This configuration improves sensitivity for detecting minor components in metabolite mixtures.

Quadrupole-TOF. Quadrupole-TOF MS instruments offer high sensitivity, mass spectral resolution, and ion mass accuracy (40). It is a sequential arrangement of a quadrupole mass analyzer, a collision cell (second quadrupole), and a TOF mass spectrometer. Molecular ions of interest (or precursor ions) are selected in the first quadrupole, and their fragmented product ions are detected in the TOF. It is very commonly used for both metabolomics and proteomics because of its high resolution, mass accuracy, and sensitivity. These instruments are capable of providing profiling data, MS/MS structure data, and accurate mass measurements that are used to determine empirical formulas.

Ion Trap-FT-ICR or Ion Trap-Orbitrap. The mass spectrometers that are capable of the highest resolution and mass accuracy are those based on FT-ICR or Orbitrap technologies. These instruments offer high resolution to resolve analytes that might have the same nominal mass but differ in their empirical formulas, and therefore their exact mass (measured out to three or four decimal places). On lower-resolution instruments, these isomers may overlap and not be distinguished. These ultrahigh resolution instruments help ensure reliable quantification of complex mixtures of metabolites from full-scan MS spectra and potentially lessen the demand on LC (elution time and chromatographic conditions) to resolve overlapping m/z peaks. It should still be emphasized that analytes with the same chemical formula will have the same exact mass, and therefore require chromatographic separation to be resolved. This is especially important for the analysis of lipids and other small-molecule metabolites; their structural isomers with the same chemical formula may be present in a tissue or cell culture sample. Linearity can cover at least two to four orders of magnitude, facilitating accurate quantification of a broad spectrum of metabolites with varying concentrations. Coverage of a broad range of molecular ions in a single LC-MS spectrum allows quantification of many metabolites over a wide m/z range, especially with isotope dilution (i.e., quantitative analyses with spiked isotope-labeled metabolites) to account for ion suppression (41).

4.2.2.3 Ionization A critical requirement for mass spectral analysis is that the analytes must be charged. Mass spectrometers operate by steering analytes in an electromagnetic field. Characteristics of their trajectory paths or kinetic rates in the electromagnetic field are exploited different mass analyzers in different ways to determine their mass. Several different mechanisms are commonly used to impart a charge on an analyte for detection. The different mechanism or energy imposed on the analytes has a profound effect on the nature of the mass spectral analysis and the chemical nature of analytes that are amenable. For example, high energy ionization techniques can cause an analyte to dissociate spontaneously into fragment ions that are not generally useful for large but chemically labile analytes; the so-called soft ionization techniques are much more gentle, and they allow an analyte to remain intact and stable through the ionization process and analysis. These soft ionization techniques are useful for a diverse range of analytes spanning from small molecules to intact proteins and nucleic acids. Because little or no fragmentation occurs with soft ionization, tandem MS techniques are generally needed to obtain structural information for these analytes.

The development of ionization techniques that are compatible with the different chromatographic separation mobile phases has been critical. Since mass analysis must occur in vacuum, interfacing the chromatographic mobile phase is a challenge. GC, which uses nitrogen or helium as a mobile phase, can be interfaced with electron impact and chemical ionization (CI) that occur within the vacuum of the mass spectrometer. Vacuum systems can adequately pump away the gas from the mobile phase introduced into the vacuum. Introducing a liquid mobile phase into the vacuum of a mass spectrometer is challenging. Liquid solvents can expand 700- to 1000-fold in volume when they change phases from liquid to gas. Early approaches of ionization under vacuum for interface with HPLC included thermospray and flow fast atom bombardment (42–44). While successful, these approaches lacked the sensitivity and robustness that were needed for high throughput and high performance operation. Atmospheric pressure ionization (API) techniques provide huge advancements that have made LC-MS a successful technology today. Ionization outside the vacuum of a mass analyzer facilitates a simple and robust interface to MS for liquid-based separation approaches, including HPLC, capillary electrophoresis (CE), and electrophoretic chromatography (EC). Effluents from LC, CE, or EC are continuously sprayed, and molecular ions are formed or extracted under atmospheric pressure before being drawn to the vacuum of MS. It can handle liquid flow rates of typical LC (<2 ml/min), and suit metabolites that are nonvolatile, of medium-to-high polarity, and thermally unstable. Commonly used API approaches include electrospray ionization (ESI), atmospheric pressure chemical ionization (APCI), and atmospheric pressure photoionization (APPI). Most commercial MS systems are often equipped with more than one option, and users can readily switch between them.

Electrospray ionization provides soft and mild ionization of a widest range of analytes, including small molecules, lipids, peptides, intact proteins, and nucleic acid oligomers. Because of this versatility, ESI is commonly used for metabolic profiling, and it preserves the integrity of metabolites allowing detection of intact molecular ions and offers superior sensitivity especially for ionic or polar metabolites soluble in aqueous solvent mixtures. Placing a high voltage (3–5 kV) on the effluent of HPLC in combination with a nebulizing gas introduced in parallel with the liquid effluent results in a spray of highly charged small droplets. As the droplets evaporate, the charge becomes concentrated and charged analytes are ejected into the gas phase. In concentrated samples, there can be competition for ionization between coeluting analytes, which can lead to suppression of ionization of some analytes. Ionization is usually related to solution chemistry and generally results in addition of a proton for positive ions and loss of a proton for negative ions. Occasionally, one-electron or radical-based ionization can occur, but this is relatively rare with ESI. Molecules with multiple sites for protonation or deprotonation can form multiply charged ions. This phenomenon is exploited in the detection of proteins and nucleic acids whose molecular weights are beyond the mass range of the mass spectrometer. Since mass spectrometers do not actually detect mass, but rather mass/charge (m/z), multiple charging of ions can readily bring the detected m/z value of a protein to within the mass range of a conventional instrument. In order to achieve the broadest detection of the widest range of metabolites, a sample should be analyzed in both positive and negative ion modes. Typical chromatography mobile phases are, for example, 0.1% aqueous formic acid, 10 mM ammonium acetate (pH 5–7), and ammonium formate (pH 4–7) with methanol or acetonitrile as a modifier (45). Lower pHs can help ionization in a positive ion mode, and higher pHs can help ionization in a negative ion mode. Electrospray ionization is applicable to most mass spectrometers, including single and triple quadrupoles, quadrupole ion trap, TOF, Q-TOF, FT-ICR, and Orbitrap.

APCI and APPI complement ESI by ionizing the samples containing less polar metabolites such as triglycerides, steroids, eicosanoids, flavonoids, and sugars. In APCI, effluents from HPLC are passed through a probe that is at high temperature (e.g., 500°C), with a nebulizing gas to evaporate the HPLC solvents. This gas phase effluent is passed by a corona discharge needle that begins the ionization process with ionization of gas molecules. A cascade of ionization and proton-transfer reactions results in either addition of a proton to analytes in a positive ion mode or abstraction of a proton in a negative ion mode. Atmospheric pressure chemical ionization generally results in singly charged ions, and is not amenable to high molecular weight analytes that require multiple charges for detection within a conventional mass range. The APCI process is less mild than ESI, but it still generally yields intact molecular ions with little in-source fragmentation. Highly thermally labile analytes can degrade with APCI, but generally, the high heat is not such a problem because the extensive evaporation actually keeps the sample relatively cool (46). Atmospheric pressure photoionization imparts a positive charge to a less polarized molecule with a photon emitted from UV source (47). Direct photoionization (PI) is often less successful due to problems that arise from solvent molecules competing with the analyte for ionization and the likely high ionization energy threshold for analytes of interest. An intermediate chemical called a *dopant* (e.g., toluene and acetone) can be added to facilitate ionization of analytes and can significantly increase ionization efficiency by charge exchange or proton transfer.

Electron impact ionization (EI) is done in the ion source that is inside the vacuum chamber of the mass spectrometer. It is commonly used with GC-MS for ionization of volatile analytes or their derivatives. Since compounds must be volatile to be separated by GC, polar analytes must generally be derivatized to increase their volatility, which adds steps to the sample preparation process of GC-MS. Derivatizations for GC-MS have been highly developed over the years, and routine procedures are available for metabolomic analyses (48). Electron ionization is a harsh ionization process that often results in extensive fragmentation of analytes. In some cases, intact molecular ions are not detected. An electron beam of 70 eV knocks off a single electron from an analyte in a gas phase to form a positive molecular ion (a one-electron process yielding a radical cation). Many molecular ions disintegrate in less than a millisecond to produce positively charged fragment ions and radicals. This EI fragmentation process is highly reproducible and robust, and the pattern and intensity of fragment ions are often unique for a given analyte. This has led to the compilation of extensive databases of EI mass spectra for many chemicals, and these spectra can serve as fingerprints to aid in compound identification. A library such as the NIST

(National Institute of Standards and Technology) GC-MS library with distinctive molecular and fragment ions can assist in metabolite identification.

Another ionization technique that is done in the vacuum chamber is chemical ionization (CI), and this is also commonly used with GC-MS. With CI, a reagent gas (e.g., methane, isobutene, or ammonia) is infused into the ion source. This reagent gas is ionized by an EI filament and initiates a cascade of reactions that result in ionization of sample analytes by addition or abstraction of a proton producing singly charged molecular ions (similar to APCI). Chemical ionization is gentler than EI and tends to yield intact molecular ions with less fragmentation. It also usually yields lower sensitivity than EI; therefore, the trade-off between reduced fragmentation and decreased sensitivity must be considered. It should be noted that ultrahigh sensitivity negative chemical ionization (NCI) methodologies are available that take advantage of the high electron capture affinity of specific derivatization moieties (49).

4.2.3 Chromatography

Analysis of complex mixtures, such as those from biological samples, involves resolving a wide range of different analytes that are present in concentrations spanning many orders of magnitudes. Although high resolution mass spectrometers can resolve many analytes based on masses, isobaric isomers (i.e., compounds with the same nominal mass and retention time that could be neither distinguished uniquely nor quantified independently) cannot be differentiated and analytes of low levels can be very difficult to identify among the more abundant analytes without chromatographic separation. The application of a chromatographic separation in front of a mass spectrometer can enhance the quality and quantity of data acquired on a sample for several important reasons. First, a chromatographic separation can be optimized to separate structural isomers that have the same chemical formula (and, thus, the same mass). Although MS/MS analyses can be used to determine chemical structures, interpretation of spectra of mixtures of (often related) isobaric compounds is complicated, at best. Second, a chromatographic separation helps reduce the number of analytes that are reaching the source and being ionized simultaneously at a given time. This helps reduce ionization suppression of low abundance analytes or those with lower ionization affinity and yield more reliable and accurate quantitative results. Chromatographic separation can help clean the sample by separating analytes from salts, polymers, or other contaminants that may be present in high concentration and affect ionization efficiency and reproducibility. Simplifying the chemical complexity at the source also helps increase the effective dynamic range, allowing low level metabolites to be more apparent and detectable among a smaller number of more abundant analytes. Third, chromatographic retention time is a key parameter that can be used for identifying analytes when the retention time of an authentic standard is obtained with the identical chromatographic conditions. Fourth, the ability to reconstruct ion chromatograms for analytes of interest, especially low level analytes, is very helpful to distinguish "real" signals of low level analytes from noise. A reconstructed ion chromatogram that shows a peak with an appropriate Gaussian peak shape is likely to be a real analyte, instead of a background ion that appears throughout the entire chromatogram. In addition, the shape of the peaks can give an indication whether a single component is present or multiple components that are only partially resolved. HPLC separations can be based on molecular polarity, hydrophobicity, or ionic properties, or combinations of these. Chemical derivatizations can also be used to improve chromatographic separations or enhance the ionization (and sensitivity) of analytes that are poorly ionizable.

For comprehensive metabolomics profiling, several chromatographic methods are often required to accommodate the wide range of chemical polarities of biological analytes. As mentioned above, the most common chromatography technologies used with MS for metabolomics are HPLC (high pressure liquid chromatography) and GC (50, 51). Capillary electrophoresis and electrochromatography are applicable but tend to be used less widely (15, 52).

HPLC can handle endogenous metabolites of a wide range of physical and chemical properties by using different columns and solvent systems. In addition, it does not require chemical derivatizations that are often required for GC. Analytes with moderate to highly hydrophobic can be separated on reversed phase HPLC (RP-HPLC), while highly polar, water-soluble molecules can be separated using hydrophilic interaction liquid chromatography (HILIC).

UPLC, or ultrahigh performance/pressure liquid chromatography (UHPLC), is a technological advancement over conventional HPLC, and it utilizes columns with packing material of very small particle size (<2 μm) for improved separation efficiency. In order to pump the solvents through these columns, pumps had to be developed that can operate at pressures as high as 15,000 psi. The chromatographic improvements include shorter run times and improved separation of analytes. The chromatographic peaks are also sharper, which increases sensitivity and chromatographic resolution. Improved resolution helps separate potential isobaric compounds. The efficient chromatographic separation with reduced peak dispersion by UPLC helps reduce ion suppression due to coeluting molecular ions during ionization (53).

One such example of taking the advantage of high resolution of UHPLC in resolving metabolites and documenting retention time for metabolite identification is give by Evans and her colleagues at Metabolon. They established an in-house LC-MS reference library based on 1500 authentic chemicals (45). Each reference compound was defined by its retention time or retention index, m/z, and MS/MS spectra including associated adducts, in-source fragments, and multimers. Narrow peaks by UHPLC made it possible to improve the identification of potential chemicals based on stringent search criteria (e.g., within 5 s retention time, and 0.4 m/z for precursor mass) and reduction of potential coeluting chemicals. Using this approach, Evans et al. showed that 339 chemicals could be identified in extracts of plasma samples with UHPLC, where only 159 were identified with standard HPLC technology. The improved resolution of chemicals on UHPLC reduced the chance of encountering coeluting isobars (i.e., isobaric compounds). In Evans' example, the nine isobars observed with standard HPLC were reduced to only one with UHPLC. Improved peak resolution with UHPLC also reduced the chance of ion suppression and improved peak shape, chromatographic reproducibility, and metabolite quantification.

Recently, C18 monolithic silica capillary columns (90 cm in length, 0.2 mm inside diameter, and 0.35 mm outside diameter) have demonstrated improved chemical resolution for many analytes (54, 55). About 200 distinct components from leaf extracts of *Arabidopsis thaliana* were resolved for identification.

Separation of water-soluble polar molecules can be difficult using reverse phase chromatography since they are often not retained without using ion pairing that can interfere with mass spectral ionization. Hydrophilic interaction liquid chromatography (HILIC) has proven to be a very useful approach for such polar compounds (56, 57). As an orthogonal approach to reversed phase chromatography, HILIC uses either silica or derivatized silica (e.g., aminopropyl silica, glycol silica) along with a mobile phase with low water content for retention on the column and higher aqueous content to elute analytes. Typical solvent systems can be similar to those used for reverse phase HPLC, except the organic (lower polarity) solvent is used to retain analytes, and aqueous content (increased polarity) is used for elution. HILIC solvents generally have a pH modifier (e.g., 0.1% aqueous formic or acetic acid with acetonitrile or methanol, or 10 mM ammonium acetate or ammonium formate in a 95 : 5 methanol/water solvent mixture). Molecular resolution relies on the stationary aqueous layer on the silica surface into which polar molecules partition from the mobile phase.

Nano-LC offers some advantages over HPLC with more conventional column diameters (e.g., 1- and 2-mm columns) in terms of increased sensitivity, decreased sample size requirements, and decreased ionization suppression; however, it does require meticulous technical attention, and lacks the robustness associated with conventional HPLC columns (58). Application of Nano-LC/MS has been reported for the analysis of polar anionic and cationic metabolites in cells, brain tissues, and CSF (59, 60) where sample size is limiting.

Capillary electrophoresis is done using an open capillary tube, and the separation is based on the differential movement of molecular ions in solution by an electric field and the electroosmotic flow (EOF) (61). The silanol groups on the wall of fused silica of a capillary have low pK_a, and they are negatively charged at pH above 3. Hydrated cations readily form a static inner layer (called the *Stern* or *Helmholtz layer*) adjacent to the negative surface of the fused silica. In the presence of an electric field, hydrated cations in a mobile outer layer (called *Gouy–Chapman layer*) that is adjacent to the static cation layer move toward the cathode under the influence of the electric field. This generates a plug flow (called electroosmatic force (EOF)) of bulk fluid in a capillary instead of laminar flow in a typical LC column under a high pressure pump. Capillary electrophoresis offers high resolution and efficiency in resolving analytes. In order to retain its high separation efficiency, the amount of sample analytes that can be loaded is limited, which undermines the sensitivity enhancements. The low flow rates (e.g., 100s nl/min) with CE require micro- or nanoionspray to couple with MS.

Soga (52) pioneered the application of CE-MS to metabolomics. The Agilent CE-MS system has an air pump to drive a flow rate of 10 μl/min. Cationic molecules were quantitatively analyzed with SIM on positive mode using 1 M formic acid as a CE electrolyte and 5 mM ammonium acetate in 50% (v/v) methanol–water as sheath liquid delivered at 10 μl/min by an LC pump, and they covered 30 protonated ions of m/z 70–1027. Anions were quantitatively analyzed in 50 mM ammonium acetate (pH 8.5) as electrolyte for CE separation, and they covered 30 deprotonated anionic molecules of m/z 70–1027. Similarly, nucleotides and CoA compounds were quantitatively analyzed in 50 mM ammonium acetate (pH 7.5), and they covered about 13 different nucleotides and CoA conjugates (62). Applying CE to *Bacillus subtilis* cell extracts resolved 1692 chemicals and facilitated the identification of 150 of them (63). Using CE-Q-TOF, Soga et al. found that ophthalmic acid (i.e., replacement of cysteine with 2-aminobutyric acid in glutathione (GSH)) was synthesized in mouse liver accompanying the depletion of reduced GSH when mice were treated with acetaminophen (64); ophthalmic acid has been considered as a candidate of oxidative stress biomarkers, indicating the depletion of hepatic reduced GSH.

GC is known to be the earliest chromatography in assisting MS to resolve chemicals (65–69). The earliest application includes clinical diagnoses of human diseases by monitoring 155 organic acids in human urine (13, 70). It was demonstrated that 50 different human diseases could be simultaneously diagnosed by monitoring urine organic acids (71). Coupled with MS, GC continues to offer a sensitive analytical method. Commonly used columns for GC include methylsiloxane or 5% phenyl polysilphenylene-siloxane stationary phases. However, in order for biological analytes to be amenable for GC, they must be derivatized chemically to increase their volatility. The derivatization is labor intensive and requires careful attention. Analytes that cannot be derivatized effectively are not generally amenable to GC. Many laboratories use GC-MS routinely as part of the metabolomics profiling platform (48). Much of this work is developed from early metabolomics work with plants, and the techniques are directly applicable to animals as well. The derivatization protocols have been refined over the years, and they must be rigorously controlled to ensure their reproducibility (72–76). The available EI spectral libraries aid in identification of analytes from GC-MS metabolite profiles (77–80).

4.2.4 Choices of LC and MS for Metabolomics

There are three primary approaches involving MS that are used to profile endogenous metabolites in complex samples. These often involve trade-off between speed, specificity, sensitivity, quantitative robustness, and the ability to detect unexpected or unknown metabolites. The first two approaches focus on unbiased or nontargeted profiling of samples using mass spectrometers that scan a wide mass range in order to detect any analytes that can be ionized in the sample, regardless of whether their structure is known or not. The first of these unbiased approaches is a direct infusion of a sample or an extract into a mass spectrometer, and it is often performed with a high resolution instrument such as a TOF,

ICR-FTMS, or Orbitrap using ESI. Scanning across a wide mass range at high resolution is often intended to detect and resolve (based on mass) as many analytes as possible. Data can be acquired in less than a minute, and averaged scans can improve S/N ratios. This approach has the benefit of speed but can be compromised in terms of specificity and quantitative accuracy. As discussed above, high resolution mass analyzers can help resolve analytes with different chemical formulas but cannot distinguish between structural isomers. MS/MS can help distinguish between such isomers, but this can still be tenuous unless authentic standards are available for all the isomers present. Ion suppression that can occur during ionization of complex mixtures can compromise quantification, and the extent of suppression can be variable across samples. With these caveats in mind, infusion experiments can be useful for an initial assessment of a sample.

The second mass scanning approach involves coupling a chromatographic separation with MS (a so-called hyphenated technique). Establishment of robust and standardized chromatography methods is essential to facilitate comparisons across samples, studies, and even laboratories. This is the most common approach used and can be done with either GC-MS or LC-MS. As mentioned above, comprehensive profiling requires more than one method to accommodate the wide range of chemical polarities that are present in complex biological samples. (In addition, multiple ways of sample preparation are generally needed to optimize the recoveries of such a broad range of chemicals. See Section 4.2.5.) While profiling can be done with a single-stage mass spectrometer (e.g., TOF, quadrupole, ICR-FTMS, or Orbitrap), the ability to perform MS/MS experiments, either as a follow-up experiment or during the profiling analysis in an automated data-dependent mode, is important. This is especially true for LC-MS with ESI where only molecular ions are observed; further structural information must be acquired from MS/MS experiments. Compounds can be identified based on mass and chromatographic retention time, but MS/MS fragmentation data can be critical to firm assignments of important analytes. Establishing a reference library based on retention time, m/z, and MS/MS is highly desirable to facilitate metabolite identification in nontargeted GC-MS or LC-MS metabolomics. Several such reference libraries or databases are being developed and are available online (METLIN: http://metlin.scripps.edu/xcms, human metabolome DB: www.hmdb.ca, Lipid maps DB: http://www.lipidmaps.org). Profiling using MS provides relative quantitative data and is not intended for absolute quantification of individual metabolites unless stable-isotope-labeled internal standards are used or calibration samples with authentic standards are analyzed. Unlike NMR, mass spectral response factors are different for different chemicals; therefore, quantitative comparison to a known amount of an authentic reference is often required for quantitative analysis using MS. Both GC and UPLC provide excellent chromatographic resolution and reproducibility to reduce potential peak overlaps and ion suppression. TOF offers mass accuracy as good as 2–5 ppm. Q-TOF provides additional molecular ion confirmation based on MS/MS of its fragmented ions and with resolution as good as full-width half-height maximum (FWHM) of 17,000 (i.e., a ratio of mass over the full width at half peak height) (81). FT-ICR and

Orbitrap technologies offer better spectral resolution (FWHM as high as 500,000), but at a higher instrument cost than Q-TOF. The resolving power obtained from coupled UPLC-FTICR/Orbitrap provides excellent metabolite profiles of often complicated metabolite mixtures in biological samples.

The third approach involves targeted profiling and is focused on a predefined list of metabolites. This approach offers the highest sensitivity, with good reproducibility and broad dynamic range (82). A triple quadrupole mass spectrometer operated in SRM mode is best suited for this analysis. Stable-isotope-labeled reference analytes can be used as internal standards in these analyses to provide absolute quantification. Since SRM is an MS/MS approach that increases specificity, the demands of chromatography to fully separate all metabolites can be lessened; therefore, shorter chromatographic run times can be accommodated for fast targeted profiling. If reference standards are run separately (the so-called external standards), care has to be exercised with the chromatography separation to minimize any analyte overlaps that can lead to ion suppression due to competitive ionization. This approach takes advantage of having a known set of analytes that have been identified as important for characterization of a particular system or process. It provides excellent sensitivity and specificity for the analytes of interest and can be especially useful for low level analytes that may be missed in scanning profiling methods. Targeted approaches have the disadvantage that they may not detect novel or unexpected metabolites, unless they happen to have the same molecular weight and form the same product ion as an analyte included in the targeted analysis.

4.2.5 Sample Preparation

There are three important aspects of sample collection and preparation that are critical to acquiring metabolomic information reliably from biological samples: sample collection, metabolic quenching, and metabolite extraction.

Sample preparation is often needed to eliminate macromolecules and inorganic salts; however, loss or degradation of endogenous metabolites must be minimized during any cleanup steps. In general, sample preparation should be convenient and fast while any residual metabolic activity is quenched and potential compound interconversion (e.g., creatine to creatinine in lyophilization) or degradation (e.g., hydrolysis of phosphate groups in ATP or oxidation of NADH or NADPH) is avoided. Treatment with cold alcohol (e.g., $<-40^{\circ}$C methanol) is simple and effective, and is widely used.

Extra care should be taken in handling and pretreating cells or tissues. Rapidly quenching cells in freezing cold methanol (e.g., -40°C or below, and 80% methanol) is recommended to lyse cells and suppress active metabolic enzymes by denaturing enzymes and precipitating proteins and lipids (83–86). For a similar reason, tissue samples are recommended to be snap-frozen with liquid nitrogen cold tongs. Subsequent tissue lysis and removal of macromolecules can be carried out in cold 80% methanol. Monitoring ATP levels and energy charge can be useful parameters to assess the quality of tissue collection and processing.

ATP tends to degrade rapidly in some tissues; therefore, measurement of ATP levels and energy charge provide an indication of the relative degradation that may have occurred during sample collection and processing (87).

Urine can be directly profiled by NMR by adding deuterated aqueous buffer for locking the magnet field and stabilizing the sample pH (26). In order to avoid clogging HPLC columns with particulates or ion suppression in MS, urine samples are often preprocessed, although they can sometimes be analyzed directly. For example, urine is mixed with a mixture of water/acetonitrile/methanol (20:72:8), and following vortexing and centrifugation, supernatants can be loaded for LC-MS analyses. Alternatively, urine samples are passed through solid phase extraction (SPE) columns, and elutes are combined from both 10 mM ammonium acetate (pH 4) and methanol washes, and loaded on LC-MS for metabolic profiling (39). With both of these approaches, care must be taken to ensure that the chromatography conditions can accommodate injection of a sample with high organic content. Urinary metabolites can be diluted by increased urine volume due, for example, to kidney toxicity or by water bottle spills; therefore, urine samples across a particular study can be normalized to equal osmolarity (e.g., dilution to similar total solute concentrations as determined using, for example, an osmometer by Advanced Instruments Inc., Norwood, MA) (88).

Plasma or serum contains high concentrations of proteins and lipids, including phospholipids. Lipoproteins are visible in NMR spectra as broad ragged peaks that obscure peaks of small molecular metabolites. Peaks of small molecules are also broadened because of interaction with lipoproteins. Editing NMR spectra with approaches such as a T_2 filter can eliminate lipoprotein peaks; however, quantitative profiling or analysis is sensitive to differential T_2 values of individual protons and can be affected by the T_2 filter. An alternative approach is physical removal of lipoproteins. There are two commonly used methods: precipitation with organic solvents (e.g., methanol or acetonitrile) and SPE using C18 SPE columns (89). Lipoproteins can be precipitated in the presence of greater than 75% ice-cold methanol or acetonitrile, and following mixing and centrifugation, supernatants can be lyophilized and reconstituted in deuterated water or buffer for NMR analyses. This approach minimizes interference from lipoproteins in metabolite profiles (83). For LC-MS analyses, the supernatants can be directly used for plasma or serum metabolomics. Alternatively, lipoproteins can be removed by passing plasma or serum through a C18 SPE column. Lipoprotein removal using SPE can be carried out using automation such as with a TOMTEC Quadra 96 (TOMTEC Inc., Hamden, CT). After eluting the retained components with water and methanol, the eluents can be lyophilized or dried under N_2 gas, and then reconstituted in deuterated buffer for NMR analyses or in a water–acetonitrile (e.g., 95 : 5) mixture for LC-MS analyses.

Cerebrospinal fluid (CSF), following centrifugation, can be analyzed directly by LC/MS or can be mixed with deuterated buffer (e.g., 40 mM phosphate buffer in 50% deuterated water) for NMR analysis. Alternatively, CSF samples can be processed using the protein removal approaches mentioned above for plasma samples.

GC-MS requires chemical derivatization to enable derivatized metabolites to be volatile enough for analysis (e.g., forming fatty acid methyl esters (FAME)). The derivatization targets labile hydrogens associated with carboxyl, hydroxyl, amine, and sulfhydryl groups for alkylation, acylation, or silylation. Various silylating reagents are commercially available to cap these functional groups with a trimethylsilyl (TMS) moiety. Commonly used trimethylsilylation reagents include BSTFA (*N,O*-bis-(trimethylsilyl)-trifluoroacetamide) and MSTFA (*N*-methyl-trimethylsilyltrifluoroacetamide) (90, 91). Both reagents are volatile and produce a highly volatile bi-product (i.e., trifluoroacetamide) that does not interfere with early elution peaks. Trimethylchlorosilane (TMCS 1%) is often added as a catalyst. Caution has to be exercised to ensure that samples are completely dry before trimethylsilylation and avoid exposing the derivatized samples to excessive moisture. Standardized derivatization strategies have been developed and published for metabolomic analyses (72, 74–76).

Chemical derivatization may not be limited to GC-MS but can also serve to enhance ionization efficiency and sensitivity with LC-MS. For example, carbonyl groups in ketones or aldehydes can be converted to oximes with hydroxylamine or alkoxylamines; and alkoxyamine introduces a functional group that can readily accept a positive charge (48). With this derivatization, care must be taken when interpreting the resulting chromatograms because oxime formation results in two (syn and anti) oxime isomers (Fig. 4.2) that are often resolved chromatographically. Other chemical derivatizations that introduce a pentafluorobenzyl group to carboxylic acids or a pentabenzoyl group to alcohols and amines can also increase sensitivity dramatically. These have been used extensively in the analysis of lipid mediators such as prostaglandins and other arachidonic acid metabolites (92–95).

4.2.6 Comparison of NMR, LC-MS, and GC-MS

NMR, LC-MS, and GC-MS are the most commonly used analytical platforms for metabolomic analyses. They are truly complementary, and the judicious choice of a particular platform depends on the intended categories of metabolites (lipids or polar compounds), concentration ranges (millimolar or nanomolar) or sample volumes, and the purpose of a study (nontargeted profiling vs targeted metabolite analyses) (96, 97). Often, the choice of an instrument is dictated by what is available in the laboratory. Any of these instruments or, even better, a combination of them can yield very useful metabolomic results (68, 69).

NMR is a relatively fast profiling approach, and it is also quite robust in reproducibility. It can be employed as a quick screening method providing quantitative

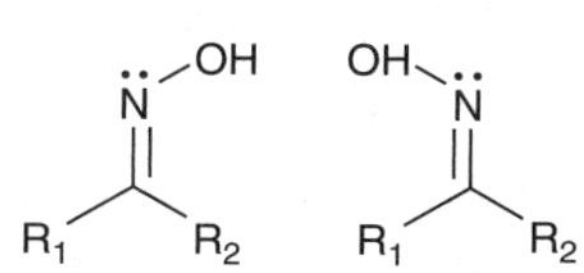

FIGURE 4.2 Two possible orientations around C═N in oximes with syn or anti isomers.

profiles for those detectable metabolites (usually micromolar or above). Although peaks often severely overlap in a 1D proton NMR spectrum, each metabolite can often have more than one signal or peak. Any one of them (or combinations of them) can assist metabolite identification and quantification. In addition, the linear additive nature of NMR signals renders overlapping peaks amenable to linear deconvolution in order to delineate and resolve individual metabolites quantitatively (98). In addition, ^{1}H NMR can provide absolute quantification with a single internal standard compound because of the fact that protons produce a near uniform response regardless of what molecule they are attached to. The signal intensity (peak area) is proportional to the number of protons that are in the sample and can be calibrated relative to the signal intensity of the internal standard for quantification. Therefore, the molar concentrations of molecules in NMR solution samples can be calculated from the proton signal intensity scaled by the number of these particular protons on the molecule of interest. The universal response factor of NMR facilitates convenient quantitative metabolite determinations (26, 98–100).

LC-MS is practical for a very wide range of analytes, including hormones, polar metabolites, cofactors, and lipids. Its capability highly depends on good chromatography for resolving chemicals and preventing ion suppression. Quantification and metabolite identification can be complicated by ion suppression, and adduct formation can result in a particular metabolite being detected as multiple species (e.g., a single analyte can be detected as a protonated ion or an ammoniated ion, and also with or without attachment of small solvent molecules such as acetonitrile). Many peaks (i.e., many m/z values at each retention time) can be readily detected by LC-MS; however, identification of metabolites can be a challenge. One solution is to establish a metabolite library of mass spectra and associated retention times for each metabolite with each particular analytical method. Metabolites can then be efficiently identified and semiquantitatively measured based on their mass and retention time, although MS/MS spectra provide more positive confirmation of the proposed structure.

GC-MS is probably the most sensitive among the three different platforms. It is limited to metabolites that can be made volatile after chemical derivatization. For example, GC-MS is often used for quantitative analyses of fatty acids as FAMEs (fatty acid methyl esters). Metabolite identification can be facilitated by matching mass spectra with a reference library such as the NIST GC-MS library. As with LC-MS, semiquantitative fold changes can be readily determined by GC-MS; however, absolute quantification requires reference standards and calibration analyses using the same or a closely related compound.

4.2.7 Flux Analysis

Flux analyses measure the rates at which metabolites are formed and consumed, rather than the transient or steady-state concentrations that are measured in the profiling experiments described above. For example, metabolic feedback may maintain a constant steady-state concentration for a highly regulated metabolite (such as glucose-6-phosphate), even though its rates of formation and consumption can be increased or decreased. A standard profiling experiment may not likely detect such rate changes in metabolites with steady net concentrations. However, stable-isotope-labeled metabolites (e.g., [1-^{13}C] glucose) can be used to study the enzymatic conversion rates (averaged over the duration of isotope incorporation into intermediate metabolites) of metabolic processes by measuring the rate of formation or disappearance of different metabolites (101). This type of study complements steady-state or transient measurements of metabolic profiling described above. Changes to enzymatic conversion rates often occur either by activation or inhibition of enzymes with chemicals intended for pharmacologic purposes or as a result of unintended and untargeted toxicity. Aided by mathematical models of either stoichiometric metabolic flux analysis (MFA) (102) or isotopomer analysis based on a ^{13}C-labeling experiment (CLE) (103), flux analyses can be used to decipher conversion rates at each step of enzymatic conversion or can track ^{13}C-labeled intermediates in a network of biochemical pathways.

A ^{13}C-labeled substrate such as [1-^{13}C]-glucose can be added to culture media or infused to animals *in vivo*. ^{13}C atoms from the glucose are distributed to intermediate metabolites through accessible metabolic pathways. NMR and MS can measure individual intermediates from lysates of *ex vivo* or *in vitro* samples to determine their rates of formation and consumption. Alternatively, *in vivo* magnetic resonance spectroscopy (MRS) can also track intermediate metabolites in related pathways at anatomical levels with ^{13}C-labeled chemicals using, for example, a hyperpolarized magnetic imaging technique (104, 105).

At a minimum, ^{13}C-based isotope flux can be used to confirm a working hypothesis that is proposed based on metabolic profiling by checking the quantitative changes of isotope-labeled chemicals and their labeling patterns in intermediate metabolites. Additional interpretation of the rates can reveal detailed kinetics of individual steps in a metabolic network; however, this involves detailed mathematical modeling and simulation (101) and requires correspondingly well-planned experimental design and sampling (103, 106–108). Interested readers should consult cited references and recent papers for more detailed information.

4.3 DATA ANALYSIS

Data processing for metabolomics can be complex and occurs at multiple stages along the analytical process. For NMR, a preliminary requirement for processing acquired 1D proton NMR spectra is phase correction; all peaks need to be adjusted to display full absorptive appearance, and any dispersive mode has to be "phased out." Software from NMR instrument vendors or any third parties can handle such phase correction; however, it is hard to find software with a robust algorithm in dealing with phase correction for spectra acquired on vastly different samples (e.g., samples with high lipoprotein) or with varying data acquisition settings (e.g., power for water suppression). Therefore, visual inspection and manual correction are still deemed necessary but are time consuming. In addition, each spectrum requires setting a reference chemical shift position (e.g., placing 0 ppm on the DSS-d_6 peak) following phase correction. After these adjustments are done, 1D proton NMR spectra for a group of study samples should be able to be properly scaled and stacked and ready for visual inspection. This provides a preliminary but direct observation of any NMR peaks that change with treatment or differentiate different treatment groups.

LC/GC-MS data are often more complicated. The files tend to be larger in size than those from NMR spectra, and they are three dimensional (time, mass, and signal intensity). Thus, simply stacking data for all samples across a study is not readily feasible. The total ion chromatogram (TIC) resembles 1D proton NMR spectra, in that it is two dimensional. It shows peak intensities displayed along chromatographic retention time. In scanning mode, a full MS spectrum is obtained approximately every second. This extends the 2D chromatogram into a third dimension, forming a data matrix (retention time $\times$ m/z $\times$ signal intensity). For identification using MS/MS, each point in the 3D matrix is then further extended to a fourth dimension (or even additional dimensions for MS^n). Clearly, data with this complexity is not amenable to simple visual inspection; therefore, software is surely needed to read and analyze such complicated multidimensional data. Many mass spectrometer vendors offer software to facilitate analysis of such complicated multidimensional data. Other third party software is also helpful in providing sophisticated statistical algorithms to aid in comparing and sorting important features and in providing quantification measurements.

4.3.1 Peak Alignment

Peak misalignments often occur in NMR spectra or LC/GC-MS chromatograms, and they complicate meaningful comparisons among treatment groups. Peak alignment is, therefore, critical for finding significant peaks that change with treatment or disease.

NMR peak chemical shift is sensitive to variation in solution properties such as pH, multivalent cations (e.g., calcium and magnesium), and metabolite concentrations. LC/GC-MS retention time can shift when chromatography conditions vary between runs because of the slow deterioration of a chromatography column, deviation of mobile phase pH or composition (for HPLC), and varying concentrations of metabolites or other components in the samples. All these factors can lead to variation of peak positions in NMR spectra or LC/GC-MS chromatograms. Peak alignment is required before data from different treatment groups can be compared. In particular, removing variance due to peak shift reduces errors in statistical analyses and facilitates automatic batch analyses for metabolite identification and quantification.

Reduction of spectral resolution, such as with bucketing, can obscure slight peak shifts (98). However, resolving power that helps minimize overlapping peaks of different chemicals with high resolution spectra or chromatograms is lost in low resolution bucketed data.

There are two categories of approaches in ameliorating variation of peak positions: experimental control and peak alignment algorithms.

There are several experimental controls for reducing variation of peak positions. For example, NMR analysis of urine samples often relies on a strong buffer (e.g., 100 mM potassium phosphate buffer, pH 7) to maintain uniform pH. The influence of multivalent cations on NMR chemical shift can be reduced by introducing chelating chemicals such as EDTA (28). For LC-MS, retention times can be readjusted with respect to a set of internal retention time markers by spiking isotope-labeled chemicals (45). These chemicals span across a chromatographic run with a short interval (e.g., every 30 s in UPLC). Retention time indices are assigned to individual markers, and thus every metabolite can be assigned an index by a linear fit based on two flanking markers. Alternatively, software-based algorithms have been developed to automate the process of spectral alignment (109, 110) (e.g., http://metlin.scripps.edu/xcms/).

Spectrum or chromatogram warping are approaches that require robust algorithms for aligning peaks. Correlation optimized warping (COW) aligns a spectrum against a target reference spectrum (e.g., any one sample spectrum) via linear stretching or compressing of spectral segments to improve correlation coefficients between the reference and aligned spectra (111). Its performance often becomes unsatisfactory in regions of crowded peaks, and it fails when peaks swap positions between spectra. The algorithm has high demand on computation time such that aligning a set of 100 spectra can take as much as several hours. Spectra processed by COW may not be amenable to metabolite quantification based on peak integration because of the stretching or compression of a spectrum or chromatogram.

Recursive segment-wise peak alignment (RSPA) is a modified version of COW (112). It roughly aligns an entire spectrum, and then moves on to smaller segments of decreasing size to fine-tune the alignment as further adjustment is required. This accommodates the demanding requirement of adjusting shifts differently for global and local variations.

Alternatively, a list of peaks, instead of varying peak positions, can be generated using generalized fuzzy Hough transformation (113). This algorithm collects a sparse peak list and assigns the correct peak correspondence even when peaks swap positions between sample spectra. It requires a peak detection algorithm; therefore, the definition of peaks in the algorithm can affect the sparse peak list.

4.3.2 Scaling and Normalization

Scaling and normalization of spectra or chromatograms are often needed before statistical analyses (e.g., principal component analysis (PCA)) can be done in order to reduce the effect of strong signals (e.g., glucose in plasma) that might otherwise dominate the spectra and inappropriately influence the comparisons across a set of data (e.g., total ion intensities in LC-MS or total peak areas in NMR spectra). The total areas under curves of spectra or chromatograms are usually first normalized to a fixed integral value (e.g., 1000).

Signal intensities at individual chemical shifts or individual retention times are often scaled to unit variance (UV). That is, the standard deviation is calculated for signal intensities at individual chemical shifts or retention times, and signal intensities across all spectra and chromatograms are scaled (i.e., divided) by a standard deviation. However, equalization of variance can lead to noise overestimation.

Pareto scaling is an alternative option to UV scaling, and it differs by scaling signal intensities with the square root of the standard deviation. It falls between no scaling and UV scaling in its impact on the data processing, and increases the importance of low concentration metabolites without significant amplification of noise (114).

Mean centering removes intensity differences between spectra or chromatograms. It subtracts averages from signal intensities at individual chemical shifts or retention times. The combination of UV scaling and mean centering is often called *autoscaling* in software such as SIMPCA (115).

4.3.3 Principal Component Analysis (PCA)

Principal component analysis reduces overwhelming and intricate experimental data (especially LC-MS chromatograms) to a data matrix of a manageable size. The original data matrix is transformed and represented by two new small matrices (i.e., score and loading matrices). The score matrix contains limited numbers of principal components with decreasing importance to capture significant changes or variances among the original data set. (Note: "principal components" refer to statistical parameters, not chemical components.) The most significant component (i.e., first principal component) accounts for the greatest amount of variability among the samples analyzed. This can be dominated by either metabolites showing the largest changes among samples or experimental artifacts (e.g., order of analysis and related changes to chromatographic conditions). Therefore, evaluation of several top principal components in a score plot helps judge which sets of principal components explain metabolite changes due to study designs (e.g., treatment, and physiological conditions). The score plot is, however, strongly influenced by scaling methods in the pretreatment of original data. Therefore, visualization of PCA by score and loading plots should be used in combination with other data analysis tools (e.g., pathway analysis tools) to aid in understanding the relevance of the changes. Although it is a valuable tool, its value when used alone for interpreting data can be limited.

The loading matrix tells the importance of variables (i.e., signal intensities of particular analytes or peaks) in causing the sample differentiation displayed in a score plot based on a particular pair of principal components.

The PCA mathematical model can also be evaluated by cross-validation that can evaluate a required number of principal components for data analyses and the validity of calculated principal components. In this approach, a few sample data points are sequentially removed from the data analysis, and PCA is used to calculate the score and loading matrices based on the remaining sample data. Comparison of data matrix constructed using the newly calculated score and loading matrices with the original sample data matrix determines the goodness of fit (R^2) and prediction (Q^2). Although goodness of fit (R^2) improves with an increasing number of principal components, the goodness of prediction (Q^2) peaks out at a certain number of principal components. This determines an appropriate number of principal components for PCA. A robust model should have $R^2 > 0.5$, $Q^2 > 0.5$, and $|R^2 - Q^2| < 0.2$–0.3 (115).

4.3.4 Projection to Latent Structures Discriminant Analysis (PLS-DA)

If sample classifications are known (e.g., dosing group or histomorphology scores), the value in correlating this information with variables (e.g., profiling data such as NMR chemical shifts or LC/GC-MS retention time–m/z) is strong in identifying metabolites that are related to these classifications. A PCA model can be oriented to emphasize the largest separation of samples according to a known classification with a process called *projection to latent structures discriminant analysis* (PLS-DA) using partial least squares. PLS is a maximum covariance model that helps maximize the linear alignment of variation in an experimental data matrix with that in the sample classification matrix. It helps unravel the variables (i.e., chemical shifts or retention time–m/z) that are responsible for sample classification or grouping.

The principal components of the data matrix can be viewed as a PCA model. The score plot of the data matrix shows the separation of samples based on the top principal components according to the sample classification. The corresponding loading plots help identify those variables (e.g., chemical shifts or retention time–m/z) that are responsible for the samples classification (i.e., sample differentiation). One way of viewing the variables utilizes a linear plot of the loadings from top principal components, and the plot can display important variables (e.g., chemical shifts or retention time–m/z) that are influential for sample differentiation. In addition, the importance of variables contributing to PLS-DA can be sorted and viewed based on the rank order of variable importance parameters (VIPs). The goodness of fit (R^2) and prediction (Q^2) can likewise be evaluated. A robust model should have $R^2 > 0.5$, $Q^2 > 0.5$, and $|R^2 - Q^2| < 0.2$–0.3 (115).

4.3.5 Metabolite Identification

Identification of metabolite chemical structures is crucial to linking metabolic profiling data with biochemical pathways for biological interpretation. Both NMR and MS have been traditionally applied iteratively and collaboratively to chemical structure elucidation; therefore, metabolomics is a perfect setting for their unsurpassed combined advantages (116). However, the complex mixture of analytes in metabolomics samples poses a new challenge for *de novo* chemical structure elucidation. The challenge arises from the presence of trace amounts of hundreds or thousands of chemicals in a biological sample. It is impossible and unthinkable to go through *de novo* chemical structure analyses on every sample or in each study. Alternatively, a reference library of NMR spectra and LC/GC-MS chromatogram/mass spectral data aided by data processing software is a necessary requirement for routine application of metabolomics.

Several NMR reference libraries are available to the public, while others are proprietary. Publically available databases include the Human Metabolome Database (HMDB, www.hmdb.ca) (117) and the Madison Metabolomics Consortium Database (MMCD, mmcd.nmrfam.wisc.edu) (118). Proprietary databases include Chenomx (100) and many other databases residing in private ownerships. The HMDB contains ^{1}H NMR spectra of 905 chemicals and ^{13}C NMR spectra of 899 chemicals (these numbers and others throughout the chapter were taken at the time the chapter was drafted). The MMCD has compiled NMR spectra for 477 compounds, and these NMR spectra include 1D-^{1}H, 1D-^{13}C, and 1D-^{13}C DEPT90, as well as spectra for 135 compounds acquired by 2D-TOCSY (total correlation spectroscopy), 2D-HSQC, and 2D-HMBC (heteronuclear multiple-bond correlation spectroscopy). Chenomx proprietary database provides 1D-^{1}H NMR spectra for 445 compounds, and these spectra are acquired at several different pHs and with five different magnetic field strengths (400, 500, 600, 700, and 800 MHz). The spectral data are stored as simulated NMR peaks based on the acquired raw spectra. An approach such as that used by Chenomx accounts for the fact that chemical shifts and peak line shape in NMR reference spectra change with solution pH, magnetic field, metabolite concentration, and molecular interaction. These reference spectra provide good coverage of the variation normally observed in spectra because of the typical experimental variables.

LC/GC-MS libraries are complicated not only for many possible MS settings (e.g., MS accuracy and resolution, and ionization) but also for different LC/GC configurations and conditions (e.g., column, mobile phase, flow rate, temperature). Retention time not only changes between different instruments but can also drift between runs. Retention time index is often adopted to minimize the problem associated with retention time drift in a given chromatographic method (45).

There are several MS databases for metabolomics. The HMDB mentioned above has MS/MS spectra for 2654 compounds and GC-MS spectra for 318 compounds. MassBank (www.massbank.jp) has 827 high resolution LC mass spectra for common metabolites. The Metlin metabolite database contains web-searchable LC-MS and MS/MS spectra for metabolites (http://metlin.scripps.edu/metabo_search.php). The Golm metabolome database (GMD) offers libraries of GC mass spectra and retention time index (MSRI) (119). Use of retention time index, instead of actual retention times, helps to eliminate problems from variability in GC retention times. The GMD displays electron impact mass spectra acquired from two different GC-MS instruments, a quadrupole GC-MS (unit mass resolution) and GC-TOF (higher mass resolution). The GMD contains 2000 mass spectra of 1089 metabolites based on these two different GC-MS platforms. Fiehn's GC-MS database (FiehnLib) contains spectra of 1000 identified metabolites of molecular weight less than 550 Da (120), and it is commercially available through Agilent Technologies as the Fiehn GC/MS Metabolomics RTL Library. The NIST offers a GC-MS library with 192,108 unique chemicals. Although some of these chemicals are biological, most of them are industrial chemicals.

4.3.6 Quantification

Both NMR and LC/GC-MS are routinely used for quantitative analyses; however, metabolite quantification by NMR and LC/GC-MS differs in, among other things, absolute versus relative quantification (depending on whether or not appropriate internal standards are used), sensitivity, and lower limit of quantification.

For NMR, using an adequate delay between repetitive pulses and signal acquisition, all nuclei (e.g., proton) in an NMR tube experience the same electromagnetic excitation and produce signals that are linearly proportional to the concentration of individual types of nuclei (e.g., three methyl protons in lactate). Quantitative NMR signals are linearly additive and independent of different functional groups on the analyte or other molecules in the sample. Such a universal response factor enables NMR quantification to rely on a single internal reference standard (e.g., internally present DSS-d_6 (26) or DMSO (121) of known concentration). This approach can be readily applied to well-resolved peaks by peak area integration.

NMR quantitative analysis of overlapping peaks can be handled in two different ways. Each metabolite often has more than one peak on an NMR spectrum; therefore, any one of those resolved peaks can be integrated for metabolite quantification. The quantitative contribution by each peak in an analyte in an NMR spectrum is determined by its stoichiometric proportions (e.g., the ratio of the peak areas of $C\mathbf{H}_3$ and $C\mathbf{H}_2$ for ethanol is 3 : 2 and of $C\mathbf{H}_3$ and C**H**OH for lactate is 3:1). If only one metabolite in a peak-overlapping region requires quantification (and its structure is known) while others (whose structures are also known) can be independently quantified from well-resolved peaks in other regions of spectra, a simple arithmetic subtraction of peak areas for each of the overlapping signals from the analytes of interest (based on peak stoichiometric ratios) can be used to decipher its concentration.

When concentrations of more than one metabolite in a peak-overlapping region are needed, linear deconvolution provides an efficient solution (98). It is widely used in metabolite concentration calculations from an *in vivo* MRS spectra (122). A robust algorithm for linear deconvolution is singular value decomposition (SVD). This approach relies on quantitative reference spectra of preidentified metabolites; therefore, chemical structures of metabolites need to be identified based on overall peak patterns in an NMR spectrum (i.e., chemical shifts, peak multiplicities, and coupling constants). Reference spectra are required to be acquired quantitatively with known concentrations of reference metabolites and an internal reference standard (e.g., DSS-d_6). NMR peaks from reference spectra also need to be aligned with respect to peaks in the sample spectrum. The SVD algorithm basically calculates the concentrations for the individual metabolites based on a linear combination of the reference spectra for each of the individual components that are overlapping in the mixture.

Relative quantification of metabolites in LC/GC-MS spectra can be determined based on the peak areas for each of the analytes (as could be determined from reconstructed ion chromatograms for each analyte; software is available from most of the major vendors to do this automatically). Absolute quantification of metabolites using LC/GC-MS requires generation of calibration curves for individual metabolites or closely related chemical structures (e.g., C17 fatty acid for even-carbon-numbered long-chain fatty acids) (95). Quantification by LC-MS is often carried out with calibration curves using external reference standards by SRM or MRM, which improves specificity, sensitivity, and dynamic range of mass m/z linear response. The concentrations of the calibration standards need to cover those potentially present in samples, or the sample concentrations need to be diluted to fall within the linear calibration ranges. An alternative to the use of calibration curves is the use of an authentic reference standard for the compound of interest at a single concentration. This is not as rigorous as the use of a full calibration curve, but it can provide a reasonable estimate of absolute quantification if the linearity of the method is known. The quantification errors due to ion suppression that occur using external standard methods can be remedied by the use of methods involving stable-isotope-labeled internal standards. Isotope-labeled metabolites used as internal references are needed in order to distinguish them from the naturally occurring analytes using a mass spectrometer. This approach may be limited by the commercial availability of isotope labeled metabolites. For cell-based metabolomics, this problem can be resolved by growing cells with stable-labeled substrates and enriching isotope-labeled endogenous metabolite products during cell harvesting. The labeled analytes from the culture can be quantified using calibration curves constructed using commercially available metabolites (without isotope labels). Then the stable-labeled cell products can be used as internal standards in subsequent profiling experiments. (123, 124).

4.3.7 Analyses of Quantitative Metabolites

The identification of important metabolites in a study and understanding their biological significance are two interactive and parallel processes. Ranking of important metabolites is often done based on statistical evaluation, while biological understanding relies on the knowledge of metabolic pathways and published literature. Synergistic integration of these two processes requires improvements over existing pathway analysis tools that have been built based on current scientific literature but are generally lacking quantitative metabolite information. The ability to make biological sense of metabolomics data based on known metabolic pathways is important for validating metabolomics results.

Instead of raw NMR or LC/GC-MS data, metabolites and their quantities can be analyzed directly by PCA to observe sample grouping and locate metabolites influencing group segregation. Other statistical analyses include the Student t test and MANOVA (multivariate analysis of variance). A simple Student t test compares the means between two treatment groups, and it can be readily carried out even in Microsoft Excel. Box plots created by statistical tools such as R (www.r-project.org) provide a good means for direct visual examination of results (http://en.wikipedia.org/wiki/Box_plot). MANOVA or its improved version such as ANOVA-simultaneous component analysis (ASCA) analyzes more than two independent variables (e.g., dose, age, and weight). It elucidates variables that carry significant influence on the results and identifies interactions between independent variables (125).

Understanding metabolite quantitative changes based on metabolic pathway analyses can help produce bona fide working hypotheses of underlying biochemical mechanisms that may link metabolomics to genomics, transcriptomics, proteomics, and external stimuli (e.g., diet, drug, and environment). Metabolomics is at a stage where it can inform about intricate biochemical processes at different levels, and can be combined with genomics, proteomics, and other biochemical data to describe metabolic phenotypes and make such integrated analyses possible and productive.

There are many pathway tools available in the public domain or in commercial packages. Most of them have evolved from gene-centric pathways. They integrate published literature results with analysis of experimental data to generate a network or pathway of related genes, proteins, and endogenous metabolites. These pathway tools help reveal the underlying relationships between genes, proteins, and metabolites, and unmask their relevant connections to clinical pathology. However, systems biology would benefit from the ability to take into account the relevance of different magnitudes of changes in genes, proteins, and metabolites. Homeostasis in a biological system can often dampen the magnitude of these changes, especially for metabolites, in spite of significant changes in gene expression or environmental perturbation.

Public biochemical pathway tools include KEGG (Kyoto Encyclopedia of Genes and Genomes, www.genome.jp/kegg), MetaCyc (www.metacyc.org), BioCyc (www.biocyc.org), and Reactome (www.reactome.org). Proprietary software includes GeneGo (www.genego.com) and Ingenuity (www.ingenuity.com). Both GeneGo and Ingenuity rely on curation biomedical information retrieved from published literature to establish a knowledge database of biochemical pathways.

KEGG is one of the most complete and widely used biochemical pathway maps. It contains 372 pathways from more than 700 organisms. Hyperlinks are embedded in the pathways with 15,000 metabolites and related enzymes.

MetaCyc is a nonredundant metabolic pathway database that has been elucidated from experimental data (126). It contains more than 1400 pathways from more than 1800 organisms that have been curated from scientific literature. BioCyc collates data from 371 Pathway/Genome Databases (PGDBs), and each database represents a description of genomic and metabolic pathways in one single organism (126).

Reactome is a pathway database encompassing human biology with inference to orthologous events in 22 nonhuman pathway systems (127, 128). It constructs pathways based on causal reactions, and those pathways include intermediate metabolism, regulatory pathways, signal transduction, and cell cycle. It contains more than 2907 reactions with 4455 literature citations (129).

GeneGo contains 700 canonical pathway maps that cover both signaling and metabolic pathways (130). The pathways and connections between molecules (genes, proteins, and metabolites) are curated by GeneGo scientists with expertise in different areas of biology, and they are of high quality in comparison to those databases that are parsed by machines. The metabolic pathways are chains of metabolic reactions that are linked to biochemical functions and cycles, and are used for interactive metabolic network generation and visualization. Ingenuity provides similar tools for systems biology analyses of genes, proteins, and metabolites based on their knowledge database that is curated by scientific experts from relevant literature information (131).

4.4 EXAMPLES OF APPLICATIONS

4.4.1 Integrated Pathway Analysis of Rat Urine Metabolic Profiles and Kidney Transcriptomic Profiles

Evaluation of metabolites as biomarkers requires extensive statistical analyses, as well as biological investigation even when the analytical methods used to identify the proposed biomarkers have been thoroughly characterized. This is due to the complexity of biological systems and difficulty in establishing a causal relationship between the biological condition and the appearance of the biomarker. Statistical evaluation ensures the biological reliability of repetitive measurements associated with particular treatments. A thorough qualification of a biomarker often requires a large set of biological samples and various treatments aimed to achieve the same biological endpoint due to complexities of biological responses. Variation in biological response can arise from internal sources (genetics, age, gender, etc.) and external perturbations (environment, diet, etc.). Well-proven and fully accepted metabolite biomarkers (such as blood glucose for diabetes and serum transaminases for liver toxicity) are often enabled by good understanding of their biological mechanisms. With current analytical technologies, an acceptable biomarker should be based on characterization of biological mechanism as assisted by integrated pathway analyses that include genes, transcripts, proteins, and metabolites. On the other hand, all of these biological data should provide invaluable support for follow-up statistical validation, and reduce time and resources by guiding improved design of subsequent *in vivo* studies and selection of samples. The biological understanding that is gained should explain not only why and how a biomarker works but also the circumstances where it *does not* work. In addition, biological understanding could yield multiple biomarkers based on a class of endogenous metabolites (e.g., neutral amino acids) instead of a single compound (e.g., leucine); this could produce panels of biomarkers that are much more robust in dealing with variability associated with living organisms from genetics, diet, and environment. In addition, knowledge of relevant biochemical mechanisms can also guide the development of diagnostic tools. For example, 2-^{18}F-2-deoxy-D-glucose is routinely used in positive electron tomography (PET) for cancer diagnosis based on the Warburg effect of elevated glycolysis in cancer cells (132).

The paper by Xu et al. (26) integrates urinary metabolomics and kidney transcriptomics to provide a reasonable biological explanation based on statistical correlation for the changes in urinary metabolites and transcription regulation of renal membrane transporters induced by the kidney toxicants cisplatin and gentamicin. The findings provide a biochemical rationalization for the elevated endogenous metabolites (including monosaccharides, amino acids, small molecules of organic carboxylic acids) observed in urine. As a result, these metabolites can serve as biomarkers for evaluating kidney toxicities based on an impaired renal reabsorption function in proximal tubules.

In this study, NMR identified and quantified about 40 endogenous metabolites in urine samples from rats dosed with a single intraperitoneal injection of cisplatin (0.0, 0.5, 3.5, and 7.0 mg/kg) or daily intraperitoneal injection of gentamicin (0, 20, 80, and 240 mg/kg). Along with NMR metabolomic analysis of urine, histopathological evaluation of necropsied kidneys provided traditional endpoints of renal toxicity, and transcriptomic profiling of extracted kidney RNA was analyzed to follow gene expression changes.

The urinary metabolites were ranked based on p values, with reference to histopathology scores from rat kidneys, using a three-way ANOVA (dose, time, and animal identification). Integrated pathway enrichment analysis was facilitated by knowledge-based canonical pathways and metabolic pathways in GeneGo's MetaCore.

The top 10 metabolic pathways that corresponded to the genes that showed the most change in the RNA profiling from both cisplatin and gentamicin data were shown to be related to the metabolism and transport of monosaccharides, amino acids, and monocarboxylic acids. One of the most significant pathways was glycolysis and glucose transport. A strong negative correlation (−0.804 to −0.959) existed between the amount of urinary glucose and sodium-dependent luminal glucose transporters encoded by the genes *SCL5A1* and *SLC5A2*. These transporters reside in the luminal membrane of renal tubular epithelial cells and help in the reabsorption of glucose that normally filters through the glomerulus. Besides usage of glucose for normal metabolism inside proximal tubule cells, extra glucose is usually transported into the capillary space and back to the blood through a facilitative glucose transporter (GLUT) family at the basolateral membrane. In contrast to sodium-dependent glucose transporters at the luminal membrane, GLUT family transporters are upregulated with cisplatin and gentamicin treatments. This seemingly contradictory down- and upregulation of glucose transporters at luminal and basolateral membranes, respectively, was consistent with the increased glucose presence in urine and the associated proximal tubule cell toxicity in kidney.

Reduced expression of glucose transporters at the luminal membrane diminished the reabsorption of glucose that was required for proximal tubule cell viability. The survival of proximal tubule cells had to rely on reverse glucose uptake from basolateral facilitative glucose transporters that work in both directions depending on glucose concentration and gradient across the membrane. Cisplatin- and gentamicin-led glucosuria was caused by the loss of glucose membrane transporters, and this resulted in an alteration in glycolysis. This was supported by synchronized perturbation of endogenous metabolites from different pathways or by different enzymatic reactions, that is, urinary lactate and alanine showed similar dose- and time-dependent profiles despite very different biochemical conversion reactions from pyruvate.

Transcription data facilitated the association of transcription regulation as a cause for the shortage of glucose transporters. Transcription of sodium-dependent luminal glucose transporters (SLC5A1 and 2) was downregulated by the reduced transcription factor HNF1α. However, increased hypoxia-induced factor HIF1α expression induced upregulation of basolateral facilitative glucose transporters (GLUT2 and 9). This appeared to be an adaptive response mechanism to sustain viability of tubular epithelial cells by deriving nutrients from blood at the basolateral side.

Similarly, urinary excretion of amino acids, lactate, acetoacetate, and 3-hydroxybutyrate was caused by downregulation of sodium-dependent luminal membrane transporters for these metabolites. Amino acid transporters SLC6A18, 19, and 20 were downregulated by collectrin that was in turn downregulated by the reduced expression of HNF1α and β. Therefore, HNF1 controlled the expression of sodium-dependent luminal membrane transporters of both glucose and amino acids.

Analyses of pathways of underlying biological mechanisms empowered better understanding of the applicability of endogenous urine metabolites in diagnosing kidney toxicity in proximal tubules. This provided functional biomarkers based on not only a particular small molecule but also a class of molecules such as neutral amino acids and monosaccharides. Understanding the biochemical pathways associated with this kidney toxicity also facilitated the use of infused labeled glucose or amino acids to monitor the function and integrity of renal proximal tubules. This could be a useful tool in circumstances where overall endogenous metabolites may actually decrease because of the reduced uptake of nutrients by animals when severe toxicity occurs suddenly and dramatically. In addition, statistical validation of these biomarkers on the basis of additional studies can be devised more effectively and efficiently by enabling design of kidney toxicity studies with a similar mechanism (e.g., proximal tubule toxicity as illustrated in this example).

4.4.2 Coordinated Modulation of Transcripts and Metabolites in Yeast

Metabolomics can help establish a quantitative relationship between the transcriptome and metabolome on the basis of the current understanding of qualitative causal relationship between transcription and transcription regulation on metabolites (133). Integration of transcriptomics and metabolomics in a systems biology approach is being developed through quantitative measurements and analyses.

One example is illustrated in the two papers by Rabinowitz, Troyanskaya, and coworkers (134, 135) using a Bayesian network analysis to elucidate the quantitative correlation between transcripts and metabolites in *Saccharomyces cerevisiae* under either nitrogen (i.e., ammonium) or carbon (i.e., glucose) deprivation. Although the examples are illustrated in yeast, the quantitative analyses can be expanded at least to *in vitro* mammalian-cell-based studies.

Yeast are grown on nitrocellulose membranes placed on top of agarose loaded with culture media to facilitate rapid modification of extracellular conditions and quenching of metabolism for parallel snapshots of the metabolome and transcriptome over time. The cells were quenched at -75°C and using liquid nitrogen for metabolome and transcriptome sampling, respectively. Metabolites were semi-quantified with LC-MS/MS (i.e., HILIC column separation and SRM detection) targeting approximately 170 water-soluble chemicals. Metabolite quantification was normalized by cell dry weight to adjust for cell growth and division. RNA was extracted with Qiagen RNEasy kit, and transcription was profiled with an Agilent Yeast Oligo Microarry with a zero-time-point sample taken as a reference. Equivalent amounts of RNA were loaded onto each array to focus on relative transcription changes instead of total changes due to nutrient starvation. Both metabolite and transcript levels were scaled by log base 2 ratios with respect to time zero values.

Comodulation of transcription and metabolism was at first studied with SVD based on temporal dynamics. The SVD extracted major components (i.e., engine vectors) by linearly factorizing complicated data sets such as temporal changes of all metabolites or transcripts. The first eigenvectors of transcripts and metabolites displayed monotonic genetic responses to the lack of nitrogen or carbon; vector signal intensities increased with time. The second eigenvectors were nutrient dependent, and they showed opposite trends relative to carbon and nitrogen deprivation. The third eigenvectors showed dynamic responses, but the biochemical meaning from these was less clear. Transcripts and metabolites responded with similar fold changes, or were closely aligned, to nutrient deprivation. This suggested that the complex interplay between genes and metabolites could be studied and may be elucidated by analyses based on pathways and rational relationships between genes and metabolites.

In addition, the second eigenvectors suggested that nitrogen or carbon deprivation had different effects. The same set of genes and metabolites could display positive or negative correlation depending on the deprivation of either nitrogen or carbon. For example, glucose phosphate showed a positive correlation to the hexokinase gene GLK1 under nitrogen starvation, while it showed negative correlation under carbon starvation; no correlation appeared when a Pearson correlation was calculated across both conditions. These correlations fit the biochemical understanding of glycolysis under normal carbon conditions (i.e., glucose) but with nitrogen starvation (i.e., without ammonium). Glucose deficiency sent signals to an intracellular sensing complex mediated by hexokinase isoenzyme 2, HXK2, that suppressed GLK1 and led to an observed inverse correlation between GLK1 and glucose phosphate under carbon starvation.

One tool for examining rational relationships between biochemical processes is Gene Ontology (GO), which covers a large scope of biological processes (such as transcription regulation, protein translation, translational modification, and cell cycles) beyond just metabolism. Metabolites are grouped in pathways such as glycolysis, TCA cycle, amino acid metabolisms, and biosynthesis. Correlation of gene–metabolite coregulation is ranked based on Bonferroni-corrected p values that are obtained from hypergeometric distribution; this is often called an *enrichment analysis*. Logically and biologically relevant relationships identified in these analyses are shown to be statistically significant for some transcripts and metabolites from specific pathways, as mentioned above, while others are less understood. For example, metabolites other than the TCA cycle's intermediary metabolites appear to show a correlation to transcripts involved in metabolism of lipids and steroids.

When pathway links between genes and metabolites are farther apart, Bayesian analysis may capture correlation patterns of genes and metabolites on the basis of existing or known experimental context, observed correlation, functional classification of metabolites, and gene–metabolite functional relationships. The functional relationship can link gene–metabolite pairs that are not linked by a single biochemical reaction. The context is parameterized in conditional probabilities as "nodes" that are connected by "edges," indicating their dependent relationships. Therefore, the probability distribution in a Bayesian network gives informative meaning that is lacking in other machine-learning algorithms. The established network of "nodes" and "edges" in terms of correlation strength and direction is leveraged with Bayesian predictions. No causality is inferred between genes and metabolites mathematically, but biological annotation based on knowledge can sometimes construct a sequence of biochemical pathways.

Prediction by Bayesian analyses can uncover novel specific gene–metabolite interactions that are physiologically relevant but are not obvious based on statistical correlation. One example provided in the paper by Rabinowitz and Troyanskaya (135) predicted a reversible glycolytic-gluconeogenic switch gene *VID24*. VID24 changed from fermentation, with the end product of ethanol, to respiratory growth with biosynthesis of glucose and carbon skeletons from ATP and ethanol. The protein Fbp1p is known to inhibit glycolysis, and VID24 switches from gluconeogenesis to glycolysis by targeting Fbp1p to vacuolar degradation. The prediction was supported by a Pearson correlation between fructose-1,6-bisphosphate (FBP) and VID24: positive correlation under nitrogen starvation, but negative correlation under carbon starvation. This fits the observation that nitrogen starvation (but with sufficient glucose) favored glycolysis, and carbon starvation favored gluconeogenesis.

The example illustrates the need to have a context-sensitive approach toward understanding intrinsic and intricate gene–metabolite relationships. A simple Pearson correlation may provide sufficient biological interpretation. However, elucidation of gene–metabolite pairs that appear to be irrelevant based on traditional analyses but are actually intricately related by causality requires thorough detection of all classes of metabolites and accurate quantification.

4.4.3 Sarcosine as a Potential Biomarker in Prostate Cancer Progression

Searching for metabolite biomarkers requires coordination with other omics technologies, as well as more conventional biochemical analyses to characterize relevant biochemical pathways. Elucidation of sarcosine as a potential prostate cancer progression biomarker illustrates the importance of biological investigation as a part of the process (including analytical and statistical evaluations) for biomarker validation (90).

4.4.3.1 Chemical Identification and Quantification. Metabolites in prostate tissue as well as urine and plasma were profiled and quantified with both LC-MS and GC-MS. Metabolites were identified based on the authors' metabolic library of 800 commercially available chemicals, and the structures of critical metabolites were confirmed by MS.

Quantification of key metabolite chemicals (urine: sarcosine, alanine; tissue: sarcosine, cysteine, thymine, glycine, and glutamic acid) was done using stable-isotope-labeled internal standards with isotope dilution GC/MS.

Samples derivatized with *t*-butyl dimethylsilyl moieties were eluted on a 15-m DB-5 capillary column (ID: 0.2 mm, 0.33 μm film thickness) and identified and quantified with an Agilent 5975 MSD mass detector using SIM.

Metabolomics analyses detected 1126 chemicals (including exogenous chemicals) in 262 human clinical samples (42 prostate-related tissues and 110 matched plasma and urine samples) (90). Among the detected chemicals, 265 (23.5%) were matched with corresponding authentic chemicals in their database. In addition, there were 28 (2.5%) isobaric candidates. These clinical samples were obtained from three different categories of patients: benign ($n = 16$), localized prostate cancers (PCA) ($n = 12$), and metastatic prostate cancers (Mets) ($n = 14$). In prostate-related tissues, 626 metabolites were quantified (28.0% were chemically identified, 3.0% were isobaric candidates, and the remaining 69.0% were chemically unidentified). Of the metabolites detected, 82.2% (i.e., 515 chemicals) were shared among the three different categories of patients. Samples from "Mets" contained 39 (i.e., 6.2%) unique metabolites, PCA samples had 3 (i.e., 0.5%) unique metabolites, and benign showed 7 (i.e., 1.12%) unique metabolites. Metabolites shared only between "PCA" and "Mets" amounted to 2.88% (i.e., 18 chemicals).

4.4.3.2 Statistical Analyses of Differential Metabolites. The differential metabolites were elucidated with a two-sided Wilcoxon rank-sum test coupled with a permutation test. Metabolites that significantly increased from benign to PCA to metastatic prostate cancers included sarcosine, uracil, kynurenine, glycerol-3-phosphate, leucine, and proline.

4.4.3.3 Pathway Analyses. Mapping differential metabolites into KEGG biochemical pathways showed an increase in amino acid metabolism and nitrogen breakdown pathways. In addition, Oncomine Concept Map (OCM) enrichment analyses supported earlier gene-based predictions of androgen-induced protein synthesis during prostate cancer development (136), as well as a strong enrichment in "methyltransferase" activity.

4.4.3.4 Sarcosine. Sarcosine is found to be increased in 42% of the PCA tissue samples and markedly increased in 79% of metastatic tissue samples, but its level was below GC-MS detection in benign tissue samples. Improved detection of sarcosine was facilitated by the development of an isotope dilution GC-MS method, with a limit of detection at 10 fmol.

Sarcosine was found to be significantly higher in urine from biopsy-positive prostate cancer patients than from biopsy-negative controls (Wilcoxon

$p < 0.0025$); however, its predictive value is modest. Sarcosine did perform somewhat better than did PSA (prostate-specific antigen) when PSA values fall in the clinical gray zone (2–10 ng/ml).

Biological relevance of the sarcosine increase in prostate cancer was evaluated in the prostate cancer cell lines VCaP, DU145, 22RV1, and LNCaP in comparison to their benign counterparts (PrEC and RWPE). Sarcosine was found to be significantly increased in prostate cancer cells ($p = 0.0218$), and it correlated with cancer invasiveness (correlation coefficient, 0.943). Overexpression of histone methyltransferase (EZH2) in benign prostate epithelial cells instigated the sarcosine increase, whereas its knockdown in DU145 prostate cells diminished sarcosine levels (4.5-fold). Addition of sarcosine to benign prostate epithelial cells exerted an invasive phenotype.

Cell invasiveness was also modulated by enzymes in the metabolic steps around sarcosine. RNA-interference-mediated knockdown of GNMT (glycine N-methyltransferase, the protein that catalyzes glycine → sarcosine) and DMGDH (dimethylglycine dehydrogenase catalyzes dimethylglycine → sarcosine) led to a reduction in cancer cell DU145 invasiveness and a decrease in sarcosine levels. On the contrary, knockdown of SARDH (sarcosine dehydrogenase catalyzes sarcosine → glycine) in benign prostate epithelial cells led to a threefold increase in endogenous sarcosine with a concomitant 3.5-fold increase in invasiveness.

Androgen-induced prostate cancer is mediated by androgen binding to its receptor on transcription factor ERG or ETV1 that increases GNMT gene expression by binding to the GNMT promoter. GNMT upregulation-induced cancer cell invasiveness was associated with a threefold increase in sarcosine. Knockdown of ERG by RNA interference resulted in a greater than threefold decrease in sarcosine and cell invasiveness. However, androgen had little influence on SARDH expression.

This example shows that endogenous metabolites (e.g., sarcosine) can be indirectly affected by the transcription factors that control regulatory enzymes (e.g., GNMT). Molecular investigation of cellular biochemistry provided solid support and validation to metabolic biomarkers such as sarcosine and a foundation to use sarcosine potentially as a universal biomarker to monitor prostate cancer prognosis. Sarcosine appears to be a useful biomarker in preclinical animal models as well as clinical settings with noninvasive detection in urine. With an understanding of the sarcosine metabolic pathways, diagnosis may be performed with improved specificity using an array of metabolites in this pathway or with stable isotope flux analysis.

4.4.4 Transcriptomics, Proteomics, and Metabolomics Response to Genetic and Environmental Perturbation in *Escherichia coli* K-12

Clearly, cells are able to respond to perturbations or stresses in different ways, maintaining their survival. Ishii and collaborators performed experiments to determine whether perturbation due to genetic mutation initiated the same adaptive mechanism as external perturbation, or do cells utilize different strategies in order to adapt. They used *Escherichia coli* K-12 and evaluated gene transcription, levels of proteins, and metabolite abundances to evaluate the mechanisms of cellular adaptation (137). The results showed that the cells have versatile mechanisms to deal with internal and external perturbations differently. The illustration with single-celled *E. coli* K-12 enhances our understanding of the ways that cells adjust to environmental changes with complementary pathways, and the genetic changes that induce robust enzyme isoforms. The cellular adaptation was achieved by maintaining stable metabolite levels in cells. The gene mutants used in Ishii's paper affected enzymes in glycolysis and the pentose phosphate pathway; the environmental perturbation was induced by altering glucose levels in glucose-limited chemostat cultures.

Although the example was based on single-celled bacteria, addressing the same question with multicelled mammalian tissues has significant implication to disease treatment and drug toxicity evaluation, especially when internal and external perturbations are intertwined. For example, the effectiveness of the same disease treatments can differ from patient to patient, and the occurrence of drug toxicity also varies between individuals. The interpretation and investigation of variable, puzzling, and often frustrating outcomes especially in clinical settings require utilization of appropriate tools (i.e., transcriptomics, proteomics, and metabolomics) at relevant time points because biological systems are often not passive but always responsive and reactive. Systems biology approaches can help guide disease treatments and assess toxicity developments.

The identification and quantification of cellular metabolites in Ishii's paper was based on capillary electrophoresis coupled to time-of-flight (CE-TOF) MS. The profiled metabolome mainly covers metabolites that are central to carbon metabolism such as glycolysis, pentose phosphate pathway, TCA cycle, amino acids, and nucleotides (52). Mixtures of available chemical standards were run in parallel for the quantification of metabolites in study samples. Peak area ratios of analytes in samples and chemical standards were used to provide absolute concentrations of more than 110 metabolites, and cellular concentrations were calculated based on single cell weight, cell dry weight analysis, and estimated volume for a single cell. Metabolic flux was performed with carbon-13-labeled glucose (1-^{13}C glucose), and analyses were done with GC-MS. Flux analyses with carbon-13-labeled glucose was used in the evaluation of pathways affected by mutation of genes such as *rpe* (ribulose-P-3-epimerase) and *zwf* (glucose-6-phosphate dehydrogenase). Quantification of genes was carried out with microarray analysis, and proteins were profiled with two-dimensional differential gel electrophoresis (2D-DIGE) and quantified by MRM LC-MS/MS. This example further emphasizes the value of identification and quantification of metabolites as well as genes, transcripts, and proteins in order to compare them meaningfully.

Quantitative comparison of changes in genes, proteins, and metabolites may be more complicated than a simple direct ratio comparison since the dynamic range of perturbation for each of these different types of molecules is different. The kinetics and timing of the changes are also different for the different classes of biomolecules. In this example, Ishii and collaborators adopted an expression index (EI). This index was based on normalization of individual measurements as a logarithm (base 2) of the ratio of measurements in specific samples relative to the median values in samples measured on the same day or within a specific series. Logarithm conversion is often taken to scale changes that span very wide ranges and also shift the distribution of changes closer to a normal distribution. Statistical evaluation of mRNAs, proteins, and metabolites of significant importance was based on one-way ANOVA.

The striking outcomes of the analyses were the difference between the responses to environmental perturbation and genetics disruption. Environmental changes with different growth rates led to gradual changes of actively regulated transcription and protein levels but little variation in metabolites. This indicated that homeostasis of metabolites was an important objective of biological systems in order to maintain their desired functionalities. The stability of metabolites was achieved at the expense of rapid and dramatic adjustment and regulation of enzymes. Both mRNAs and enzymes were induced or repressed to maintain stable metabolite concentrations following an external perturbation. When growth rates increased, glucose transporter proteins (PtsH, especially PtsI) were induced to facilitate increased glucose uptake. Increased glucose uptake was welcomed by upregulation of genes coding enzymes for glycolysis (such as pfkA and pfkB for 6-phosphofructose-1-kinase, fbaA for fructose bisphosphate aldolase, and gapA for glyceraldehyde-P dehydrogenase). In order to closely couple glycolysis with the mitochondrial TCA cycle for bioenergetic production, proteins in the glyoxylate shunt were repressed. The anaplerotic glyoxylate pathways were often activated when four-carbon TCA cycle intermediates (such as malate and succinate) were needed, with availability of only two-carbon metabolites such as acetate. Fast growth rates observed with glucose as a substrate resulted in reduced levels of proteins such as PckA (phosphoenolpyruvate carboxykinase), GlcB (malate synthase G), AceB (malate synthase A), and AceA (isocitrate lyase).

On the other hand, metabolites were not passive bystanders. The concentration changes of metabolites and small-molecule enzyme cofactors (such as ATP) were indicative of the direction and magnitude of an enzymatic activity in terms of thermodynamic Gibbs free energy. Changes of metabolites due to external perturbation of nutrients such as glucose cascaded to enzymatic adjustments and transcription regulation of enzyme expression.

Although some genetic mutations can be lethal (illustrated with mutation to fbaB: fructose bisphosphate aldolase, and gapC: glyceraldehyde-P dehydrogenase in *E. coli*), other genetic mutations could be circumvented by redundancies in enzyme activities in cells. Cellular robustness offers isozymes and/or alternative metabolic pathways to make the same metabolites and maintain homeostasis. Mutation of one isoform of the *pfkA* (6-phosphofructose-1-kinase) gene was compensated by more than 13-fold increases of another isoform pfkB (6-phosphofructose-1-kinase). The availability of two different isoforms of the protein involved in a key rate-limiting step of glycolysis offered robustness to this bioenergetics pathway.

The deleterious effect on metabolites in the pentose phosphate pathway by mutation to genes encoding enzymes in the pathways was compensated by operating necessary pathways in reverse to produce the missing metabolites. For example, mutation to rpe (ribulose-P-3-epimerase) disrupted the formation of D-xylose 5-phosphate from D-ribulose 5-phosphate. Instead, transketolases (TktA, and especially TktB) were upregulated at the transcription level to make D-xylose 5-phosphate from the glycolysis intermediate glyceraldehyde-3-phosphate, which was a reversal of the usual direction of this reaction. If a mutation upstream of rpe occurred at glucose-6-phosphate dehydrogenase (zwf) disabling the formation of D-6-phosphate-glucono-1,5-lactone from α-glucose-6-phosphate, then essential metabolites could be formed from the reversed pentose phosphate pathway mainly with the help of transketolases (TktA and TktB). Interestingly, the transcription levels of genes encoding enzymes other than the mutated one remained similar to those in wild type.

Adjustments at the gene transcription level and the resulting changes in protein amounts, or pathway reroutes are designed to maintain the stability of the metabolome despite environmental perturbation and genetic mutations. All these adjustments serve the ultimate goal of keeping cellular functions at homeostasis by sustaining pathways that are essential for bioenergetics, proliferation, and homeostasis.

4.4.5 Metabolic Profiles Associated with Cellular Mitochondrial Toxicity and *In Vivo* Mitochondrial Diseases

Despite different etiologies of toxicity and disease, they can have similarities especially in the metabolites that might be produced by pathways that are perturbed similarly. Mitochondrial disease and mitochondrial toxicity share a common problem in ATP production from oxidative reactions in mitochondria. Comparing these two seemingly different problems side by side can help researchers in both areas of disease therapeutics and assessment of drug toxicity.

Cellular models have been valuable tools to investigate the underlying intricacies of biochemistry in areas such as genetics, diseases, and toxicities. Similarly, cellular metabolic profiling provides an approach to observe changes to metabolic profiles both inside the cell and in culture media. Although some small-molecule metabolites continue to be useful as robust biomarkers for *in vivo* diagnoses (e.g., glucose for diabetics, cholesterol for coronary artery disease, sorbitol for cataracts), many informative metabolites often elude our attention because of dynamic processes such as systemic circulation, active transport, and transient metabolism to other metabolites. An approach to biomarker discovery that starts with cell models may help establish an understanding of relevant biochemical pathways to facilitate identification of a robust biomarker(s) *in vivo*.

Using metabolomics to search for robust metabolite biomarkers of mitochondrial disease in plasma provided an example of utilizing both *in vitro* models of disease or toxicity, and *in vivo* human subjects (138). Shaham et al. started the search for accessible mitochondrial toxicity biomarkers in myotube cells differentiated from myoblast C2C13 cells in the presence of 2% horse serum for 4 days. Mitochondrial toxicity was induced by either 0.1 μM rotenone or

0.5 μM antimycin. After 8 h incubation, the medium was removed by aspiration, and the cells were quickly washed with ice-cold phosphate buffered saline (PBS) before quenching and extracting with 80% methanol at dry ice temperature. Targeted metabolite quantification of the medium and cell extracts was carried out by MRM using LC-MS/MS, and unbiased metabolite profiling was carried out by proton NMR. LC-MS/MS detected 117 intracellular endogenous metabolites and 84 metabolites in culture media. The corresponding NMR analysis identified 43 intracellular metabolites and 30 metabolites in cell culture (139). Three-quarters of the intracellular metabolites and two-thirds of the extracellular metabolites identified by NMR matched concordantly with LC-MS/MS analyses. LC-MS/MS had the advantage of greater sensitivity to detect metabolites present at lower concentrations. NMR provided quick and direct metabolite quantification with a single quantification reference standard. Therefore, these two complementary analyses ensured expedient and broad coverage of the metabolome.

Rotenone is known to inhibit NADH dehydrogenase by binding to the ubiquinone binding site of complex I of the electron transport chain in mitochondria (ETC, alternatively called the *respiratory chain* (RC)). Intimate coupling of NADH dehydrogenase in complex I and malate dehydrogenase in the TCA cycle facilitated detection of the inhibition to complex I by accumulation of malate and fumarate (Fig. 4.3). Antimycin inhibits succinate dehydrogenase by binding to coenzyme Q and interrupting electron transfer in ETC at complex III (Fig. 4.3). This resulted in accumulation of succinate.

Metabolic profiles in cells showed increases in malate and fumarate with rotenone, and in succinate with antimycin. In addition, succinate also appeared in culture media. The appearance of succinate outside the cells was likely facilitated by its membrane transporter. Uridine was formed as a downstream product in metabolic pathways where dihydroorotate was metabolized by dihydroorotate dehydrogenase (DHODH) that was coupled to coenzyme Q. Succinate dehydrogenase was also coupled to coenzyme Q and uridine decreased when succinate accumulated in the presence of antimycin. Uridine production was not impaired by rotenone inhibition of NADH dehydrogenase. In addition, nonoxidative glucose metabolism as a result of impaired mitochondrial function was reflected by high glucose consumption and increased production of lactate and alanine. Decreased biosynthesis of proteins and phospholipids was indicated by reduced uptake of amino acids and choline from culture media as well as alteration of other metabolites in cells and media.

Phosphocreatine is the energy reservoir for high energy phosphoryl groups to sustain a steady supply of ATP in muscle cells. With mitochondrial toxicity due to rotenone and antimycin, phosphocreatine was depleted (i.e., from 50–60% to about 20% in the pool of phosphocreatine and creatine) with a corresponding

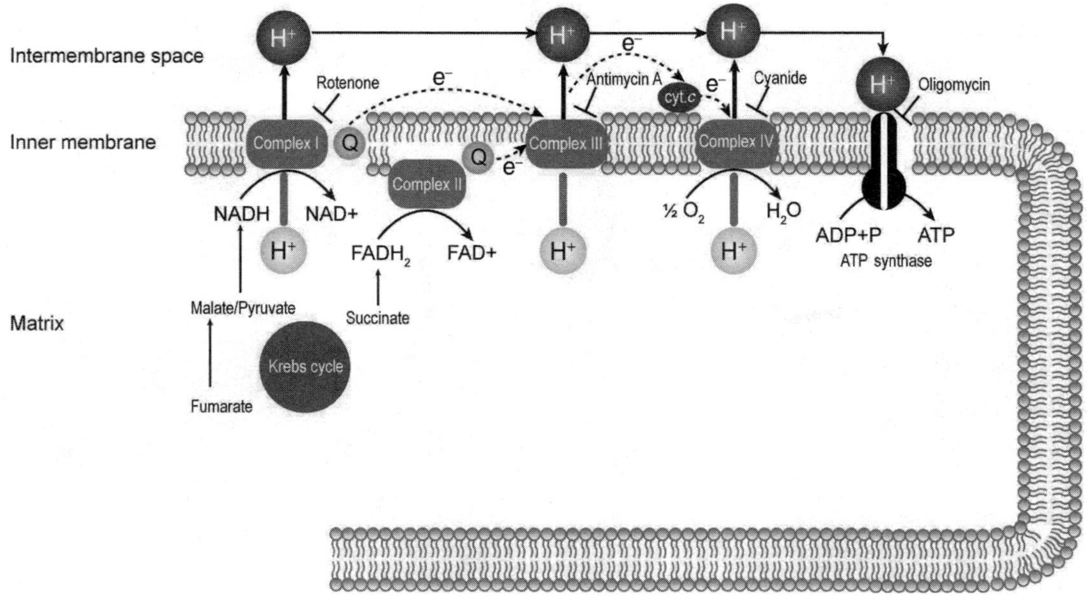

FIGURE 4.3 Illustration of the coupling of the electron transfer chain (ETC) and Krebs cycle (alternatively called *citric acid cycle* or *tricarboxylic acid cycle* (TCA)) metabolites in the mitochondrial inner membrane. Rotenone, antimycin A, cyanide, and oligomycin are known mitochondrial toxicants inhibiting at sites of complex I, III, IV, and ATP synthase, respectively. Krebs cycle metabolites that rely on NADH and $FADH_2$ for enzymatic conversion change intracellularly and in culture media in response to inhibition in the ETC.

accumulation of creatine (Fig. 4.4). LC-MS/MS detected increased creatine in culture media treated with rotenone.

Defects in genes coding for proteins that are necessary for normal electron transfer and oxidative phosphorylation (OXPHOS) in mitochondria in humans can result in disorders in mitochondrial OXPHOS or inherited respiratory chain diseases (RCDs). These represent some of the most common inborn errors of metabolism in humans (140, 141). Profiling plasma metabolites from two independent cohorts of RCD patients who have either a pathogenic mutation or

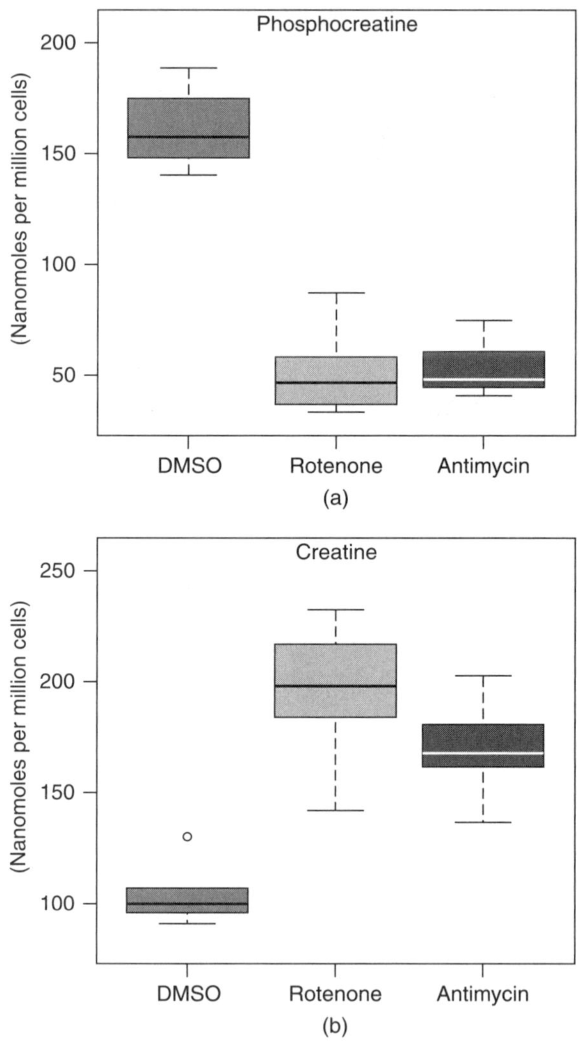

FIGURE 4.4 (a) Depletion of phosphocreatine and (b) accumulation of creatine in C2C12-differentiated myotube cells treated with 0.1 μM rotenone or 0.5 μM antimycin for 8 h. The measurements were carried out using NMR on cell lysates.

abnormally low RC enzymes and myopathy revealed a total of 68 metabolites. Of the 32 metabolites that changed in RCD patients relative to control human subjects, 26 were also found to change in cell culture media. Besides two of the commonly changed metabolites, lactate and alanine that are considered to be RCD plasma biomarkers, creatine increased by 201–233%, while the change in uridine was marginal in the two independent cohorts. Among the three marker metabolites (i.e., lactate, alanine, and creatine) in plasma, creatine exhibited the largest and most significant difference between RCD patients and controls.

This example illustrates the synergistic benefit of analyzing metabolic profiles from both *in vitro* cellular and *in vivo* models (i.e., animals and human). Proceeding to metabolite biomarker elucidation, these two systems help check, confirm, and validate the results seen in both approaches by revealing the pathways associated with common underlying biochemical mechanisms. Unlike individual metabolites or a single family of metabolites, metabolic profiles spanning several different pathways can lead to a panel of effective biomarkers that offers increased specificity to a diagnosis. Quantitative variation in biomarker responses can be an indication of subtle differences of mechanistic or biochemical effects.

4.5 FINAL THOUGHTS

Improvements in analytical technologies and pathway analysis tools will in no doubt continue to serve metabolomics for a wide variety of applications. Metabolomics may not be a fixed set of procedures, but its actual roles should be malleable and flexible, which allows it to address many problems encountered in basic research, safety assessment, clinical research, and pharmaceutical research and development. The infancy of metabolomics in comparison to other omics suggests that creative thinking and close interaction among collaborators are as important as the technology itself to drive its development. The strength of metabolomics is its holistic approach in understanding basic biochemical mechanisms that is complementary to other biochemical and molecular methods (e.g., genomics, transcriptomics, proteomics, and biochemical assays). The commonality in metabolite chemical structures and metabolic pathways between different species and organisms is also a clear benefit (e.g., the structure of ATP is the same in all species, and it is a ubiquitous energy molecule). Proposing biomarker candidates based on well-understood biochemical mechanisms helps build more confidence in the relevance to the biomarkers, and will help biomarkers to reach a wider acceptance. At the analytical level, molecular identification and quantification are indispensable to drive biological interpretation for metabolomic and biochemical results.

What is not covered here is another great potential end point of metabolomics based on *in vivo* MRS. Metabolic profiling, especially ^{1}H NMR, provides a direct and noninvasive link between chemical analysis and imaging (MRI) for *in vivo* diagnoses. Metabolic profiling using *in vivo* NMR, or MRS, can offer a three-dimensional molecular view of metabolism in tissues in real time, thus reflecting disease or toxicity. Metabolic profiling is being improved and offers tremendous potential for future medical diagnoses (142). NMR metabolomics serves as a tool to lay the ground work for the assessment and translation of candidate metabolite biomarkers for *in vivo* MRS diagnoses. Prescreens by NMR metabolomics in tissues can help select candidate metabolites on the basis of basic biochemical understanding. Both proton and carbon NMR analyses can help evaluate the feasibility of monitoring particular metabolites for *in vivo* quantitative MRS diagnoses. Recent developments of hyperpolarized ^{13}C NMR or MRS offer the potential of high resolution and great sensitivity that are often challenging in traditional ^{1}H NMR analyses (143). Interested readers can read literature cited here and current publications on *in vivo* MRS.

ACKNOWLEDGMENTS

We would like to acknowledge Dr. Frank Sistare's critical reading or this manuscript, and thank him for his insightful suggestions to this chapter.

REFERENCES

1. Fiehn O. Metabolomics—the link between genotypes and phenotypes. Plant Mol Biol 2002;48:155–171.
2. Kind T, Scholz M, Fiehn O. How large is the metabolome? A critical analysis of data exchange practices in chemistry. PLoS One 2009;4:e5440.
3. Robertson DG. Metabonomics in toxicology: a review. Toxicol Sci 2005;85: 809–822.
4. Freedman DS, Otvos JD, Jeyarajah EJ, Barboriak JJ, Anderson AJ, Walker JA. Relation of lipoprotein subclasses as measured by proton nuclear magnetic resonance spectroscopy to coronary artery disease. Arterioscler Thromb Vasc Biol 1998;18:1046–1053.
5. Gibney MJ, Walsh M, Brennan L, Roche HM, German B, van Ommen B. Metabolomics in human nutrition: opportunities and challenges. Am J Clin Nutr 2005;82:497–503.
6. Memelink J. Tailoring the plant metabolome without a loose stitch. Trends Plant Sci 2005;10:305–307.
7. Lu W, Kimball E, Rabinowitz JD. A high-performance liquid chromatography-tandem mass spectrometry method for quantitation of nitrogen-containing intracellular metabolites. J Am Soc Mass Spectrom 2006;17:37–50.
8. Yuan J, Doucette CD, Fowler WU, Feng XJ, Piazza M, Rabitz HA, Wingreen NS, Rabinowitz JD. Metabolomics-driven quantitative analysis of ammonia assimilation in E. coli. Mol Syst Biol 2009;5:302.
9. Wu L, Mashego MR, van Dam JC, Proell AM, Vinke JL, Ras C, van Winden WA, van Gulik WM, Heijnen JJ. Quantitative analysis of the microbial metabolome by isotope dilution mass spectrometry using uniformly 13C-labeled cell extracts as internal standards. Anal Biochem 2005;336:164–171.

10. Markuszewski MJ, Szczykowska M, Siluk D, Kaliszan R. Human red blood cells targeted metabolome analysis of glycolysis cycle metabolites by capillary electrophoresis using an indirect photometric detection method. J Pharm Biomed Anal 2005;39:636–642.

11. Mortishire-Smith RJ, Skiles GL, Lawrence JW, Spence S, Nicholls AW, Johnson BA, Nicholson JK. Use of metabonomics to identify impaired fatty acid metabolism as the mechanism of a drug-induced toxicity. Chem Res Toxicol 2004;17:165–173.

12. Tanaka K, Ikeda Y. Hypoglycin and Jamaican vomiting sickness. Prog Clin Biol Res 1990;321:167–184.

13. Tanaka K, Hine DG, West-Dull A, Lynn TB. Gas-chromatographic method of analysis for urinary organic acids. I. Retention indices of 155 metabolically important compounds. Clin Chem 1980;26:1839–1846.

14. Sweatman BC, Farrant RD, Holmes E, Ghauri FY, Nicholson JK, Lindon JC. 600MHz 1H-NMR spectroscopy of human cerebrospinal fluid: effects of sample manipulation and assignment of resonances. J Pharm Biomed Anal 1993;11: 651–664.

15. Dettmer K, Aronov PA, Hammock BD. Mass spectrometry-based metabolomics. Mass Spectrom Rev 2007;26:51–78.

16. Bennett BD, Kimball EH, Gao M, Osterhout R, Van Dien SJ, Rabinowitz JD. Absolute metabolite concentrations and implied enzyme active site occupancy in Escherichia coli. Nat Chem Biol 2009;5:593–599.

17. Wilson I, Nicholson JK. Metabonomics and global systems biology. In: Griffin WJ, editor. Metabolomics, metabonomics and metabolite profiling. RSC Publishing, (Cambridge, UK) 2008;295–314.

18. Sands CJ, Coen M, Ebbels TM, Holmes E, Lindon JC, Nicholson JK. Data-driven approach for metabolite relationship recovery in biological (1)h NMR data sets using iterative statistical total correlation spectroscopy. Anal Chem 2011;83:2075–2082.

19. Bollard ME, Contel NR, Ebbels TM, Smith L, Beckonert O, Cantor GH, Lehman-McKeeman L, Holmes EC, Lindon JC, Nicholson JK, Keun HC. NMR-based metabolic profiling identifies biomarkers of liver regeneration following partial hepatectomy in the rat. J Proteome Res 2010;9:59–69.

20. Lindon JC, Nicholson JK. Analytical technologies for metabonomics and metabolomics, and multi-omic information recovery. Trends Anal Chem 2008;27: 194–204.

21. Bodenhausen G, Ruben DJ. Natural abundance nitrogen-15 NMR by enhanced heteronuclear spectroscopy. Chem Phys Lett 1980;69:185–189.

22. Muller L. Sensitivity enhanced detection of weak nuclei using heteronuclear multiple quantim coherence. J Am Chem Soc 1979;101:4481–4484.

23. Styles P, Soffe NF, Scott CA, Cragg DA, white DJ, White PCJ. A high-resolution NMR probe in which the coil and preamplifier are cooled with liquid helium. J Magn Reson 1984;60:397–404.

24. Kelly AE, Ou HD, Withers R, Dotsch V. Low-conductivity buffers for high-sensitivity NMR measurements. J Am Chem Soc 2002;124:12013–12019.

25. Robosky LC, Reily MD, Avizonis D. Improving NMR sensitivity by use of salt-tolerant cryogenically cooled probes. Anal Bioanal Chem 2007;387:529–532.

26. Xu EY, Perlina A, Vu H, Troth SP, Brennan RJ, Aslamkhan AG, Xu Q. Integrated pathway analysis of rat urine metabolic profiles and kidney transcriptomic profiles to elucidate the systems toxicology of model nephrotoxicants. Chem Res Toxicol 2008;21:1548–1561.
27. Reily MD, Robosky LC, Manning ML, Butler A, Baker JD, Winters RT. DFTMP, an NMR reagent for assessing the near-neutral pH of biological samples. J Am Chem Soc 2006;128:12360–12361.
28. Asiago V, Gowda GA, Zhang S, Shanaiah N, Clark J, Raftery D. Use of EDTA to minimize ionic strength dependent frequency shifts in the 1H NMR spectra of urine. Metabolomics 2008;4:328–336.
29. Robosky LC, Wade K, Woolson D, Baker JD, Manning ML, Gage DA, Reily MD. Quantitative evaluation of sebum lipid components with nuclear magnetic resonance. J Lipid Res 2008;49:686–692.
30. Ogg RJ, Kinsley PB, Taylor JS. WET: a T1-B1-insensitive water suppression method for in vivo localized 1H NMR spectroscopy. J Magn Reson B 1994;104:1–10.
31. Pratt EL. The growth of microorganisms on media exposed to ultraviolet radiations. J Bacteriol 1936;32:613–619.
32. Glish GL, Burinsky DJ. Hybrid mass spectrometers for tandem mass spectrometry. J Am Soc Mass Spectrom 2008;19:161–172.
33. Tolstikov VV, Fiehn O. Analysis of highly polar compounds of plant origin: combination of hydrophilic interaction chromatography and electrospray ion trap mass spectrometry. Anal Biochem 2002;301:298–307.
34. Potterat O, Wagner K, Haag H. Liquid chromatography-electrospray time-of-flight mass spectrometry for on-line accurate mass determination and identification of cyclodepsipeptides in a crude extract of the fungus Metarrhizium anisopliae. J Chromatogr A 2000;872:85–90.
35. Guilhaus M, Selby D, Mlynski V. Orthogonal acceleration time-of-flight mass spectrometry. Mass Spectrom Rev 2000;19:65–107.
36. Schaub TM, Hendrickson CL, Horning S, Quinn JP, Senko MW, Marshall AG. High-performance mass spectrometry: Fourier transform ion cyclotron resonance at 14.5 Tesla. Anal Chem 2008;80:3985–3990.
37. Makarov A, Denisov E, Kholomeev A, Balschun W, Lange O, Strupat K, Horning S. Performance evaluation of a hybrid linear ion trap/orbitrap mass spectrometer. Anal Chem 2006;78:2113–2120.
38. Makarov A, Denisov E, Lange O, Horning S. Dynamic range of mass accuracy in LTQ Orbitrap hybrid mass spectrometer. J Am Soc Mass Spectrom 2006;17:977–982.
39. Idborg-Bjorkman H, Edlund PO, Kvalheim OM, Schuppe-Koistinen I, Jacobsson SP. Screening of biomarkers in rat urine using LC/electrospray ionization-MS and two-way data analysis. Anal Chem 2003;75:4784–4792.
40. Plumb RS, Stumpf CL, Gorenstein MV, Castro-Perez JM, Dear GJ, Anthony M, Sweatman BC, Connor SC, Haselden JN. Metabonomics: the use of electrospray mass spectrometry coupled to reversed-phase liquid chromatography shows potential for the screening of rat urine in drug development. Rapid Commun Mass Spectrom 2002;16:1991–1996.

41. Kiefer P, Portais JC, Vorholt JA. Quantitative metabolome analysis using liquid chromatography-high-resolution mass spectrometry. Anal Biochem 2008;382: 94–100.
42. Voyksner RD, Williams FP, Hines JW. Analysis of candidate anticancer drugs by thermospray high-performance liquid chromatography-mass spectrometry and by high-resolution mass spectrometry. J Chromatogr 1985;347:137–146.
43. Barber M, Bardoli RS, Sedgwick RD, Tyler AN. Fast atom bombardment of solids as an ion source in mass spectrometry. Nature 1981;293:270–275.
44. Barber M, Bordoli RS, Garner GV, Gordon DB, Sedgwick RD, Tetler LW, Tyler AN. Fast-atom-bombardment mass spectra of enkephalins. Biochem J 1981;197:401–404.
45. Evans AM, Dehaven CD, Barrett T, Mitchell M, Milgram E. Integrated, nontargeted ultrahigh performance liquid chromatography/Electrospray ionization tandem mass spectrometry platform for the identification and relative quantification of the small-molecule complement of biological systems. Anal Chem 2009;81:6656–6667.
46. Rosenberg E. The potential of organic (electrospray- and atmospheric pressure chemical ionisation) mass spectrometric techniques coupled to liquid-phase separation for speciation analysis. J Chromatogr A 2003;1000:841–889.
47. Raffaelli A, Saba A. Atmospheric pressure photoionization mass spectrometry. Mass Spectrom Rev 2003;22:318–331.
48. Fiehn O, Kopka J, Trethewey RN, Willmitzer L. Identification of uncommon plant metabolites based on calculation of elemental compositions using gas chromatography and quadrupole mass spectrometry. Anal Chem 2000;72:3573–3580.
49. Lewy AJ, Markey SP. Analysis of melatonin in human plasma by gas chromatography negative chemical ionization mass spectrometry. Science 1978;201:741–743.
50. Yanes O, Tautenhahn R, Patti GJ, Siuzdak G. Expanding coverage of the metabolome for global metabolite profiling. Anal Chem 2011;83:2152–2161.
51. Kind T, Fiehn O. Advances in structure elucidation of small molecules using mass spectrometry. Bioanal Rev 2010;2:23–60.
52. Soga T. Capillary electrophoresis-mass spectrometry for metabolomics. Methods Mol Biol 2007;358:129–137.
53. Sieber M, Wagner S, Rached E, Amberg A, Mally A, Dekant W. Metabonomic study of ochratoxin a toxicity in rats after repeated administration: phenotypic anchoring enhances the ability for biomarker discovery. Chem Res Toxicol 2009;22:1221–1231.
54. Tolstikov VV, Fiehn O, Tanaka N. Application of liquid chromatography-mass spectrometry analysis in metabolomics: reversed-phase monolithic capillary chromatography and hydrophilic chromatography coupled to electrospray ionization-mass spectrometry. Methods Mol Biol 2007;358:141–155.
55. Tolstikov VV, Lommen A, Nakanishi K, Tanaka N, Fiehn O. Monolithic silica-based capillary reversed-phase liquid chromatography/electrospray mass spectrometry for plant metabolomics. Anal Chem 2003;75:6737–6740.
56. Naidong W. Bioanalytical liquid chromatography tandem mass spectrometry methods on underivatized silica columns with aqueous/organic mobile phases. J Chromatogr B Analyt Technol Biomed Life Sci 2003;796:209–224.
57. Cubbon S, Antonio C, Wilson J, Thomas-Oates J. Metabolomic applications of HILIC-LC-MS. Mass Spectrom Rev 2010;29:671–684.

58. Noga M, Sucharski F, Suder P, Silberring J. A practical guide to nano-LC troubleshooting. J Sep Sci 2007;30:2179–2189.
59. Myint KT, Uehara T, Aoshima K, Oda Y. Polar anionic metabolome analysis by nano-LC/MS with a metal chelating agent. Anal Chem 2009;81:7766–7772.
60. Myint KT, Aoshima K, Tanaka S, Nakamura T, Oda Y. Quantitative profiling of polar cationic metabolites in human cerebrospinal fluid by reversed-phase nanoliquid chromatography/mass spectrometry. Anal Chem 2009;81:1121–1129.
61. Cunico RL, Gooding KM, Wehr T. Capillary Electrophoresis. Basic HPLC and CE of biomolecules. Bay Bioanalytical Laboratory (Richmond, CA); 1988. p. 388.
62. Soga T, Ueno Y, Naraoka H, Matsuda K, Tomita M, Nishioka T. Pressure-assisted capillary electrophoresis electrospray ionization mass spectrometry for analysis of multivalent anions. Anal Chem 2002;74:6224–6229.
63. Soga T, Ohashi Y, Ueno Y, Naraoka H, Tomita M, Nishioka T. Quantitative metabolome analysis using capillary electrophoresis mass spectrometry. J Proteome Res 2003;2:488–494.
64. Soga T, Baran R, Suematsu M, Ueno Y, Ikeda S, Sakurakawa T, Kakazu Y, Ishikawa T, Robert M, Nishioka T, Tomita M. Differential metabolomics reveals ophthalmic acid as an oxidative stress biomarker indicating hepatic glutathione consumption. J Biol Chem 2006;281:16768–16776.
65. Gates SC, Sweeley CC, Krivit W, DeWitt D, Blaisdell BE. Automated metabolic profiling of organic acids in human urine. II. Analysis of urine samples from "healthy" adults, sick children, and children with neuroblastoma. Clin Chem 1978;24:1680–1689.
66. Gates SC, Dendramis N, Sweeley CC. Automated metabolic profiling of organic acids in human urine. I. Description of methods. Clin Chem 1978;24:1674–1679.
67. Gates SC, Sweeley CC. Quantitative metabolic profiling based on gas chromatography. Clin Chem 1978;24:1663–1673.
68. Bedair M, Sumner LW. Current and emerging mass-spectrometry technologies for metabolomics. Trends Anal Chem 2008;27:238–250.
69. Scalbert A, Brennan L, Fiehn O, Hankemeier T, Kristal B, van Ommen B, Pujos-Guillot E, Verheij E, Wishart D, Wopereis S. Mass-spectrometry-based metabolomics: limitations and recommendations for future progress with particular focus on nutrition research. Metabolomics 2009;5:435–458.
70. Tanaka K, West-Dull A, Hine DG, Lynn TB, Lowe T. Gas-chromatographic method of analysis for urinary organic acids. II. Description of the procedure, and its application to diagnosis of patients with organic acidurias. Clin Chem 1980;26:1847–1853.
71. Jellum E, Kvittingen EA, Stokke O. Mass spectrometry in diagnosis of metabolic disorders. Biomed Environ Mass Spectrom 1988;16:57–62.
72. Roessner U, Wagner C, Kopka J, Trethewey RN, Willmitzer L. Technical advance: simultaneous analysis of metabolites in potato tuber by gas chromatography-mass spectrometry. Plant J 2000;23:131–142.
73. Roessner U, Luedemann A, Brust D, Fiehn O, Linke T, Willmitzer L, Fernie A. Metabolic profiling allows comprehensive phenotyping of genetically or environmentally modified plant systems. Plant Cell 2001;13:11–29.
74. Adams MA, Chen Z, Landman P, Colmer TD. Simultaneous determination by capillary gas chromatography of organic acids, sugars, and sugar alcohols in plant tissue extracts as their trimethylsilyl derivatives. Anal Biochem 1999;266:77–84.

75. Fiehn O, Kopka J, Dormann P, Altmann T, Trethewey RN, Willmitzer L. Metabolite profiling for plant functional genomics. Nat Biotechnol 2000;18:1157–1161.
76. Katona ZF, Sass P, Molnár-Perl I. Simultaneous determination of sugars, sugar alcohols, acids and amino acids in apricots by gas chromatography—mass spectrometry. J Chromatogr A 1999;847:91–102.
77. McLafferty FW, Zhang MY, Stauffer DB, Loh SY. Comparison of algorithms and databases for matching unknown mass spectra. J Am Soc Mass Spectrom 1998;9:92–95.
78. McLafferty FW, Stauffer DA, Loh SY, Wesdemiotis C. Unknown identification using reference mass spectra. Quality evaluation of databases. J Am Soc Mass Spectrom 1999;32:438–443.
79. Pool WG, Leeuw JW, Van der Graaf BJ. Automated extraction of pure mass spectra from gas chromatographic/mass spectrometric data. J Mass Spectrom 1997;32:438–443.
80. Halket JM, Przyborowska A, Stein SE, Mallard WG, Down S, Chalmers RA. Deconvolution gas chromatography/mass spectrometry of urinary organic acids—potential for pattern recognition and automated identification of metabolic disorders. Rapid Commun Mass Spectrom 1999;13:279–284.
81. Balogh MP. Debating resolution and mass accuracy. LC-GC Europe 2004;17: 152–159.
82. Lu W, Bennett BD, Rabinowitz JD. Analytical strategies for LC-MS-based targeted metabolomics. J Chromatogr B Analyt Technol Biomed Life Sci 2008;871:236–242.
83. Want EJ, O'Maille G, Smith CA, Brandon TR, Uritboonthai W, Qin C, Trauger SA, Siuzdak G. Solvent-dependent metabolite distribution, clustering, and protein extraction for serum profiling with mass spectrometry. Anal Chem 2006;78:743–752.
84. Lu W, Kwon YK, Rabinowitz JD. Isotope ratio-based profiling of microbial folates. J Am Soc Mass Spectrom 2007;18:898–909.
85. Winder CL, Dunn WB, Schuler S, Broadhurst D, Jarvis R, Stephens GM, Goodacre R. Global metabolic profiling of Escherichia coli cultures: an evaluation of methods for quenching and extraction of intracellular metabolites. Anal Chem 2008;80:2939–2948.
86. Lin CY, Wu H, Tjeerdema RS, Viant MR. Evaluation of metabolite extraction strategies from tissue samples using NMR metabolomics. Metabolomics 2007;3:55–67.
87. Faijes M, Mars AE, Smid EJ. Comparison of quenching and extraction methodologies for metabolome analysis of Lactobacillus plantarum. Microb Cell Fact 2007;6:27.
88. Warrack BM, Hnatyshyn S, Ott KH, Reily MD, Sanders M, Zhang H, Drexler DM. Normalization strategies for metabonomic analysis of urine samples. J Chromatogr B Analyt Technol Biomed Life Sci 2009;877:547–552.
89. Michopoulos F, Lai L, Gika H, Theodoridis G, Wilson I. UPLC-MS-based analysis of human plasma for metabonomics using solvent precipitation or solid phase extraction. J Proteome Res 2009;8:2114–2121.
90. Sreekumar A, Poisson LM, Rajendiran TM, Khan AP, Cao Q, Yu J, Laxman B, Mehra R, Lonigro RJ, Li Y, Nyati MK, Ahsan A, Kalyana-Sundaram S, Han B, Cao X, Byun J, Omenn GS, Ghosh D, Pennathur S, Alexander DC, Berger A, Shuster JR, Wei JT, Varambally S, Beecher C, Chinnaiyan AM. Metabolomic profiles delineate potential role for sarcosine in prostate cancer progression. Nature 2009;457:910–914.

91. Gullberg J, Jonsson P, Nordstrom A, Sjostrom M, Moritz T. Design of experiments: an efficient strategy to identify factors influencing extraction and derivatization of Arabidopsis thaliana samples in metabolomic studies with gas chromatography/mass spectrometry. Anal Biochem 2004;331:283–295.
92. Junier MP, Dray F, Blair I, Capdevila J, Dishman E, Falck JR, Ojeda SR. Epoxygenase products of arachidonic acid are endogenous constituents of the hypothalamus involved in D2 receptor-mediated, dopamine-induced release of somatostatin. Endocrinology 1990;126:1534–1540.
93. Lee SH, Blair IA. Targeted chiral lipidomics analysis by liquid chromatography electron capture atmospheric pressure chemical ionization mass spectrometry (LC-ECAPCI/MS). Methods Enzymol 2007;433:159–174.
94. Mesaros C, Lee SH, Blair IA. Analysis of epoxyeicosatrienoic acids by chiral liquid chromatography/electron capture atmospheric pressure chemical ionization mass spectrometry using [C-13]-analog internal standards. Rapid Commun Mass Spectrom 2010;24:3237–3247.
95. Ciccimaro E, Blair IA. Stable-isotope dilution LC-MS for quantitative biomarker analysis. Bioanalysis 2010;2:311–341.
96. Lenz EM, Wilson ID. Analytical strategies in metabonomics. J Proteome Res 2007;6:443–458.
97. Atherton HJ, Bailey NJ, Zhang W, Taylor J, Major H, Shockcor J, Clarke K, Griffin JL. A combined 1H-NMR spectroscopy- and mass spectrometry-based metabolomic study of the PPAR-alpha null mutant mouse defines profound systemic changes in metabolism linked to the metabolic syndrome. Physiol Genomics 2006;27:178–186.
98. Xu Q, Sachs JR, Wang TC, Schaefer WH. Quantification and identification of components in solution mixtures from 1D proton NMR spectra using singular value decomposition. Anal Chem 2006;78:7175–7185.
99. Xu EY, Schaefer WH, Xu Q. Metabolomics in pharmaceutical research and development: metabolites, mechanisms and pathways. Curr Opin Drug Discov Devel 2009;12:40–52.
100. Weljie AM, Newton J, Mercier P, Carlson E, Slupsky CM. Targeted profiling: quantitative analysis of 1H NMR metabolomics data. Anal Chem 2006;78:4430–4442.
101. Wiechert W. 13C metabolic flux analysis. Metab Eng 2001;3:195–206.
102. Varma A, Palsson BO. Stoichiometric flux balance models quantitatively predict growth and metabolic by-product secretion in wild-type Escherichia coli W3110. Appl Environ Microbiol 1994;60:3724–3731.
103. Marx A, de Graaf AA, Wiechert W, Eggeling L, Sahm H. Determination of the fluxes in the central metabolism of Corynebacterium glutamicum by nuclear magnetic resonance spectroscopy combined with metabolite balancing. Biotechnol Bioeng 1996;49:111–129.
104. Golman K, Olsson LE, Axelsson O, Mansson S, Karlsson M, Petersson JS. Molecular imaging using hyperpolarized 13C. Br J Radiol 2003;76(2): S118–S127.
105. Golman K, Zandt RI, Lerche M, Pehrson R, Ardenkjaer-Larsen JH. Metabolic imaging by hyperpolarized 13C magnetic resonance imaging for in vivo tumor diagnosis. Cancer Res 2006;66:10855–10860.
106. Christensen B, Nielsen J. Isotopomer analysis using GC-MS. Metab Eng 1999;1: 282–290.

107. de Graaf AA, Mahle M, Mollney M, Wiechert W, Stahmann P, Sahm H. Determination of full 13C isotopomer distributions for metabolic flux analysis using heteronuclear spin echo difference NMR spectroscopy. J Biotechnol 2000;77:25–35.
108. Kleijn RJ, Geertman JM, Nfor BK, Ras C, Schipper D, Pronk JT, Heijnen JJ, van Maris AJ, van Winden WA. Metabolic flux analysis of a glycerol-overproducing Saccharomyces cerevisiae strain based on GC-MS, LC-MS and NMR-derived C-labelling data. FEMS Yeast Res 2007;7:216–231.
109. Smith CA, Want EJ, O'Maille G, Abagyan R, Siuzdak G. XCMS: processing mass spectrometry data for metabolite profiling using nonlinear peak alignment, matching, and identification. Anal Chem 2006;78:779–787.
110. Tautenhahn R, Patti GJ, Kalisiak E, Miyamoto T, Schmidt M, Fan YL, McBee J, Baliga NS, Siuzdak G. metaXCMS: second-order analysis of untargeted metabolomics data. Anal Chem 2011;83:696–700.
111. Nielsen N-PV, Carstensen JM, Smedsgaard J. Aligning of single and multiple wavelength chromatographic profiles for chemometric data analysis using correlation optimized warping. J Chromatogr A 1998;805:17–35.
112. Veselkov KA, Lindon JC, Ebbels TM, Crockford D, Volynkin VV, Holmes E, Davies DB, Nicholson JK. Recursive segment-wise peak alignment of biological (1)h NMR spectra for improved metabolic biomarker recovery. Anal Chem 2009;81:56–66.
113. Alm E, Torgrip RJ, Aberg KM, Schuppe-Koistinen I, Lindberg J. A solution to the 1D NMR alignment problem using an extended generalized fuzzy Hough transform and mode support. Anal Bioanal Chem 2009;395:213–223.
114. van den Berg RA, Hoefsloot HC, Westerhuis JA, Smilde AK, van der Werf MJ. Centering, scaling, and transformations: improving the biological information content of metabolomics data. BMC Genomics 2006;7:142.
115. Eriksson L, Johansson E, Kettaneh-Wold N, Trygg J, Wikstrom C, Wold S. Volume 1, Multi- and megavariate data analysis. Umetrics Academy (Umeå, Sweden); 2006.
116. Appiah-Amponsah E, Shanaiah N, Nagana Gowda GA, Owusu-Sarfo K, Ye T, Raftery D. Identification of 4-deoxythreonic acid present in human urine using HPLC and NMR techniques. J Pharm Biomed Anal 2009;50:878–885.
117. Wishart DS, Tzur D, Knox C, Eisner R, Guo AC, Young N, Cheng D, Jewell K, Arndt D, Sawhney S, Fung C, Nikolai L, Lewis M, Coutouly MA, Forsythe I, Tang P, Shrivastava S, Jeroncic K, Stothard P, Amegbey G, Block D, Hau DD, Wagner J, Miniaci J, Clements M, Gebremedhin M, Guo N, Zhang Y, Duggan GE, Macinnis GD, Weljie AM, Dowlatabadi R, Bamforth F, Clive D, Greiner R, Li L, Marrie T, Sykes BD, Vogel HJ, Querengesser L. HMDB: the human metabolome database. Nucleic Acids Res 2007;35:D521–D526.
118. Cui Q, Lewis IA, Hegeman AD, Anderson ME, Li J, Schulte CF, Westler WM, Eghbalnia HR, Sussman MR, Markley JL. Metabolite identification via the Madison metabolomics consortium database. Nat Biotechnol 2008;26:162–164.
119. Kopka J, Schauer N, Krueger S, Birkemeyer C, Usadel B, Bergmuller E, Dormann P, Weckwerth W, Gibon Y, Stitt M, Willmitzer L, Fernie AR, Steinhauser D. GMD@CSB.DB: the golm metabolome database. Bioinformatics 2005;21:1635–1638.
120. Kind T, Wohlgemuth G, Lee DY, Lu Y, Palazoglu M, Shahbaz S, Fiehn O. FiehnLib: mass spectral and retention index libraries for metabolomics based on quadrupole and time-of-flight gas chromatography/mass spectrometry. Anal Chem 2009;81:10038–10048.

121. Xu Q, Klees J, Teyral J, Capen R, Huang M, Sturgess AW, Hennessey JP Jr., Washabaugh M, Sitrin R, Abeygunawardana C. Quantitative nuclear magnetic resonance analysis and characterization of the derivatized Haemophilus influenzae type b polysaccharide intermediate for PedvaxHIB. Anal Biochem 2005;337:235–245.

122. Provencher SW. Estimation of metabolite concentrations from localized in vivo proton NMR spectra. Magn Reson Med 1993;30:672–679.

123. Bennett BD, Yuan J, Kimball EH, Rabinowitz JD. Absolute quantitation of intracellular metabolite concentrations by an isotope ratio-based approach. Nat Protoc 2008;3:1299–1311.

124. Basu SS, Mesaros C, Gelhaus SL, Blair IA. Stable isotope labeling by essential nutrients in cell culture for preparation of labeled coenzyme A and its thioesters. Anal Chem 2011;83:1363–1369.

125. Smilde AK, Jansen JJ, Hoefsloot HC, Lamers RJ, van der Greef J, Timmerman ME. ANOVA-simultaneous component analysis (ASCA): a new tool for analyzing designed metabolomics data. Bioinformatics 2005;21:3043–3048.

126. Caspi R, Foerster H, Fulcher CA, Kaipa P, Krummenacker M, Latendresse M, Paley S, Rhee SY, Shearer AG, Tissier C, Walk TC, Zhang P, Karp PD. The MetaCyc database of metabolic pathways and enzymes and the BioCyc collection of Pathway/Genome Databases. Nucleic Acids Res 2008;36:D623–D631.

127. Joshi-Tope G, Gillespie M, Vastrik I, D'Eustachio P, Schmidt E, de Bono B, Jassal B, Gopinath GR, Wu GR, Matthews L, Lewis S, Birney E, Stein L. Reactome: a knowledgebase of biological pathways. Nucleic Acids Res 2005;33:D428–D432.

128. Vastrik I, D'Eustachio P, Schmidt E, Gopinath G, Croft D, de Bono B, Gillespie M, Jassal B, Lewis S, Matthews L, Wu G, Birney E, Stein L. Reactome: a knowledge base of biologic pathways and processes. Genome Biol 2007;8:R39.

129. Matthews L, Gopinath G, Gillespie M, Caudy M, Croft D, de Bono B, Garapati P, Hemish J, Hermjakob H, Jassal B, Kanapin A, Lewis S, Mahajan S, May B, Schmidt E, Vastrik I, Wu G, Birney E, Stein L, D'Eustachio P. Reactome knowledgebase of human biological pathways and processes. Nucleic Acids Res 2009;37:D619–D622.

130. Nikolsky Y, Kirillov E, Zuev R, Rakhmatulin E, Nikolskaya T. Functional analysis of OMICs data and small molecule compounds in an integrated "knowledge-based" platform. Methods Mol Biol 2009;563:177–196.

131. Ganter B, Zidek N, Hewitt PR, Muller D, Vladimirova A. Pathway analysis tools and toxicogenomics reference databases for risk assessment. Pharmacogenomics 2008;9:35–54.

132. Busk M, Horsman MR, Jakobsen S, Bussink J, van der Kogel A, Overgaard J. Cellular uptake of PET tracers of glucose metabolism and hypoxia and their linkage. Eur J Nucl Med Mol Imaging 2008;35:2294–2303.

133. Desvergne B, Michalik L, Wahli W. Transcriptional regulation of metabolism. Physiol Rev 2006;86:465–514.

134. Brauer MJ, Yuan J, Bennett BD, Lu W, Kimball E, Botstein D, Rabinowitz JD. Conservation of the metabolomic response to starvation across two divergent microbes. Proc Natl Acad Sci USA 2006;103:19302–19307.

135. Bradley PH, Brauer MJ, Rabinowitz JD, Troyanskaya OG. Coordinated concentration changes of transcripts and metabolites in Saccharomyces cerevisiae. PLoS Comput Biol 2009;5:e1000270.

136. Tomlins SA, Mehra R, Rhodes DR, Cao X, Wang L, Dhanasekaran SM, Kalyana-Sundaram S, Wei JT, Rubin MA, Pienta KJ, Shah RB, Chinnaiyan AM. Integrative molecular concept modeling of prostate cancer progression. Nat Genet 2007;39:41–51.

137. Ishii N, Nakahigashi K, Baba T, Robert M, Soga T, Kanai A, Hirasawa T, Naba M, Hirai K, Hoque A, Ho PJ, Kakazu Y, Sugawara K, Igarashi S, Harada S, Masuda T, Sugiyama N, Togashi T, Hasegawa M, Takai Y, Yugi K, Arakawa K, Iwata N, Toya Y, Nakayama Y, Nishioka T, Shimizu K, Mori H, Tomita M. Multiple high-throughput analyses monitor the response of E. coli to perturbations. Science 2007;316:593–597.

138. Shaham O, Slate NG, Goldberger O, Xu Q, Ramanathan A, Souza AL, Clish CB, Sims KB, Mootha VK. A plasma signature of human mitochondrial disease revealed through metabolic profiling of spent media from cultured muscle cells. Proc Natl Acad Sci USA 2010;107:1571–1575.

139. Xu Q, Vu H, Liu L, Wang TC, Schaefer WH. Metabolic profiles show specific mitochondrial toxicities in vitro in myotube cells. J Biomol NMR 2011;49:207–219.

140. Thorburn DR. Diverse powerhouses. Nat Genet 2004;36:13–14.

141. DiMauro S, Schon EA. Mitochondrial respiratory-chain diseases. N Engl J Med 2003;348:2656–2668.

142. Shulman RG, Rothman DL. Metabolomics by in vivo NMR. Wiley (West Sussex, England); 2005.

143. Kurhanewicz J, Bok R, Nelson SJ, Vigneron DB. Current and potential applications of clinical 13C MR spectroscopy. J Nucl Med 2008;49:341–344.

PART II

CLINICAL APPLICATION OF BIOMARKERS

5

VASCULAR BIOMARKERS AND IMAGING STUDIES

Karin Potthoff, Ulrike Fiedler, and Joachim Drevs

5.1 TUMOR ANGIOGENESIS

Angiogenesis, the growth of new blood vessels from preexisting vasculature, is a key process that physiologically occurs during embryonic and adult life, but it is also involved in several pathological conditions including tumor growth (1). There are several natural regulators of angiogenesis, that is, angiogenic and antiangiogenic growth factors, that normally are well balanced. Vascular endothelial growth factor (VEGF) and members of the basic fibroblast growth factor (bFGF) family are, among others, positive regulators of angiogenesis, whereas thrombospondin, angiostatin, endostatin, and IL-12 are inhibitors of angiogenesis (2). Angiopoietins are a family of molecules that, like VEGF, are specific for endothelial cells. These ligands work in conjunction with VEGF in promoting angiogenesis and vascular remodeling. Angiopoietin-1 (Ang-1) and angiopoietin-2 (Ang-2) have been identified as ligands with mostly opposing functions of their endothelial-cell-specific Tie-2 receptor, also a receptor tyrosine kinase, regulating endothelial cell survival and vascular maturation. Ang-1 is located in the extracellular matrix and acts in a paracrine agonistic manner to maintain and stabilize mature vessels by promoting interactions between endothelial cells and smooth muscle support cells, whereas Ang-2 appears to act primarily as an autocrine antagonistic regulator in controlling endothelial cell quiescence and responsiveness to VEGF (3), but context dependently, it might act as an agonist for the Tie-2 receptor with pleiotropic effects on tumor vessel structure, perfusion, oxygenation, and apoptosis (4–6).

Predictive Approaches in Drug Discovery and Development: Biomarkers and In Vitro/In Vivo Correlations, First Edition. Edited by J. Andrew Williams, Jeffrey R. Koup, Richard Lalonde, and David D. Christ.

In 1971, Judah Folkman hypothesized that solid tumors are angiogenesis dependent and need blood supply from surrounding vessels to grow over a certain size of a few millimeters in diameter (7). Since then, numerous reports have pointed to the crucial role of angiogenesis in the primary establishment of solid tumors and secondary site metastases, and much research has been devoted to determining the impact of angiogenesis on tumor development and progression and the reciprocal influences of tumor products on the microvasculature (8). Up to 2- to 3-mm^3 solid tumors are able to grow without blood vessel supply, but above this size, diffusion becomes insufficient and they require new blood vessels for growth and survival (1, 9). The so-called angiogenic switch depends on a net balance of positive and negative angiogenic factors in the tumor and occurs if there is a misbalance in favor for proangiogenic stimuli that lead to neovascularization. Thus, these new blood vessels provide nutrients and growth factors for tumor progression and, at the same time, allow tumor cells to disseminate and undergo metastasis in distant organs. Most human tumors arise and remain *in situ* without angiogenesis for months to years before switching to an angiogenic phenotype. Thereafter, however, new vessels perfuse the tumor and endothelial cells release growth factors and matrix degrading proteinases that facilitate invasion. Interestingly, different types of tumor cells use different molecular strategies in the angiogenic switch. VEGFs and bFGF (also known as *FGF-2)* are the most potent endothelial-cell-specific angiogenic factors that play a key role in promoting tumor angiogenesis and lymphangiogenesis by activating vascular endothelial growth factor receptor (VEGFR) tyrosine kinases in endothelial cells. VEGF binds to one of three tyrosine kinase receptors, that is, VEGF receptor-1 (VEGFR-1), VEGFR-2, and VEGFR-3. While VEGFR-1 is associated with vascular development and may have a function in quiescent endothelium, VEGFR-2 activation seems to be both necessary and sufficient to mediate VEGFR-dependent angiogenesis (10). VEGFR-3, via Notch regulation, is present on endothelial tip cells and is critical to sprouting angiogenesis (11, 12); besides, it plays an important role in the regulation of lymphangiogenesis (13).

5.2 ANTIANGIOGENESIS

Today, antiangiogenesis is one of the most attractive areas for drug development. Inhibitors of angiogenesis block the development of new blood vessels and halt new blood vessel growth, thereby starving tumors of essential nutrients and oxygen by cutting off their blood supply. They broadly fall into three categories, acting by preventing new blood vessels from sprouting, by attacking a tumor's established blood supply, and by attacking both the cancer cells and the blood vessel cells. In early 2004, a major landmark in medical oncology has been achieved with the first FDA-approved angiogenesis inhibitor Avastin (bevacizumab; Genentech, Inc., San Francisco, CA), a humanized monoclonal antibody blocking VEGF itself. Today antiangiogenic treatment strategies are well established for treatment of cancer patients with various tumor types including colorectal cancer, breast cancer, lung cancer, and renal cell carcinoma. Bevacizumab is FDA approved for the treatment of metastatic colorectal cancer (14), metastatic non-small-cell lung cancer (15), and advanced breast cancer (16) in combination with chemotherapy. In addition, small-molecule receptor tyrosine kinase inhibitors targeting the VEGFR, that is, the multitargeted inhibitors sunitinib and sorafenib, have been approved for treatment of advanced renal cell carcinoma. Sunitinib inhibits VEGFRs, PDGFR (platelet-derived growth factor receptor), c-kit, and Flt-3 and has also been approved for imatinib-resistant gastrointestinal stromal tumors (17). Other VEGF antagonists that are still at various stages of clinical development include small-molecule oral antiangiogenesis agents and small-molecule receptor tyrosine kinase inhibitors targeting the VEGFR (i.e., vandetanib, cediranib, pazopanib, axitinib, AMG 706, XL647, enzastaurin, and others), as well as antiangiogenic agents with different mechanisms of action, including direct endothelial cell antagonists (angiostatin, endostatin, thrombospondin-1, troponin I); antagonists of endothelial cell migration (ECM), such as inhibitors of ECM modeling; AMG 386 that neutralizes Ang/Tie-2 interaction; and VEGF Trap, a chimeric molecule that combines extracellular portions of VEGFR-1 and VEGFR-2 with the Fc portion of immunoglobulin G1 to form a molecule that binds and "traps" VEGF (18, 19).

Consistent with previous reports on growth and recurrence of experimental tumors during VEGF blockade, it has become evident that many patients treated with inhibitors of VEGF or VEGFR will ultimately develop progressive disease (20). Patients frequently develop resistances after 6–12 months of treatment. It is currently not well understood why patients do develop resistances or show very different responses. The reason for this might be (i) the activation of compensatory pathways in the tumor (21); (ii) organ- and tissue-specific endothelium structures, expression profiles, and responses (22); or (iii) cytogenetic abnormalities in tumor endothelial cells (23). Thus, the endothelium, especially the tumor endothelium, seems to be more complex than expected.

5.3 BIOMARKERS IN ANTIANGIOGENIC DRUG DEVELOPMENT

5.3.1 Rationale

Effective biomarker research is becoming one of the cornerstones of efficient drug discovery and early clinical drug development because, in contrast to cytotoxic agents, antiangiogenic agents might not necessarily be applied in maximal tolerable doses (MTDs) but in optimal biological doses (BODs). Especially, soluble biomarkers and imaging biomarkers are increasingly being applied to understand mechanisms of disease and to predict outcome of novel antiangiogenic agents. Soluble biomarkers are most attractive because of the easy access to blood or other accessible biological fluid levels by an available and validated method *in vivo*. In recent years, imaging techniques have been increasingly used as biomarkers to demonstrate the efficacy of novel antiangiogenic agents by imaging of the tumor and its blood supply. While existing biomarkers such as tumor markers are used related to tumor growth in general and are widely used over the whole development process of a new agent, the biomarkers described here are more related to the mode of action of the agent and are mainly used in early stages of the drug development, identifying the optimal dose used later in the development of a certain drug. The overall goal of the integration of vascular and imaging biomarkers in the development of new antiangiogenic agents is to monitor biochemical coverage and to stratify patient populations, thereby fulfilling the aim to deliver the "right antiangiogenic agent" at the "right dose" to the "right patient."

We provide a brief overview on the current status of soluble biomarkers and different imaging techniques in cancer in general and of diagnostic and biological predictive biomarkers for antiangiogenic agents in particular, with focus on translational investigations and the recent progresses made in targeting the major regulators of angiogenesis. In addition, we give some preclinical insights that allow identification of prospects for future translational research projects.

5.3.2 Strategy and Goals

There are different strategies and goals for discovery, development, and implementation of biomarkers in antiangiogenic drug development (i) to make earlier, better informed decisions on dosing and application schedule during drug development; (ii) to use biomarkers as surrogate endpoints for regulatory approval; and (iii) to routinely use a clinical biomarker to associate a treatment to a particular patient subpopulation that has demonstrated a substantial clinical response (24). Stages of biomarker assay development are well described by Severino et al. (25) to identify a disease target, to develop a new drug that is specific for the biomarker, to correlate the biomarker with the outcome of treatment, and finally, to use biomarkers to select patients for specific treatment.

During development of novel antiangiogenic agents, the key question that arises is why do drugs fail and others succeed. In traditional drug development, drugs fail because of lack of efficacy, lack of safety, poor bioavailability, or problems in drug supply, and therefore, in early clinical development, drug safety and drug bioavailability have to be established quickly and it is of great importance to get a rapid measure of response. There is evolving evidence that while going toward "personalized medicine," the identification and use of novel biomarkers to accurately assess and individualize therapy, that is, to stratify the patients, to diagnose the exact type of disease in each patient and the individual patient's drug sensitivity to side effects will play an increasingly important role. This "new paradigm" is being addressed by the "FDA Critical Path Initiative," integrating biomarker assay technologies for safety and effectiveness in preclinical and early clinical, that is, exploratory phase, drug development programs to improve predictability and efficiency along the path from laboratory concept to commercial products (26).

Clinically meaningful biomarkers may be based on genotypes, proteins, metabolic patterns, histology, or imaging techniques (24), and they ideally are suitable for early diagnosis, monitoring, and prediction of therapeutic response. Some patients do not respond to antiangiogenic therapy at all (27). Therefore, biomarkers in the clinical setting are urgently needed (i) to stratify patients before treatment, (ii) to monitor the patients' response to the therapy early during therapy, (iii) to classify patients into responders and nonresponders, and (iv) to determine the OBD for the drug.

A biomarker may also become a surrogate endpoint in clinical trials if it can predict clinical benefit, harm, or lack of benefit or harm (28). In the future, it is likely that a combination approach to simultaneously measure different markers such as soluble biomarkers and imaging markers would be most successful in detecting early disease to identify patients with early stage of disease, who are likely to develop advanced disease and would benefit from additional therapies or to predict patients' outcomes based on prognostic significance independent of initial tumor burden, lymph node metastases, and other common prognostic indicators.

5.4 TISSUE MARKERS

Immunohistological analysis of microvessel densities (MVDs) in tumor biopsies is widely used to determine tumor angiogenesis. Throughout the literature, the MVD in tumor sections is determined by three different pan-endothelial cell markers, CD31, CD34, and vWF (29). These markers are reliable markers for blood vessels and widely used to determine the effect of an antiangiogenic or antitumor therapy on the vascular bed. Surprisingly, meta-analyses of MVDs reported in the literature show tremendous variations between the studies in dependence of the markers used for the histochemical analysis, even if the same tumor type was analyzed (30–32). Besides, blood vessels are lymphatic vessels and are very important for tumor growth and dissemination (33, 34). During the last years, antibodies became publicly available, allowing the detection of lymphatic vessels in tumors in mice and humans. In tumors, lymphatic vessels are mainly found in the tumor margin, and more recently, intratumoral lymphatic vessels have been identified too (35). During the last years, the lymphatic MVD could be correlated with lymph node metastasis and might serve as a prognostic factor (34, 35).

MVDs of blood and lymph vessels were widely determined in tumor biopsies to correlate the MVD with the prognosis for the patients. Very rarely, the MVD was determined in the same patient during an antiangiogenic therapy to monitor the patient's response. One example is a small clinical trial with the anti-VEGF antibody bevacizumab in patients with rectal cancers. Upon treatment, in five out of six patients, the MVD decreased after 12 days (36). However, these invasive methodologies make them unsuitable for analysis on large cohorts of patients in clinical trials. Moreover, the determination of MVD does not predict the patients' response to antiangiogenic therapy or give information of the functionality of the vessels.

Besides the analysis of the tumor endothelium by histological markers, more quantitative methods have been developed to monitor the effectiveness of a drug in patients. In a very elegant study proving the target inhibition of sorafinib, ERK(extracellular signal-regulated kinases)-phosphorylation was monitored in peripheral blood cells that were isolated from patients before, during, and after therapy. These cells did not respond to phorbol ester stimulation if the patients received sorafinib (37). This indicates that peripheral blood cells can serve as surrogates for the response of tumor cells or stromal cells to an antiangiogenic or antitumor therapy.

5.5 BLOOD MARKERS

5.5.1 Circulating Endothelial Cells and Progenitor Cells

The response of a patient to an antiangiogenic therapy can also be monitored by the detection of peripheral blood cells (38). These cells are easily accessible and traceable and can be sorted by density gradients or fluorescence-activated cell sorting (FACS) analysis using different cell surface markers. Besides leukocytes and monocytes, the pool of peripheral blood cells also contains circulating endothelial cells (CECs) and circulating endothelial cell progenitor cells (CEPCs). CECs and CEPCs are interesting cell populations because it could be shown in mice that the mobilization of CEPCs correlates with induction of angiogenesis and that the levels of CEPCs decrease in response to antiangiogenic therapy (39). Moreover, it has been shown in patients with rectal cancer that therapy with bevacizumab reduces the levels of viable CECs and CEPCs (36). Further studies will be needed to assess the clinical value of CECs and CEPCs as biomarkers of angiogenesis. This would be mandatory since the expression of the surface markers has been shown to be variable (CECs express the cell surface markers CD31 and CD146 but do not express the leukocyte marker CD45 and the stem cell marker CD133; CEPCs do express CD146, CD31, and the stem cell marker CD133) and the methodologies for sampling are not yet standardized. Moreover, the use of different brands of flow cytometers results in variations of the results (38). Besides CECs or CEPCs, circulating Tie-2-expressing mononuclear cells have been identified, which promote tumor angiogenesis (40). The impact of these cells as biomarkers is not yet evaluated but might be of great importance in the future. Another interesting cell population has been identified recently in mice with tumors that were refractory to an anti-VEGF therapy. These cells are CD11b+Gr1+ myeloid cells that promote angiogenesis and are mobilized by the secreted protein Bv8 (41). Although nothing is known about these cells in humans, it will be interesting to see in the future whether an increase of CD11b+Gr1+ myeloid cells is found in patients who do not respond to antiangiogenic therapy or whether Bv8 could be detected in plasma of patients that are refractory to anti-VEGF therapies.

5.5.2 Soluble Blood Markers

Blood is the most favorable source for the identification of novel biomarkers since it is easily accessible and contains, besides cellular constituents, soluble circulating cytokines, chemokines, protein fragments, and other soluble factors. In clinical trials with antiangiogenic drugs, many of the soluble receptor fragment levels and cytokine and chemokine levels have been determined, and it could be shown that the factors can serve as surrogates for the biological activity of antiangiogenic drugs (42, 43). These surrogate markers are on one hand soluble cytokines that mediate angiogenesis, such as VEGF-A, VEGF-C, FGF-2, placental growth factor (PLGF), Ang-1, Ang-2, the interleukins (IL-6 and IL-8), stromal cell-derived factor1 (SDF-1), hepatocyte growth factor (HGF), or osteopontin, and on the other hand, soluble fragments of cell surface receptors such as the soluble VEGF receptors (sVEGFR-1, sVEGFR-2, sVEGFR-3), sTie-2, sE-selectin, sSCFR (soluble c-kit), or soluble intercellular adhesion molecule-1 (sICAM-1). Most of the cytokines are expressed by the tumor or the tumor stroma and the upregulation of their expression is frequently directly correlated with tumorigenesis and tumor angiogenesis. The soluble receptor fragments are frequently shed forms of receptor tyrosine kinases that are released from the activated angiogenic or inflammatory endothelium (43). This suggests that the markers do not exclusively correlate with tumor angiogenesis but also with inflammatory responses. Indeed, most of the molecules are also involved in inflammatory diseases such as arthritis and sepsis as well as in diabetic diseases such as retinopathy (44). In this respect, increased Ang-2 levels were detected in serum or plasma of not only patients with severe sepsis or critically ill patients but also tumor patients with, for example, acute myeloid leukemia or lung cancer (45–48).

The key critical molecule for the induction of angiogenesis or tumor angiogenesis VEGF-A also serves as a powerful surrogate marker in antiangiogenic therapies. Blood VEGF-A levels are on average two- to fourfold higher in cancer patients than those in healthy individuals (49) and have been shown to change in response to antiangiogenic therapies in multiple clinical trials (42, 50, 51). There are many clinical trials in which changes of the above-mentioned markers have been determined as the response to an antiangiogenic therapy. In a phase I/II escalation study with PTK787/ZK 22584 in patients with colorectal cancers, it could be shown that plasma VEGF-A and FGF-2 levels changed in relation to the biological activity of the drug. Moreover, the changes in plasma VEGF-A and FGF-2 correlated with the clinical outcome (50). For another antiangiogenic compound that potently inhibits all VEGFRs, AZD2171, it could be shown that sVEGFR-2 levels decrease depending on time and dose (51). In a glioblastoma study with the same compound, normalization of the tumor vasculature was observed. Interestingly, it could be shown in this study that FGF-2, SDF-1α, and the number of viable CECs increased when tumors escaped treatment and that the number of CEPCs increased when tumors progressed after drug interruption (52). In a clinical trial with sunitinib, similar changes in soluble blood marker levels were observed during therapy; for example, in a clinical trial with metastatic renal carcinoma patients, it could be shown that PLGF and VEGF-A levels increased during therapy, whereas the plasma levels of VEGF-C and sVEGFR-3 decreased (53). In a similar trial with metastatic breast cancer patients, it could be shown that plasma levels of VEGF-A increased and the levels of the sVEGFRs and of soluble c-KIT decrease during treatment with sunitinib (54).

Collectively, the measurements of soluble blood markers proved that the markers are powerful surrogates to monitor efficacy of antiangiogenic drugs.

Interestingly, not much is known about the source of the surrogate markers in plasma or serum of tumor patients. Most factors are overexpressed in tumor tissue or in the tumor stroma, but not all factors are freely diffusible and can easily end up in blood. It is more likely that an antiangiogenic therapy is not specifically targeting the tumor endothelium, which is apparent by the observed side effects (55) and suggests that the soluble markers are not exclusively released from the tumor compartment. In this respect, it could be shown for VEGF-A that VEGF-A in the tumor tissue contributes only 0.3–16% per 1 kg tumor to the total VEGF-A in men and that most VEGF is stored in the muscle. The main source of VEGF in the blood compartment is the platelets and leukocytes, which indicates that elevated blood levels of VEGF are caused by the release of VEGF from these compartments (49). This is confirmed by a study on non-tumor-bearing mice treated with sunitinib. In these mice, a dose-dependent modulation of VEGF-A, PLGF, SDF-1, stem cell factor (SCF), sVEGFR-2, granulocyte colony-stimulating factor (G-CSF), sTie-2, and osteopontin plasma levels was observed; moreover, this dose-dependent modulation was consistent with the OBD in patients (56).

Collectively, this shows that many parameters reflect specific aspects of angiogenesis and indicates that more than one parameter needs to be studies to monitor the effect of an antiangiogenic drug. Thus, the combination of different approaches is needed to get a robust quantification of angiogenesis; for example, the combination of multiple surrogate markers, circulating peripheral blood cells and CECs, and imaging techniques.

5.6 IMAGING TECHNIQUES

5.6.1 Introduction

Imaging can be used as a biomarker both in early development to assess drug delivery to and interaction with a target and later in development to monitor disease response. Thus, in the field of angiogenesis, the use of different imaging techniques plays an important role in facilitating the translation of many advances made on the bench to the bedside (57) and to monitor disease during the course of antiangiogenic therapy. While conventional imaging technologies such as CT (computed tomography) and MRI (magnetic resonance imaging) require a well-defined anatomical lesion to be viewed and rely on the measurement of the reduction of tumor size, for novel antiangiogenic agents, anatomical imaging has become inappropriate and functional studies, for example, dynamic contrast-enhanced MRI, are necessary in order to monitor therapeutic effects.

Modern imaging techniques permit the observation of structural, cellular, molecular, and functional characteristics of biological systems with minimal invasiveness and provide a proven source of highly useful biomarkers. Molecular imaging, for instance, has recently generated considerable excitement as key component of twenty-first century cancer management for being used as biomarker in early antiangiogenic drug development (58). By using modern imaging tools, the need for periodic biopsies may be alleviated in clinical trials, which is of great benefit especially in patient populations where single or multiple tumor biopsies are not possible or even prohibitive (e.g., in glioblastoma patients) during the course of treatment. Thus, imaging is seen as a key technology for assessing and accelerating development as well as guiding use of new antiangiogenic agents. Besides, advanced development of imaging technologies and rapid incorporation of imaging tools as biomarkers and surrogate markers for response evaluation is on the forefront of the Critical Path Initiative of the FDA (26). Imaging allows the measurement of many very useful biological parameters of angiogenesis and antiangiogenic approaches, which could previously only be measured with invasive techniques (58). Several imaging modalities have been used to assess the status of tumor neovasculature *in vivo* including ultrasound (US), MRI, CT, photon emission imaging (positron emission tomography (PET) and single photon emission computed tomography (SPECT)), and optical, that is, near-infrared absorption and scattering, techniques. However, there is a great need to establish standards for imaging biomarkers and to standardize data collection and analysis for the different imaging methodologies used.

5.6.2 Ultrasound

Ultrasound is one of the preferred techniques for imaging angiogenesis *in vivo* due to its wide use in routine clinical practice and its risk-free application in terms of absence of patient exposure to radiation and radionuclides. Furthermore, US technology can often be repeated and is mobile and flexible in use. High frequency sound waves are utilized to produce images, and the time delay between sound wave emission and detection correlates with the depths of the reflected surface (59). For imaging of (neo)vasculature, Doppler US can be used by detecting moving reflectors that represent the surfaces of red blood cells in flowing blood. The classical Doppler techniques such as color Doppler ultrasound (CDUS) and power sonography are today the routine measurements of pathological changes in large feeding vessels in tumors and can be used as an adjunct in the detection of malignant lesions, for example, in breast cancer, or in prostate lesions (60, 61) to reveal the vascular anatomy of the lesions, branching pattern, shunts, and even blind ending vessels according to a visual scoring system (9). However, flow sensitivity limits the ability of Doppler US to identify vessels smaller than 50 μm.

Modern US techniques using contrast-enhanced US permit the measurement of tissue perfusion irrespective of vessel size or flow velocity (62). One of those advanced US methods detecting smaller vessels than those amenable to conventional CDUS is the method of microbubble-enhanced CDUS (63). Microbubble contrast agents work by replenishment kinetics of microbubbles. They resonate to the pressure changes of the sound wave and rapidly contract and expand in reply, providing a comprehensive quantification of tissue perfusion. This technique enhances not only Doppler flow but also US images in different grayscale images.

Importantly, microbubbles, in contrast to agents used in contrast-enhanced MRI and CT, are strictly located intravascularly and, thus, predispose for monitoring intravascular flow. Their sensitivity in detecting focal liver lesions is comparable to that of other imaging modalities such as CT or MRI, and it provides a high accuracy in lesion characterization. Furthermore, CDUS is expected to play a major role in estimating the tumor burden of involved lymphatic tissue (63).

Another modern technology described is the contrast-enhanced US using cadence contrast pulse sequencing (CPS) technology. Using CPS enables excellent visualization of the microvasculature associated with prostate cancer and can improve the detection of prostate cancer compared with systematic biopsy (64). The most exciting advantage of microbubbles may be their great potential as carriers in site-specific gene therapy and their potential as drug-loading microbubbles, releasing small molecules into the cytoplasm after microbubble-induced permeabilization of cell membranes in combination with low intensity US, as has been shown in *in vivo* experiments with xenografts (65, 66). Interestingly, US in terms of therapeutic US utilizing low intensities and noncontinuous waves has recently emerged as a method to safely deliver genes to cells and tissues as has been shown *in vitro* and *in vivo* for genes encoding angiogenesis inhibitors in prostate cancer cells (67).

The future role of US as well as contrast-enhanced CDUS and therapeutic US in antiangiogenic therapy will be defined by addressing the limitations of operator-dependent reproducibility and repeatability. Thus, standard methods are needed for quantification of tumor vascular metrics irrespective of the examining physician. Furthermore, tumors within certain anatomic regions, that is, lung, bone, are less amenable to US interrogation (59); they require imaging techniques such as CT or MRI.

5.6.3 Dynamic Contrast-Enhanced MRI

At present, dynamic contrast-enhanced magnetic resonance imaging (DCE-MRI) is used as gold standard in clinical trials with antiangiogenic agents to interrogate the status of tumor neovasculature (68). It was first described preclinically in a murine renal cell carcinoma model using the VEGFR inhibitor PTK787/ZK 222584 (9). Dynamic contrast-enhanced MRI is performed with a small-molecular-weight gadolinium-based paramagnetic contrast agent that is indirectly imaged based on its magnetic effect on neighboring water molecules. It is a useful technique to study the pathophysiology with particular emphasis on tumor perfusion and tumor vascular permeability. Besides, intratumoral hypoxia can be predicted using DCE-MRI, as shown by Newbold et al. (69) in patients with head and neck cancer.

The key to DCE-MRI analysis is to determine the kinetic delivery of gadolinium contrast into the tumor. Gadolinium does not enter into intact cells, Its diffusion occurs passively across the capillary endothelial membrane into the extravascular, extracellular space, that is, the interstitial space. After rapid administration of gadolinium chelate, the changes of signal intensity in the tumor tissue can rapidly be measured during the early first-pass bolus and subsequent equilibrium phases of contrast entry into the tumor interstitium. A kinetic term, K^{trans} (values measured amount of contrast agent per minute in region of interest), represents this first-order kinetic constant of gadolinium passage from plasma to tumor interstitium and back again. In tumors with high vessel permeability, K^{trans} is equivalent to tumor perfusion. In general, K^{trans} presents a mixture of tumor perfusion and tumor vascular permeability (59). Imaging techniques in DCE-MRI are identical to contrast-enhanced T1-weighted imaging used routinely in contrast-enhanced MRI, but aspects of DCE-MRI examination differ from those of routine MRI (70, 71); for example, an advanced planning is required before the patient is scanned. General guidelines for DCE-MRI examination have been published by Evelhoch et al. (72) and Leach et al. (73). Nevertheless, many questions have arisen in regard to the reproducibility of DCE-MRI tumor metrics. Therefore, in many clinical trials, central measurement is indicated if DCE-MRI is used, especially in multicenter clinical trials. Kinetic modeling in DCE-MRI is still under investigation (74). Other MRI techniques such as diffusion-weighted magnetic resonance imaging (DW-MRI) and magnetic resonance spectroscopy (MRS) are also capable of detecting changes in cell density and metabolite content within tumors. They are still being evaluated for response evaluation of antiangiogenic agents.

Dynamic contrast-enhanced MRI is a promising biomarker for assessing antiangiogenic treatment effects *in vivo*. Changes in the physiologic state of tumor vascularity after vascular-targeted therapy, that is, variables related to tumor blood flow and microvessel permeability, can be quantified to evaluate vascular response. Since these responses, for example, the reduction of vascular permeability, may precede morphologic tumor shrinkage, DCE-MRI might serve as a noninvasive means of monitoring early tumor response to vascular-targeted therapy (75). In a phase I study of PTK/ZK in liver metastases, changes in tumor vascular status as assessed by DCE-MRI were used to define the minimal biologically active dose (76). Dynamic contrast-enhanced MRI can also be used to demonstrate a dose–response relationship, providing useful information for phase II dosing, as has been demonstrated in a phase I study of the oral angiogenesis inhibitor AG-013736 in solid tumors (77). Recently, it has been found that DCE-MRI allows complete discrimination of malignant lesions from benign lesions in patients with breast lesions, as validated by comparison with gold standard pathology analyses of subsequent biopsy tissue samples (78).

It appears that novel DCE-MRI protocols and future developments, such as combination PET-MRI, might be able to address those and other issues on early tumor response to antiangiogenic agents in the near future (79). Monitoring antitumor effects by functional MRI and correlating them with existing validated biomarker has been proven to be a very useful tool in very early stages of drug development. Prospective studies are needed to further test standardized DCE-MRI methodology; to evaluate novel, that is, more specific contrast agents; and to establish MRI as a biomarker itself.

5.6.4 DCE-CT

As with MRI, dynamic contrast-enhanced computed tomography (DCE-CT) can be used to assess the status of tumor neovasculature in a noninvasive manner. Blood flow, blood volume, and capillary permeability, as well as MVD, may be evaluated. The advantage of functional CT scans is the possibility to implement them into routine CT examination for using them as biomarkers of angiogenesis in tumor tissue *in vivo*. CTs are widely available, and furthermore, quantification of tumor response by dynamic CT is quite simple. Although there is a lack of experience due to a limited number of clinical studies with dynamic CT in tracking tumor response to antiangiogenic agents, it has been demonstrated that perfusion values obtained with first-pass DCE T2-weighted MRI and CT may be used interchangeably in squamous cell carcinomas in the upper aerodigestive tract (80). In clinical trials, prediction of responsiveness to antiangiogenic agents has been addressed successfully by using CT perfusion metrics, suggesting a potential role of DCE-CT in patient stratification (81, 82). However, several factors limit the utility of CT perfusion outside areas such as the brain and the head and neck region, that is, the fixed axial plane of CT scanning that may hinder CT perfusion studies in moving areas of the body, such as liver and lung, or the relatively high radiation dosages that may limit the number of high quality images that can be obtained repetitively (59).

Nevertheless, radiologists must be convinced that by using DCE-CT as a promising approach, it is becoming possible to gain functional information during routine tumor imaging. The most promising future role for CT in detecting angiogenic changes will be functional CT with advanced multislice systems incorporated in PET scanners (9).

5.6.5 Positron Emission Tomography

PET and SPECT rely on the injection of radioisotopes, whose short lifetime requires access to a cyclotron for their production. SPECT images derive from the emission of a single photon by a decaying isotope, whereas PET images rely on the creation of coincident γ rays originating from the annihilation between the emitted positron and an adjacent electron. The PET image is being constituted due to localization of the coincident γ rays to the origin of photons within the patient's body. PET has become the modality of choice for functional tissue assessment because of its greater sensitivity compared with SPECT (83).

PET is the most useful of the imaging modalities employed for biomarker discovery and detection so far due to the opportunity to visualize not only the anatomical structure of a tumor but its biological activity as well as physiological reactions. Thus, PET can be used as an ideal imaging biomarker measuring a biological effect of anticancer treatment in the affected tissue of interest.

PET relies on the injection of radioisotopes, whose short lifetime requires access to a cyclotron for their production. The radiotracers are compounds labeled with proton-rich atomic isotopes being internalized after injection or inhalation and decay by emission of a positively charged electron, that is, the positron (9).

The tracer concentrations are typically very small, on the order of picomolar, and are able to track *in vivo* physiology without altering the function of the body system. Tracer uptake is indicated by the standardized uptake value (SUV), a measure of mean tissue tracer concentration relative to the mean whole-body tracer concentration. ^{18}F-Fluorodeoxyglucose (FDG) is the best studied PET radiotracer. FDG-PET images are based on the increased glucose metabolism by cancer cells (84). FDG-PET shows the relative glucose uptake and metabolism of various tissues in the body, and thus, besides tumor diagnosis, it is increasingly clinically relevant for monitoring tumor therapy. Glucose uptake serves as response surrogate. Interestingly, Bos et al. (85) found that in breast cancer, FDG uptake correlates not only with glucose transporter and hexokinase expression but also with MVD. However, it is not clear that alterations in tumor neovasculature will be easily correlated to changes in tumor FDG-PET.

While FDG-PET remains the standard of routine clinical PET imaging in oncology, today, many other tracers are available visualizing various biologic or physiologic aspects of a tumor (e.g., tissue- and blood-based markers). PET, according to the tracer that is being used, can be applied to directly evaluate parameters of angiogenesis, that is, hemodynamic parameters such as blood flow (i.e., ^{15}O-labeled water) and blood volume of distribution (i.e., labeling of red blood cells); tissue properties such as glucose metabolism and hypoxia; tissue expression of specific markers of angiogenesis, for example, VEGF or integrins (59); or even cell proliferation and apoptosis. ^{15}O-labeled H_2O enables measurement of blood flow because of its free diffusion capability, its short half-life of 2 min, and its dosimetric properties. It has been applied in several tumor types including brain tumors and breast cancer (86–88). In breast lesions, for instance, blood flow and volume of distribution measured with ^{15}O-labeled H_2O were shown to be significantly different between malignant and normal tissue (88). Several investigators have evaluated PET to monitor tumor changes in antiangiogenic therapies. Interestingly, in a randomized phase II study evaluating effects of chemotherapy in patients with breast cancer, Miller et al. (89) showed that post treatment, tumor blood volume remained stable in responding patients but tended to increase after therapy in patients without pathologic complete response (pCR). Herbst et al. (90) found in a phase I trial that ^{15}O-H_2O-PET demonstrated decreases in tumor blood flow in the absence of morphologic tumor regression, and these changes have gone along with decreased metabolic activity.

In several tumor types, hypoxia is associated with a poor prognosis and may be a predictor of response to anticancer therapy too. The interplay between tumor perfusion, tumor neovascular maturation, and tumor hypoxia is an important metric for determining prognosis and tumor responsiveness. A noninvasive technique for imaging oxygen distribution in tumors is using PET imaging with radiotracers for hypoxia, for example, tracers that target viable, but hypoxic cells such as 18-F-fluoromisonidazole, 18-F-EF5, and copper (II) diacetyl-bis(*N*4)-methylthiosemicarbazone (91). A novel approach in imaging hypoxia has been to evaluate its effects on cell markers. Hypoxia inducible factor α (HIF1α) mediates a complex cascade of events in tissue with low partial pressures of oxygen. Carbonic anhydrase 9 (CA9) is upregulated via HIF1α. cG250 is a chimeric antibody against CA9. Recently, ^{124}I-cG250 PET was shown to have a sensitivity of 94% and specificity of 100% in the detection of clear cell phenotype in patients with renal cell carcinoma (92). Therefore, quantification of CA9 expression represents a promising strategy to interrogate tumor hypoxia before and during treatment with novel targeted agents. Furthermore, molecular imaging methodologies using PET technique have recently generated considerable excitement. Quantitative PET imaging of VEGFR will, for example, facilitate the planning of whether, and when, to start antiangiogenic treatment and enable more robust and effective monitoring of such treatment. Recently, a preclinical study has been published where VEGF(121) was conjugated with DOTA (1,4,7,10-tetra-azacylododecane N,N',N'',N''-tetraacetic acid) and then labeled with (64) Cu for PET imaging of mice bearing different-sized human glioblastoma. PET imaging showed good linear correlation with the relative tumor tissue VEGFR-2 expression as measured by Western blot. The tumor uptake of (64) Cu-DOTA-VEGF(121) measured by small-animal PET imaging reflected tumor VEGFR-2 expression level *in vivo* (93). Given these early results, targeting VEGF and VEGFR expression as a means of monitoring the effects of antiangiogenic therapy could be a very valuable tool for evaluation of patients with a variety of malignancies and to monitor those patients undergoing antiangiogenic therapies that block VEGF/VEGFR-2 function. Quantitative VEGF/VEGFR targeting, however, has not been evaluated as a marker of tumor angiogenic activity in human clinical trials so far. Great attention lies on the development of VEGFR-2-specific PET tracers to noninvasively measure VEGFR-2 expression. Wang et al. in a preclinical model demonstrated efficacy of the VEGFR-2-specific PET tracer, (64)Cu-DOTA-VEGF(DEE). It has comparable tumor targeting efficacy to (64)Cu-DOTA-VEGF(121) but much reduced renal toxicity. Thus, this tracer may be translated into the clinic for imaging tumor angiogenesis and monitoring antiangiogenic treatment efficacy (94).

A new molecule discovered by General Electric is a radiolabeled small peptide in a configuration that allows high affinity binding of the peptide to specific integrin receptors including $\alpha_v\beta_3$. Integrin $\alpha_v\beta_3$ is highly expressed on endothelial cells during angiogenesis and is associated with endothelial cell differentiation, proliferation, and migration, whereas its expression in normal vasculature and other organs is weak. Radiolabeled RGD (argenine-glycine-aspartic acid) peptide imaging is one of the promising strategies to evaluate integrin expression *in vivo* and assess tumor neovasculature noninvasively. PET imaging with 18F-galacto-RGD has been used to evaluate angiogenesis in patients with metastatic melanoma (95). Recently, a significant correlation between radiotracer uptake and tumor $\alpha_v\beta_3$ integrin expression by immunohistochemistry was found in man (96). [18F]Gal-RGD-PET, thus, is a promising method to image angiogenesis in patients (97).

In summary, PET can be used as a biomarker due to its capability to directly measure a biological process in the tissue of interest and not only a plasma concentration level. Therefore, PET is an ideal imaging tool for investigating molecular processes during the design of new compounds. Even PET systems for small-animal work are becoming increasingly useful for critical preclinical studies of novel antiangiogenic compounds.

As seen in DCE-MRI and CT technologies, standardization of quantification also is an issue for FDG-PET. Baseline and follow-up scans have to be acquired according to the same protocol, and the individual patient, then, serves as own control.

5.6.6 PET-CT

CT provides a ready source of structural information and has successfully been combined with PET instrumentation to offer combined structural/molecular information. Although PET-CT still is not available at many centers, it represents the future for molecular and angiogenic imaging. PET-CT has proved to be superior to PET or CT alone in both diagnosing cancer and evaluating therapeutic effects. Combining anatomic and functional results works not only additively but also synergistically. A higher acceptance by the nonradiological medical personnel and the better "image-in-mind" effect too should not be underestimated (9). One very valuable advantage is the potential to standardize PET-CT testing protocols, for example, quantification of tracer uptake. Tracer uptake analysis is a very promising tool for refining the evaluation of treatment response and determining prognostic factors of disease. Measuring the total tumor FDG uptake defined as [mean intratumoral FDG concentration] × [tumor volume in CT] is superior to the measurement of FDG uptake alone.

5.7 SUMMARY

In summary, angiogenesis is and will remain an important therapeutic anticancer target that is suitable for complex and personalized treatment regimens. Biomarkers are finding increased application in clinical diagnostics, driven by visions for personalized medicine and therapeutics. Their promise is particularly strong for antiangiogenic agents. Vascular biomarkers and modern imaging techniques, including molecular imaging of VEGFRs, as well as multimodality approaches, will facilitate progress made from bench to bedside. They can be applied in treatment selection, that is, *in vitro* diagnostics and imaging as well as during treatment monitoring. Dynamic contrast-enhanced MRI represents today's gold standard in the assessment of the OBD for antiangiogenic agents. A combination of multiple imaging modalities may yield complementary information and offer synergistic advantages over any modality alone. Thus, imaging techniques such as PET and PET-CT or even PET-MRI will provide additional advantages for early drug development in the future.

Combination of noninvasive surrogate parameters such as soluble biomarkers and quantitative imaging techniques, that is, VEGF/VEGFR expression, may aid in patient stratification, new drug development and validation, treatment monitoring, and dose optimization, and it may help in deciding when or whether to start antiangiogenic treatment. In the future, quantitative PET molecular imaging coupled with selective labeled biomarkers may facilitate *in vivo* antiangiogenic drug efficacy by noninvasively assessing the expression of parameters of tumor neovascularization, guiding dose and regime by measuring target drug binding and receptor occupancy as well as potentially detecting the existence of mutations leading to either drug interaction or failure of treatment. Owing to the variety of antiangiogenic treatment approaches, the most optimal biomarker method will be defined by the drug's mode of action. In the era of personalized medicine, discovering new prognostic or predictive biomarkers and fast clinical translation and incorporation of soluble biomarkers and novel imaging techniques such as integrin imaging into anticancer clinical trials will be critical for the maximum benefit of cancer patients.

REFERENCES

1. Carmeliet P, Jain RK. Angiogenesis in cancer and other diseases. Nature 2000; 407:249–257.
2. Shojaei F, Ferrara N. Antiangiogenesis to treat cancer and intraocular neovascular disorders. Lab Invest 2007;87:227–230.
3. Fiedler U, Scharpfenecker M, Koidl S, Hegen A, Grunow V, Schmidt JM, Kriz W, Thurston G, Augustin HG. The Tie-2 ligand Angiopoietin-2 is stored in and rapidly released upon stimulation from endothelial cell Weibel-Palade bodies. Blood 2004;103:4150–4156.
4. Papapetropoulos A, Fulton D, Mahboubi K, Kalb RG, O'Connor DS, Li F, Altieri DC, Sessa WC. Angiopoietin-1 inhibits endothelial cell apoptosis via the Akt/survivin pathway. J Biol Chem 2000;275:9102–9105.
5. Oliner J, Min H, Leal J, Yu D, Rao S, You E, Tang X, Kim H, Meyer S, Han SJ, Hawkins N, Rosenfeld R, Davy E, Graham K, Jacobsen F, Stevenson S, Ho J, Chen Q, Hartmann T, Michaels M, Kelley M, Li L, Sitney K, Martin F, Sun JR, Zhang N, Lu J, Estrada J, Kumar R, Coxon A, Kaufman S, Pretorius J, Scully S, Cattley R, Payton M, Coats S, Nguyen L, Desilva B, Ndifor A, Hayward I, Radinsky R, Boone T, Kendall R. Suppression of angiogenesis and tumor growth by selective inhibition of angiopoietin-2. Cancer Cell 2004;6:507–516.
6. Cao Y, Sonveaux P, Liu S, Zhao Y, Mi J, Clary BM, Li CY, Kontos CD, Dewhirst MW. Systemic overexpression of angiopoietin-2 promotes tumor microvessel regression and inhibits angiogenesis and tumor growth. Cancer Res 2007;67:3835–3844.
7. Folkman J. Tumor angiogenesis: therapeutic implications. N Engl J Med 1971;285:1182–1186.
8. Bergers G, Benjamin LE. Tumorigenesis and the angiogenic switch. Nat Rev Cancer 2003;3:401–410.
9. Drevs J, Schneider V. The use of vascular biomarkers and imaging studies in the early clinical development of anti-tumour agents targeting angiogenesis. J Intern Med 2006;260:517–529.

10. Meyer M, Clauss M, Lepple-Wienhues A, Waltenberger J, Augustin HG, Ziche M, Lanz C, Büttner M, Rziha HJ, Dehio C. A novel vascular endothelial growth factor encoded by Orf virus, VEGF-E, mediates angiogenesis via signalling through VEGFR-2 (KDR) but not VEGFR-1 (Flt-1) receptor tyrosine kinases. EMBO J. 1999; 18:363–74.
11. Tammela T, Zarkada G, Wallgard E, Murtomäki A, Suchting S, Wirzenius M, Waltari M, Hellström M, Schomber T, Peltonen R, Freitas C, Duarte A, Isoniemi H, Laakkonen P, Christofori G, Ylä-Herttuala S, Shibuya M, Pytowski B, Eichmann A, Betsholtz C, Alitalo K. Blocking VEGFR-3 suppresses angiogenic sprouting and vascular network formation. Nature 2008;454:656–660.
12. Padera TP, Jain RK. VEGFR3: a new target for antiangiogenesis therapy? Dev Cell 2008;15:178–179.
13. Laakkonen P, Waltari M, Holopainen T, Takahashi T, Pytowski B, Steiner P, Hicklin D, Persaud K, Tonra JR, Witte L, Alitalo K. Vascular endothelial growth factor receptor 3 is involved in tumor angiogenesis and growth. Cancer Res 2007;67:593–599.
14. Hurwitz, H. Integrating the anti-VEGF-A humanized monoclonal antibody bevacizumab with chemotherapy in advanced colorectal cancer. Clin Colorectal Cancer. 2004;Suppl 2:S62–8.
15. Sandler A, Gray R, Perry MC, Brahmer J, Schiller JH, Dowlati A, Lilenbaum R, Johnson DH. Paclitaxel-carboplatin alone or with bevacizumab for non-small-cell lung cancer. N Engl J Med. 2006;355:2542–50.
16. Miller K, Wang M, Gralow J, Dickler M, Cobleigh M, Perez EA, Shenkier T, Cella D, Davidson NE. Paclitaxel plus bevacizumab versus paclitaxel alone for metastatic breast cancer. N Engl J Med 2007;357:2666–2676.
17. Cabebe E, Wakelee H. Sunitinib: a newly approved small-molecule inhibitor of angiogenesis. Drugs Today. 2006;42:387–98.
18. Deleted in Proof.
19. Cabebe E, Wakelee H. Role of anti-aniogenesis agents in treating NSCLC: focus on bevacizumab and VEGFR tyrosine kinase inhibitors. Curr Treat Options Oncol 2007;8:15–27.
20. Huang J, Bae JO, Tsai JP, Kadenhe-Chiweshe A, Papa J, Lee A, Zeng S, Kornfeld ZN, Ullner P, Zaghloul N, Ioffe E, Nandor S, Burova E, Holash J, Thurston G, Rudge J, Yancopoulos GD, Yamashiro DJ, Kandel JJ. Angiopoietin-1/Tie-2 activation contributes to vascular survival and tumor growth during VEGF blockade. Int J Oncol 2009;34:79–87.
21. Bhatt RS, Seth P, Sukhatme VP. Biomarkers for monitoring antiangiogenic therapy. Clin Cancer Res 2007;13:777s–780s.
22. Cines DB, Pollak ES, Buck CA, Loscalzo J, Zimmerman GA, McEver RP, Pober JS, Wick TM, Konkle BA, Schwartz BS, Barnathan ES, McCrae KR, Hug BA, Schmidt AM, Stern DM. Endothelial cells in physiology and in the pathophysiology of vascular disorders. Blood 1998;91:3527–3561.
23. Hida K, Hida Y, Shindoh M. Understanding tumor endothelial cell abnormalities to develop ideal anti-angiogenic therapies. Cancer Sci 2008;99:459–466.
24. Trusheim MR, Berndt ER, Douglas FL. Stratified medicine: strategic and economic implications of combining drugs and clinical biomarkers. Nature Rev Drug Discov 2007;6:287–293.

25. Severino MF, DuBose RF, Patterson SD. A strategic view on the use of pharmacodynamic biomarkers in early clinical drug development. IDrugs 2006;9:849–853.

26. FDA. 2008. Available at http://www.fda.gov/oc/initiatives/criticalpath/.

27. Bergers G, Hanahan D. Modes of resistance to anti-angiogenic therapy. Nat Rev Cancer 2008;8:592–603.

28. Atkinson AJ, Colburn WA, DeGruttola VG, DeMets DL, Downing GJ, Hoth DF, Oates JA, Peck CC, Schooley RT, Spilker BA, Woodcock J, Zeger SL. Biomarkers and surrogate endpoints: Preferred definitions and conceptual framework Clin Pharmacol Ther 2001;69:89–95.

29. Weidner N. Intratumor microvessel density as a prognostic factor in cancer. Am J Pathol 1995;147:9–19.

30. Mineo TC, Ambrogi V, Baldi A, Rabitti C, Bollero P, Vincenzi B, Tonini G. Prognostic impact of VEGF, CD31, CD34, and CD105 expression and tumour vessel invasion after radical surgery for IB-IIA non-small cell lung cancer. J Clin Pathol 2004;57:591–597.

31. Uzzan B, Nicolas P, Cucherat M, Perret GY. Microvessel density as a prognostic factor in women with breast cancer: a systematic review of the literature and meta-analysis. Cancer Res 2004;64:2941–2955.

32. Des Guetz G, Uzzan B, Nicolas P, Cucherat M, Morere JF, Benamouzig R, Breau JL, Perret GY. Microvessel density and VEGF expression are prognostic factors in colorectal cancer. Meta-analysis of the literature. Br J Cancer 2006;94:1823–1832.

33. Achen MG, Stacker SA. Molecular control of lymphatic metastasis. Ann N Y Acad Sci 2008;1131:225–234.

34. Tobler NE, Detmar M. Tumor and lymph node lymphangiogenesis—impact on cancer metastasis. J Leukoc Biol 2006;80:691–696.

35. El-Gohary YM, Metwally G, Saad RS, Robinson MJ, Mesko T, Poppiti RJ. Prognostic significance of intratumoral and peritumoral lymphatic density and blood vessel density in invasive breast carcinomas. Am J Clin Pathol 2008;129:578–586.

36. Willett CG, Boucher Y, di Tomaso E, Duda DG, Munn LL, Tong RT, Chung DC, Sahani DV, Kalva SP, Kozin SV, Mino M, Cohen KS, Scadden DT, Hartford AC, Fischman AJ, Clark JW, Ryan DP, Zhu AX, Blaszkowsky LS, Chen HX, Shellito PC, Lauwers GY, Jain RK. Direct evidence that the VEGF-specific antibody bevacizumab has antivascular effects in human rectal cancer. Nat Med 2004;10:145–147, Epub 2004 Jan 25. Erratum in: Nat Med 10:649.

37. Strumberg D, Richly H, Hilger RA, Schleucher N, Korfee S, Tewes M, Faghih M, Brendel E, Voliotis D, Haase CG, Schwartz B, Awada A, Voigtmann R, Scheulen ME, Seeber S. Phase I clinical and pharmacokinetic study of the Novel Raf kinase and vascular endothelial growth factor receptor inhibitor BAY 43–9006 in patients with advanced refractory solid tumors. J Clin Oncol 2005;23:965–972.

38. Bertolini F, Mancuso P, Shaked Y, Kerbel RS. Molecular and cellular biomarkers for angiogenesis in clinical oncology. Drug Discov Today 2007;12:806–812.

39. Mancuso P, Colleoni M, Calleri A, Orlando L, Maisonneuve P, Pruneri G, Agliano A, Goldhirsch A, Shaked Y, Kerbel RS, Bertolini F. Circulating endothelial-cell kinetics and viability predict survival in breast cancer patients receiving metronomic chemotherapy. Blood 2006;108:452–459.

40. Venneri MA, De Palma M, Ponzoni M, Pucci F, Scielzo C, Zonari E, Mazzieri R, Doglioni C, Naldini L. Identification of proangiogenic TIE2-expressing monocytes (TEMs) in human peripheral blood and cancer. Blood 2007;109:5276–5285.
41. Shojaei F, Ferrara N. Refractoriness to antivascular endothelial growth factor treatment: role of myeloid cells. Cancer Res 2008;68:5501–5504.
42. DePrimo SE, Bello C. Surrogate biomarkers in evaluating response to anti-angiogenic agents: focus on sunitinib. Ann Oncol 2007;18 Suppl 10:11–19.
43. Sassa C, Guibal A, Del Conte G, Ruegg C. Biomarkers of angiogenesis for the development of antiangiogenic therapies in oncology~: tools or decorations~? Nat Clinc Pract 2008;5:378–391.
44. Carmeliet P. Angiogenesis in life, disease and medicine. Nature 2005;438:932–936.
45. Kümpers P, Lukasz A, David S, Horn R, Hafer C, Faulhaber-Walter R, Fliser D, Haller H, Kielstein JT. Excess circulating angiopoietin-2 is a strong predictor of mortality in critically ill medical patients. Crit Care 2008;12:R147.
46. Orfanos SE, Kotanidou A, Glynos C, Athanasiou C, Tsigkos S, Dimopoulou I, Sotiropoulou C, Zakynthinos S, Armaganidis A, Papapetropoulos A, Roussos C. Angiopoietin-2 is increased in severe sepsis: correlation with inflammatory mediators. Crit Care Med 2007;35:199–206, Erratum in: Crit Care Med 2007;35:1224.
47. Schliemann C, Bieker R, Thoennissen N, Gerss J, Liersch R, Kessler T, Büchner T, Berdel WE, Mesters RM. Circulating angiopoietin-2 is a strong prognostic factor in acute myeloid leukemia. Leukemia 2007;21:1901–1906.
48. Park JH, Park KJ, Kim YS, Sheen SS, Lee KS, Lee HN, Oh YJ, Hwang SC. Serum angiopoietin-2 as a clinical marker for lung cancer. Chest 2007;132:200–206.
49. Kut C, Mac-Gabhann F, Popel AS. Where is VEGF in the body? A meta-analysis of VEGF distribution in cancer. Br J Cancer 2007;97:978–985.
50. Drevs J, Zirrgiebel U, Schmidt-Gersbach CI, Mross K, Medinger M, Lee L, Pinheiro J, Wood J, Thomas AL, Unger C, Henry A, Steward WP, Laurent D, Lebwohl D, Dugan M, Marmé D. Soluble markers for the assessment of biological activity with PTK787/ZK 222584 (PTK/ZK), a vascular endothelial growth factor receptor (VEGFR) tyrosine kinase inhibitor in patients with advanced colorectal cancer from two phase I trials. Ann Oncol 2005;16:558–565.
51. Drevs J, Siegert P, Medinger M, Mross K, Strecker R, Zirrgiebel U, Harder J, Blum H, Robertson J, Jürgensmeier JM, Puchalski TA, Young H, Saunders O, Unger C. Phase I clinical study of AZD2171, an oral vascular endothelial growth factor signaling inhibitor, in patients with advanced solid tumors. J Clin Oncol 2007;25:3045–3054.
52. Batchelor TT, Sorensen AG, di Tomaso E, Zhang WT, Duda DG, Cohen KS, Kozak KR, Cahill DP, Chen PJ, Zhu M, Ancukiewicz M, Mrugala MM, Plotkin S, Drappatz J, Louis DN, Ivy P, Scadden DT, Benner T, Loeffler JS, Wen PY, Jain RK. AZD2171, a pan-VEGF receptor tyrosine kinase inhibitor, normalizes tumor vasculature and alleviates edema in glioblastoma patients. Cancer Cell 2007;11:83–95.
53. Rini BI, Michaelson MD, Rosenberg JE, Bukowski RM, Sosman JA, Stadler WM, Hutson TE, Margolin K, Harmon CS, DePrimo SE, Kim ST, Chen I, George DJ. Antitumor activity and biomarker analysis of sunitinib in patients with bevacizumab-refractory metastatic renal cell carcinoma. J Clin Oncol 2008;26:3743–3748.
54. Burstein HJ, Elias AD, Rugo HS, Cobleigh MA, Wolff AC, Eisenberg PD, Lehman M, Adams BJ, Bello CL, DePrimo SE, Baum CM, Miller KD. Phase II study of sunitinib malate, an oral multitargeted tyrosine kinase inhibitor, in patients with metastatic

breast cancer previously treated with an anthracycline and a taxane. J Clin Oncol 2008;26:1810–1816.

55. Verheul HM, Pinedo HM. Possible molecular mechanisms involved in the toxicity of angiogenesis inhibition. Nat Rev Cancer 2007;7:475–485.
56. Ebos JM, Lee CR, Christensen JG, Mutsaers AJ, Kerbel RS. Multiple circulating proangiogenic factors induced by sunitinib malate are tumor-independent and correlate with antitumor efficacy. Proc Natl Acad Sci USA 2007;104:17069–17074.
57. Li KCP. Angiogenesis imaging in the post-genomic era. Br J Radiol 2003;S1–S2.
58. Jain RK. Lessons from multidisciplinary translational trials on anti-angiogenic therapy of cancer. Nat Rev Cancer 2008;8:309–316.
59. Perini R, Choe R, Yodh AG, Sehgal C, Divgi CR, Rosen MA. Non-invasive assessment of tumor neovasculature: techniques and clinical applications. Cancer Metastasis Rev 2008;27:615–630.
60. Arger PH, Malkowicz SB, VanArsdalen KN, Sehgal CM, Holzer A, Schultz SM. Color and power Doppler sonography in the diagnosis of prostate cancer: comparison between vascular density and total vascularity. J Ultrasound Med 2004;23:623–630.
61. Yang WT, Tse GMK, Lam PKW, Metreweli C, Chang J. Correlation between color power Doppler sonographic measurement of breast tumor vasculature and immunohistochemical analysis of microvessel density for the quantitation of angiogenesis. J Ultrasound Med 2002;21:1227–1235.
62. Delorme S, Krix M. Contrast-enhanced ultrasound for examining tumor biology. Cancer Imaging 2006;6:148–152.
63. Weskott HP. Emerging roles for contrast-enhanced ultrasound. Clin Hemorheol Microcirc 2008;40:51–71.
64. Aigner F, Pallwein L, Mitterberger M, Pinggera GM, Mikuz G, Horninger W, Frauscher F. Contrast-enhanced ultrasonography using cadence-contrast pulse sequencing technology for targeted biopsy of the prostate. BJU Int 2009;103: 458–463.
65. Watanabe Y, Aoi A, Horie S, Tomita N, Mori S, Morikawa H, Matsumura Y, Vassaux G, Kodama T. Low-intensity ultrasound and microbubbles enhance the antitumor effect of cisplatin. Cancer Sci 2008;99:2525–2531.
66. Tinkov S, Bekeredjian R, Winter G, Coester C. Microbubbles as ultrasound triggered drug carriers. J Pharm Sci. 2009;98:1935–1961.
67. Duvshani-Eshet M, Benny O, Morgenstern A, Machluf M. Therapeutic ultrasound facilitates antiangiogenic gene delivery and inhibits prostate tumor growth. Mol Cancer Ther 2007;6:2371–2382.
68. Hylton N. Dynamic contrast-enhanced magnetic resonance imaging as an imaging biomarker. J Clin Oncol 2006;24:3293–3298.
69. Newbold K, Castellano I, Charles-Edwards E, Mears D, Sohaib A, Leach M, Rhys-Evans P, Clarke P, Fisher C, Harrington K, Nutting C. An exploratory study into the role of dynamic contrast-enhanced magnetic resonance imaging or perfusion computed tomography for detection of intratumoral hypoxia in head-and-neck cancer. Int J Radiat Oncol Biol Phys. 2009;74:29, 37.
70. Padhani AR. Where are we with imaging oxygenation in human tumors. Cancer Imaging 2005;5:128–130.

71. O'Connor JP, Jackson A, Parker GJ, Jayson GC. DCE-MRI biomarkers in the clinical evaluation of antiangiogenic and vascular disrupting agents. Brit J Cancer 2007;96:189–195.
72. Evelhoch J, Garwood M, Vigneron D, Knopp M, Sullivan D, Menkens A. Expanding the use of magnetic resonance in the assessment of tumor response to therapy: workshop report. Cancer Res 2005;65:7041–7044.
73. Leach MO, Brindle KM, Evelhoch J, Griffiths JR, Horsman MR, Jackson A, Jayson GC, Judson IR, Knopp MV, Maxwell RJ, McIntyre D, Padhani AR, Price P, Rathbone R, Rustin GJ, Tofts PS, Tozer GM, Vennart W, Waterton JC, Williams SR, Workman P. The assessment of antiangiogenic and antivascular therapies in early-stage clinical trials using magnetic resonance imaging: issues and recommendations. Brit J Cancer 2005;92:1599–1610.
74. Schmid VJ, Whitcher B, Padhani AR, Taylor NJ, Yang GZ. A Bayesian hierarchical model for the analysis of a longitudinal dynamic contrast-enhanced MRI oncology study. Magn Reson Med 2009;61:163–174.
75. Rosen MA, Schnall MD. Dynamic contrast-enhanced magnetic resonance imaging for assessing tumor vascularity and vascular effects of targeted therapies in renal cell carcinoma. Clin Cancer Res 2007;13:770s–776s.
76. Mross K, Drevs J, Müller M, Medinger M, Marmé D, Hennig J, Morgan B, Lebwohl D, Masson E, Ho YY, Günther C, Laurent D, Unger C. Phase I clinical and pharmacokinetic study of PTK/ZK, a multiple VEGF receptor inhibitor, in patients with liver metastases from solid tumours. Eur J Cancer 2005;23:1291–1299.
77. Liu G, Rugo HS, Wilding G, McShane TM, Evelhoch JL, Ng C, Jackson E, Kelcz F, Yeh BM, Lee FT Jr, Charnsangavej C, Park JW, Ashton EA, Steinfeldt HM, Pithavala YK, Reich SD, Herbst RS. Dynamic contrast-enhanced magnetic resonance imaging as a pharmacodynamic measure of response after acute dosing of AG-013736, an oral angiogenesis inhibitor, in patients with advanced solid tumors: results from a phase I study. J Clin Oncol 2005;23:5464–5473.
78. Huang W, Li X, Morris EA, Tudorica LA, Seshan VE, Rooney WD, Tagge I, Wang Y, Xu J, Springer CS. The magnetic resonance shutter speed discriminates vascular properties of malignant and benign breast tumors *in vivo*. Proc Natl Acad Sci USA 2008;105:17943–17948.
79. Sorensen AG, Batchelor TT, Wen PY, Zhang WT, Jain RK. Response criteria for glioma. Nat Clin Pract Oncol 2008;5:634–644.
80. Bisdas S, Medov L, Baghi M, Konstantinou GN, Wagenblast J, Thng CH, Vogl TJ, Koh TS. A comparison of tumour perfusion assessed by deconvolution-based analysis of dynamic contrast-enhanced CT and MR imaging in patients with squamous cell carcinoma of the upper aerodigestive tract. Eur Radiol 2008;18:843–850.
81. Makari Y, Yasuda T, Doki Y, Miyata H, Fujiwara Y, Takiguchi S, Matsuyama J, Yamasaki M, Hirao T, Koyama MK, Nakamuara H, Monden M. Correlation between tumor blood flow assessed by perfusion CT and effect of neoadjuvant therapy in advanced esophageal cancers. J Surgical Oncol 2007;96:220–229.
82. Zima A, Carlos R, Gandhi D, Case I, Teknos T, Mukherji SK. Can pretreatment CT perfusion predict response of advanced squamous cell carcinoma of the upper aerodigestive tract treated with induction chemotherapy. Am J Neuroradiol 2007;28:328–334.
83. Schnall MD, Rosen MA. Primer on imaging technologies for Cancer. J Clin Oncol 2006;24:3225–3233.

84. Warburg O. On the origin of cancer cells. Science 1956;123:309–314.

85. Bos R, van Der Hoeven JJ, van Der Wall E, van Der Groep P, van Diest PJ, Comans EF, Joshi U, Semenza GL, Hoekstra OS, Lammertsma AA, Molthoff CF. Biologic correlates of (18) fluorodeoxyglucose uptake in human breast cancer measured by positron emission tomography. J Clin Oncol 2002;20:379–387.

86. Herscovitch P, Markham J, Raichle ME. Brain blood flow measured with intravenous H2(15)O. I. Theory and error analysis. J Nucl Med 1983;24:782–789.

87. Haberkorn U, Strauss LG, Reisser C, Haag D, Dimitrakopoulou A, Ziegler S, Oberdorfer F, Rudat V, van Kaick G. Glucose uptake, perfusion, and cell proliferation in head and neck tumors: relation of positron emission tomography to flow cytometry. J Nucl Med 1991;32:1548–1555.

88. Wilson CB, Lammertsma AA, McKenzie CG, Sikora K, Jones T. Measurements of blood flow and exchanging water space in breast tumors using positron emission tomography: a rapid and noninvasive dynamic method. Cancer Res 1992;52:1592–1597.

89. Miller KD, Soule SE, Calley C, Emerson RE, Hutchins GD, Kopecky K. Randomized phase II trial of the anti-angiogenic potential of doxorubicin and docetaxel; primary chemotherapy as Biomarker Discovery Laboratory. Breast Cancer Res Treat 2005;8:187–197.

90. Herbst RS, Mullani NA, Davis DW, Hess KR, McConkey DJ, Charnsangavej C, O'Reilly MS, Kim HW, Baker C, Roach J, Ellis LM, Rashid A, Pluda J, Bucana C, Madden TL, Tran HT, Abbruzzese JL. Development of biologic markers of response and assessment of antiangiogenic activity in a clinical trial of human recombinant endostatin. J Clin Oncol 2002;20:3804–3814.

91. Padhani AR, Krohn KA, Lewis JS, Alber M. Imaging oxygenation of human tumours. Eur Radiol 2007;17:861–872.

92. Divgi CR, Pandit-Taskar N, Jungbluth AA, Reuter VE, Gönen M, Ruan S, et al. Preoperative characterization of clear-cell renal carcinoma using iodine-124-labelled antibody chimeric G250 (124IcG250) and PET in patients with renal masses: a phase I trial. Lancet Oncol 2007;8:304–310.

93. Chen K, Cai W, Li ZB, Wang H, Chen X. Quantitative PET imaging of VEGF receptor expression. Mol Imaging Biol. 2009;11:15–22.

94. Wang H, Cai W, Chen K, Li ZB, Kashefi A, He L, Chen X. A new PET tracer specific for vascular endothelial growth factor receptor 2. Eur J Nucl Med Mol Imaging. 2007;34:2001–2010.

95. Sivolapenko GB, Skarlos D, Pectasides D, Stathopoulou E, Milonakis A, Sirmalis G, Stuttle A, Courtenay-Luck NS, Konstantinides K, Epenetos AA. Imaging of metastatic melanoma utilising a technetium-99m- labelled RGD-containing synthetic peptide. Euro J Nucl Med 1998;25:1383–1389.

96. Beer AJ, Haubner R, Goebel M, Luderschmidt S, Spilker ME, Wester HJ, Weber WA, Schwaiger M. Biodistribution and pharmacokinetics of the alphavbeta3-selective tracer 18F-galacto-RGD in cancer patients. J Nucl Med 2005;46:1333–1341.

97. Beer AJ, Haubner R, Sarbia M, Goebel M, Luderschmidt S, Grosu AL, Schnell O, Niemeyer M, Kessler H, Wester HJ, Weber WA, Schwaiger M. Positron emission tomography using [18F] Galacto-RGD identifies the level of integrin alpha(v)beta3 expression in man. Nat Rev Cancer 2003; 3:401–410.

6

CARDIOVASCULAR BIOMARKERS AS EXAMPLES OF SUCCESS AND FAILURE IN PREDICTING SAFETY IN HUMANS

SIMON AUTHIER, MICHAEL K. PUGSLEY, ERIC TRONCY, AND MICHAEL J. CURTIS

Biomarker: A characteristic that is objectively measured and evaluated as an indicator of normal biological processes, pathogenic processes, or pharmacologic responses to a therapeutic intervention (1)

—(National Institute of Health Biomarker Definitions Working Group).

A surrogate endpoint or marker is a laboratory measurement or physical sign that is used in therapeutic trials as a substitute for a clinically meaningful endpoint that is a direct measure of how a patient feels, functions or survives and is expected to predict the effect of the therapy (2).

6.1 INTRODUCTION

Drug discovery begins with a hypothesis. This hypothesis leads to development of a new chemical entity (NCE) that is designed to alter a particular target. However, before clinical efficacy can be tested, proof of concept requires identification of a likelihood of efficacy. This is interrogated using animal models of disease and usually involves use of biomarkers. Biomarkers are also used as surrogate indicators of the state of a disease (for diagnosis and prognosis). Biomarkers

Predictive Approaches in Drug Discovery and Development: Biomarkers and In Vitro/In Vivo Correlations, First Edition. Edited by J. Andrew Williams, Jeffrey R. Koup, Richard Lalonde, and David D. Christ.

(those used for safety assessment purposes as well as efficacy) are therefore cornerstones of medical research. In their central role, biomarkers reflect the presumed link between conceptual understanding of pathophysiological processes and their modulation for therapeutic purposes. In this unique position, biomarkers serve a role in translational medicine, where research advances are converted into practical decision algorithms for clinicians or potential new therapies for development by the pharmaceutical industry. While this concept may appear simple, it depends on the predictivity of the biomarker. Given the close relatedness of biomarkers used in drug discovery (nonclinical) and clinical applications, both perspectives will be presented and discussed.

For clinicians, biomarkers are useful for diagnosis, prognosis, or to help select and guide treatment for a given patient. In drug development, biomarkers are the scientific endpoints that orient study design and decision making. In clinical trials, a biomarker may be elevated to the status of "surrogate endpoint," where it serves as a quantitative measure of the effectiveness or the potential safety of a treatment. A surrogate endpoint is therefore a biomarker that (to be of value) has been validated for a given application as a predictive measure of the true clinical outcome. The controversial process by which a biomarker is validated to become a surrogate marker will be discussed throughout the chapter.

Biomarkers are used in all areas of medicine and many well-known examples include the use of bone density as a surrogate of fractures in osteoporosis (3, 4), CD4 lymphocyte count and quantitative measurement of viral load and proviral DNA in human immunodeficiency virus (HIV) (5, 6), or albuminemia in chronic renal disease (7).

6.2 THE INTERDEPENDENCY OF NONCLINICAL RESEARCH, CLINICAL TRIALS, AND THERAPEUTICS

To set the basis for discussion, the context of drug development will be outlined. Drug development can be divided into clinical and nonclinical areas of investigation. Clinical investigation includes research activities commensurate with study in healthy volunteers or patients, while nonclinical investigation encompasses research activities for drug development that include a diverse spectrum of *in vitro* and *in silico* assays and animal models. Nonclinical research is initiated before first in human (FIH) administration (preclinical) but can continue during clinical trials. Nonclinical testing requirements increase as the process proceeds and the NCE advances in development; however, note that data received in clinical trials may also prompt additional nonclinical testing. The different requirements of nonclinical testing for drug development in relation to clinical trials and drug approval will be presented later. Drug development research and clinical diagnosis share the same final interest—the patient. As a result, a significant proportion of biomarkers used in nonclinical research and early clinical trials are also used in the clinic during the conduct of phases 3 (i.e., randomized controlled multicenter trials on large patient populations) and 4 (postmarketing safety surveillance) studies.

There is a growing consensus among regulatory authorities and the pharmaceutical industry regarding the need for biomarkers that are common to nonclinical research and clinical trials and which can eventually be used by clinicians. This is important because nonclinical data can be used to inform clinical decision making when issues of efficacy and (more commonly) safety arise. For example, a biomarker used in nonclinical research to quantify liver toxicity could later be used to monitor signs of possible hepatic toxicity in clinical trials. Thus, common clinical and nonclinical biomarkers offer the promise of greater coherence and ease of decision making. However, for this to work, a question about the process of biomarker development must be addressed.

Biomarkers are developed by integrating the outcomes of clinical trials, clinical research (independent of ongoing evaluation of therapeutic interventions), and nonclinical research. Drugs with known clinical effects are most useful for development and validation of biomarkers in nonclinical models, where nonclinical biomarkers are assessed for their ability to predict a confirmed clinical outcome. Clinical research excluding therapeutic interventions may identify biomarkers for diagnostic or prognostic purposes. However, establishing causality between change in the biomarker and improvement of clinical outcome due to the treatment (drug) is often no more than a "leap of faith" as will be explained later using high density lipoproteins (HDLs) for cardiovascular disease as an example.

6.3 EVOLUTION OF BIOMARKER DEVELOPMENT

The development of biomarkers and their role in drug development have evolved rapidly leveraged primarily by advances in life science technologies. At a time when biomarkers carry great hope for medical advances, the history of biomarkers may have a lesson to teach the modern medical world. Considerable efforts have been invested to characterize the predictive value of each current biomarker as a surrogate endpoint. The iconic Framingham heart study (started in 1947) was among the pioneer initiatives of the era of prospective epidemiological clinical studies to undertake systematic investigation of causes of cardiovascular disease and risk factors. Findings from the Framingham study allowed for an assessment of biomarker validation for cardiovascular disease resulting in the utilization and subsequent adoption of serum cholesterol (8) as a primary biomarker for cardiovascular health status. As described in the initial study outline by Meadors (1947): "this project is designed to study the expression of coronary artery disease in a normal or unselected population and to determine the factors predisposing to the development of the disease through clinical and laboratory examination and long-term follow-up of such a group" (8). While serum cholesterol can be used as a biomarker to establish the general health status of the cardiovascular system, the troponins (T and I), recognized as highly sensitive and specific markers of myocardial damage, illustrate biomarkers that have been developed to provide direct evidence of disease (9). The use of biomarkers to assess disease risk factors is common in clinical diagnosis (e.g., for identification of signs of malignancy by histology of tumor biopsy) and also in clinical trials (e.g., from assessment of QT prolongation), whereas the effect of a treatment on a biomarker may be used to predict efficacy or safety (e.g., troponin T and I, serum level of low density lipoprotein (LDL), or glomerular filtration rate).

However, the use of unvalidated biomarkers (i.e., characteristics that have not yet been determined to be reliable) is potentially hazardous. First, it is recognized that a treatment effect on a surrogate endpoint does not necessarily guarantee correct inference of the treatment effect on the relevant clinical endpoint (10–12). The concept of biomarkers as risk factors is intimately related to validation of surrogate endpoints. Surrogate endpoints may include biomarkers that represent direct evidence of disease as illustrated previously with troponins or biomarkers validated as predictive of clinical outcome exemplified by the QT interval that is widely used to assess the risk of the syndrome torsades de pointes (TdP). The QT interval, which represents the interval between the start of ventricular depolarization and the end of repolarization, is recognized by the scientific community (13, 14) and regulatory agencies (15) as the most convenient biomarker to assess the risk of developing TdP. Consequently, most of the attention from both the scientific community and regulatory agencies [Food and Drug Administration (FDA), European Medicines Agency (EMEA), and the Ministry of Health, Labour and Welfare (MHLW)] has been directed toward QT prolongation as a risk factor for drug-induced TdP. Sensitivity and specificity limitations of QT prolongation have been reported by several groups (16, 17) and there has been criticism of overreliance on the use of the QT interval (18). Some drugs such as amiodarone and pentobarbital induce QT prolongation but have no reported ability to cause TdP. The use of QT prolongation as a surrogate for TdP in drug development may lead to discontinuation of valuable treatments. On the other hand, while increasing evidence is emerging that QT *shortening* predisposes to ventricular fibrillation (19), regulatory guidelines on QT interval have yet to address this possible concern. As a result, a widely accepted and validated biomarker used as risk factor for a potentially fatal condition relies on questionable and evolving foundations. Despite the limitations of QT prolongation, ethical and economical considerations prevent use of the true clinical endpoint (TdP in patients) to assess the safety of new treatments. Considerable efforts have been made to refine and validate the use of QT as a surrogate marker for TdP (20–23) but this has tended to serve only to emphasize its limitations. Increasingly, QT is seen as just one part of an integrated risk assessment (24).

6.4 COMPOSITE ENDPOINTS

As one might expect, given the complex nature of most diseases, if there is no single definitive biomarker for a given condition, a combination of biomarkers is normally used to forecast potential clinical outcome (mortality, morbidity, and quality of life). Medicine has always aimed at improving the predictive value of biomarkers. Selection and validation processes have evolved into an organized framework, where evidence-based medicine (EBM) benefits from meta-analyses of the medical literature, risk–benefit assessment, and randomized controlled trials to weigh the predictive value of biomarker combinations. The use of an integrated approach combining more than one biomarker to increase the predictive value is noted in the clinic where prognostic indexes using multiple biomarkers have been developed in major areas of medicine, including, but not limited to, cardiology (25, 26), oncology (27, 28), and neurology (29). Similar approaches have been developed and are now utilized in nonclinical drug development where an integrated risk assessment is used to estimate the sensitivity and specificity of a combination of nonclinical models. Pollard et al. (30) recently assessed the predictive value of a combination of nonclinical assays to quantify TdP risk potentials. When combining *in vitro* (**h**uman ***E**ther-á-go-go*-**R**elated **G**ene; hERG) and *in vivo* QT data, the predictive value to man was reported to be greater than 80%. This may seem high, but in fact, it implies a 20% failure rate to predict a potentially life-threatening condition. Initiatives to assess the integrated predictive value of drug development screening platforms may have a long-term impact on development of new therapies where selection and timing of the various assays is traditionally based on experience of the research groups rather than on calculated and proven predictive value. With calculated predictive value, one could reassess the construct of a drug development program and optimize timeline and resource allocation.

Among the disciplines using multiple factor analysis, genomic and proteomic approaches offer potentially one of the best hopes for rapid medical progress. With the increasing availability of microarray technologies, genomics and proteomics give rise to a new paradigm in biomarker development. The quest to establish a relationship between biomarkers and clinical outcome has challenged medical research for the past century. The modern medical world is now faced with a unique challenge: determination of whether the correlations identified have genuine predictive value. Genomics and proteomics are particularly affected by this since although they allow extensive characterization of chromosome and protein expression, so much data are generated by these powerful screening technologies that correlation of one or more biomarker with an experimental variable is inevitable. It then becomes necessary to interrogate the relevance of the correlation. This validation of biomarker candidates is an area of intensive activity. The imperative to confirm the scientific value of "discoveries" from high output microarray technologies requires novel approaches to data analysis supported by bioinformatics (31, 32). At bed side, patient genome screening is now commercially available and can be used to evaluate multiple single-nucleotide polymorphisms (SNPs) for disease susceptibility. Genome profiling is a start point for personalized preventive medicine and targeted therapies (33). In spite of recognized potential for improved diagnosis, the clinical utility of SNPs remains limited given the lack of controlled clinical trials to evaluate the clinical value of genetic biomarker screening (34).

6.5 CONSIDERATIONS FOR THE USE OF BIOMARKERS: IS VALIDATION ACHIEVABLE?

Validation requires value as well as validity. A biomarker is useful only if it is sufficiently accurate *and* a therapeutically useful drug with a good risk/benefit ratio is available (i.e., the biomarker can be used to usefully inform therapeutic decision making). Prostate-specific antigen (PSA; also known as *kallikrein III* or *P30 antigen*) is a prostate-specific protein that is usually present in minute quantities in the serum of normal men but which is elevated in prostate cancer (35). The measurement of PSA for use in prostate cancer assessment began commercially in 1982—yet more than 25 years later, its value as a routine screening diagnostic tool is still debated (36) partly due to a relatively high rate of false negatives (reported to be as high as 27%) (37). The psychological consequences of a false positive in the case of a cancer biomarker may outweigh the biomarker's diagnostic value as a routine screening tool.

This emphasizes the key driver in validation: the patient is the primary focus. New generations of biomarkers succeed older generations on the basis of improved sensitivity, specificity, or other considerations such as economical and psychological impacts. Thus, lactate dehydrogenase (LDH), a marker of cardiac ischemia (38), has been largely replaced by troponin T, and troponin T is now challenged by a more sensitive marker (Heart fatty acid binding protein; H-FABP) for early detection of myocardial ischemia (39, 40).

Biomarkers used as surrogate endpoints evolve in a regulated environment where generic validation for a clinical condition takes priority over validation for a given drug or treatment (41). In other words, an ideal validation would demonstrate the predictive value of a surrogate endpoint across different drug classes to treat a given clinical indication (42). However, even widely accepted biomarkers struggle to comply with such stringent validation requirements, as will be discussed below, but first the regulatory context of biomarker validation will be presented.

6.6 REGULATORY CONSIDERATIONS

In the pharmaceutical industry, guidelines that have been provided by regulatory authorities serve as a start point for nonclinical and clinical study designs. Regulatory approval is usually based on the manifest effects of the treatment on survival or on the symptoms of the disease (41). Approval is based "... upon a determination that the product has an effect on a clinical endpoint or on a surrogate endpoint that is reasonably likely to predict clinical benefit." Examples of surrogate endpoints that were accepted by the US FDA (www.fda.gov) and the EMEA (http://www.emea.europa.eu) include blood pressure and cholesterol for heart attacks, stroke, and death.

Validation of biomarkers is recognized as a process that needs to be independent from drug submission review (43). Several initiatives by regulatory authorities have provided for a better understanding of biomarkers and their use in the regulatory approval of investigational drugs. A pilot group structure was developed by the FDA around the Interdisciplinary Pharmacogenomic Review Group (IPRG). Although the primary mission of the IPRG was to establish a scientific and regulatory framework for reviewing genomic data, it was also logical to allow the contributors from this group to aid in the qualification of new biomarkers for the evaluation of new drugs. This subsequent initiative comprised FDA experts from the Center for Drug Evaluation and Research (CDER), Center for Biologicals Evaluation and Research (CBER), Center for Devices and Radiological Health, and National Center for Toxicological Research and is known as the *Biomarker Qualification Review Team*. This team is mandated to coordinate the evaluation of data submitted as related to the qualification of novel biomarkers of drug safety using clinical, nonclinical, and statistical methodology (43). Coordination initially involves a review of the intended context of use of the biomarker utilizing data submitted from the applicant. The context of use is a critical component of qualification, since a biomarker may be relevant in more than one particular clinical setting. Thus, once the context of use has been reviewed, the biomarker qualification study strategy is devised and, in an iterative process, a consensus can be sought between the regulatory authority and the sponsor. After completion of the qualification study, the Biomarker Qualification Review Team will decide on approval or rejection of the new biomarker based on the study results.

While the urge to develop biomarkers is recognized by regulatory authorities, motives to prioritize the task differ from one clinical area to another. In some cases, new technologies provide unmatched opportunities for biomarker development such as imaging for lung tumors (44). In other cases, such as cardiovascular disease, the use of clinical outcome as a primary endpoint may not be feasible and the use of biomarkers such as LDL-cholesterol level (a biomarker for atherosclerosis) provides a reasonably validated (albeit not wholly definitive) surrogate endpoint.

Other initiatives to facilitate biomarker development include scientific forums such as the *Cardiovascular Biomarkers and Surrogate Endpoints Symposium* held annually since 2003. The symposium includes international experts and FDA representatives in a collaboration designed to address issues surrounding biomarker and surrogate endpoint application for the assessment of cardiovascular disease along with evaluation of the development of novel diagnostics. The symposium has been marked by controversies regarding the value of biomarkers in cardiovascular (and diabetes) drug approvals. In spite of significant efforts to increase the use of biomarkers, the debate on the validity of surrogate endpoints remains unresolved. Rosiglitazone (Avandia®), a thiazolidinedione agonist of the nuclear peroxisome proliferation-activation receptor gamma (PPAR-γ), was fast-tracked and received regulatory approval in 2000 for the treatment of type II diabetes mellitus based on its ability to improve glycemic control by improving insulin sensitivity. However, a meta-analysis revealed that rosiglitazone increases the risk of myocardial ischemic events (45).

6.7 PROBLEMS ARISING FROM THE USE OF UNVALIDATED BIOMARKERS

The use of biomarkers as surrogate endpoints may lead to grossly inaccurate predictions as illustrated by the Cardiac Arrhythmia Suppression Trial (CAST). Before conduct of this clinical trial, an increased risk for cardiovascular death was believed to correlate with the incidence of ventricular premature beats (VPBs) (46). Encainide and flecainide, class I sodium channel blocking antiarrhythmic drugs, effectively suppressed VPB incidence and were consequently approved by the FDA for life-threatening and symptomatic ventricular arrhythmias. The CAST confirmed a reduction in the incidence of VPBs in survivors of myocardial infarction treated with encainide and flecainide. However, encainide and flecainide also caused a higher rate of lethal ventricular arrhythmia incidence and total mortality (47). Paradoxically, it was accepted at the time that VPBs were only hypothetically a surrogate for lethal arrhythmias, and the CAST study was in fact billed as an attempt to test the "cardiac arrhythmia suppression hypothesis"—the notion that if a drug suppresses VPBs acutely in hospital, it will improve long-term survival (48). This illustrates the danger of using an unvalidated surrogate to direct drug discovery.

Aside from the risk of lack of causality with clinical outcome, using a surrogate endpoint in a clinical trial is unlikely to identify off target adverse effects or beneficial effects mediated through biomarker unrelated mechanisms. The presence of unexpected toxicity or lack of efficacy sometime requires large clinical trials to be confirmed. The use of biomarkers to assess efficacy and safety of drugs warrants a conservative position both for regulators and the pharmaceutical (and biopharmaceutical?) industry. In spite of biomarker drawbacks, they remain a key component in drug development and present definitive advantages in clinical trials, allowing completion of the trials in a much shorter time (49).

6.8 BIOMARKERS AND DRUG DEVELOPMENT STAGNATION

In March 2004, the US FDA launched the Critical Path Initiative in response to a marked decrease in the number of innovative medical products submitted for regulatory approval. In its report, the FDA highlighted the difficulties of medical product development and called for joint efforts (by both industry and regulators) to improve the use of scientific tools, including validated biomarkers. In 2006, the FDA released the Critical Path Opportunities List, which provided a list of greatest opportunities in which genomics, proteomics, imaging, and bioinformatics were recognized as valuable components of medical/medicinal product development. In 2007, three years after the Critical Path Initiative was launched, only a total of 17 new molecular entities (NMEs) and 2 biologic license applications (BLAs) that were submitted were approved by the US FDA—a record low since 1983 (50). While the urge to accelerate drug development is keenly felt by all, appropriate solutions still await to enhance approval yet provide a greater degree of drug safety. One explanation for the drug development stagnation resides in the approval of drugs in newer therapeutic areas (i.e., "first in class" drugs), where regulators require more comprehensive clinical (and preclinical) data due to the limited experience with the therapeutic target and developed drugs for that target. Thus, the risk–benefit assessments by regulators become obscured by the increased volume of data on the NCE.

Another possible explanation for drug development stagnation is the absence of biomarkers for predicting short-term efficacy in some degenerative diseases such as osteoarthritis (OA). OA is one of the most prevalent disease in North America with a patient population of 27 million in the United States alone (51), representing 3.5% of home care patients (52). While disease-modifying drugs such as interleukin-6 (IL-6) (53) and tumor necrosis factor-a (TNF-a) (54) blockers are developed for the treatment of rheumatoid arthritis (DMARDs), OA remains a condition with scarce drug candidate development potential (DMOAD). Yet approved treatments are mainly limited to drugs that alleviate pain and symptoms (55) but have limited to no impact on actual disease progression.

Drug development in this field suffers from the general perception of limited predictive value of small animal research models (hence low research output). In recent studies, the Pond-Nuki dog model has, however, been proven to be predictive as positive results with diacerhein (56), doxycycline (57), and licofelone (58) were all reproduced in clinical trials (59–61). When large animal models present the only real predictive *in vivo* model, financial and ethical considerations become a significant burden to high throughput screening of drug candidates.

The reduction in the progression of structural changes in OA induced by drugs does not always translate into clinical benefits. Absence of symptom improvement with DMOADs could be related to a number of factors such as the selection of patient population (e.g., inadequate stage of the disease and limited alteration in biomechanics), slow disease progression and chronic disability rendering the evaluation of the efficacy of treatment difficult, or simply due to the absence of a validated surrogate endpoint. OA involves all synovial joint tissue components and the emphasis on the loss of cartilage in the evaluation of efficacy of new treatments was likely misleading. OA is a painful disorder. Since articular cartilage is not innervated, the link between cartilage and pain severity may be due to other aspects of OA disease pathology, involving a potential role for cartilage. Hence, there is no reason to expect that pain or discomfort would accompany cartilage loss. In contrast, longitudinal magnetic resonance imaging (MRI) studies correlating pain with cartilage loss found no association between the severity of pain and the severity of cartilage morphology alterations (62, 63). The above-indicated points have all been barriers to drug development and are consequences of technical and measurement limitations that new approaches will have to overcome. Indeed, a recent study of Raynauld et al. (61) using quantitative MRI (qMRI—Fig. 6.1) allowed to link structural joint benefits and pain relief of a DMOAD (licofelone) in a multicenter trial of knee OA patients. These results suggest that technology advances are likely to hold part of the solution to the drug development stagnation through refinement of biomarkers. Further

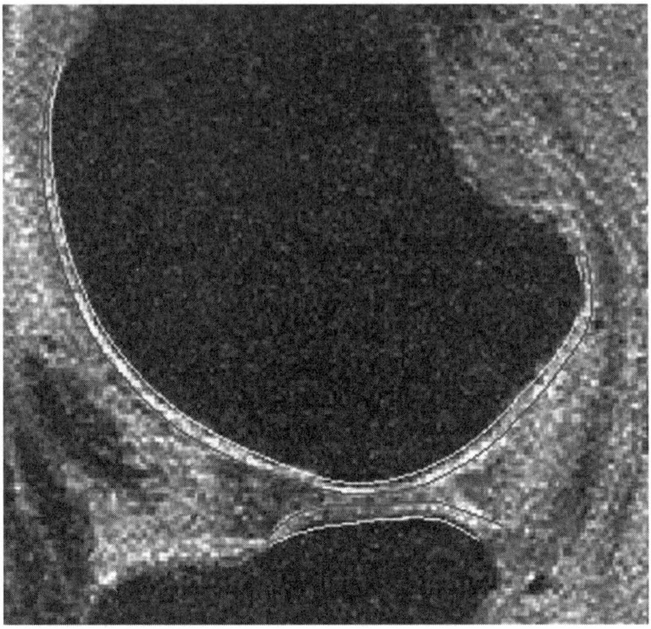

(a)

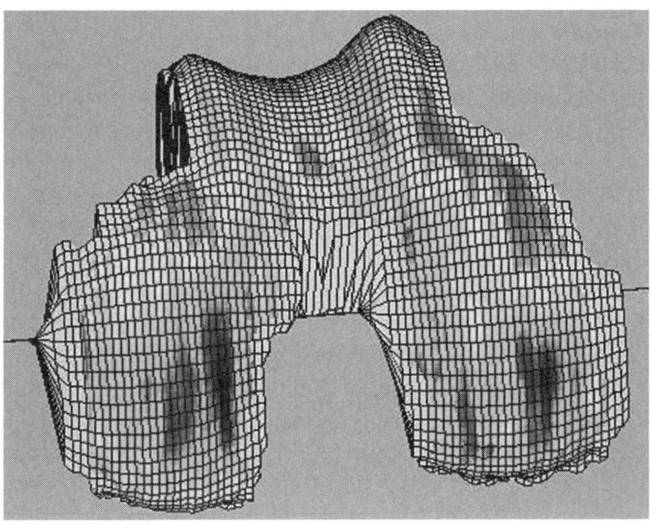

(b)

FIGURE 6.1 Cartilage quantitative evaluation of knee joint using magnetic resonance imaging (MRI). From the MRI picture (a), the cartilage is mapped and reconstructed in 3D (b,c). The cartilage mapping on MRI is useful to track cartilage defects (d) and validate them to macroscopic defects on this dog model (e). Quantitative MRI could also allow computerized measurement of cartilage volumetry (f). *Source:* Pictures are courtesy of ArthroLab/ArthroVision, Montreal (QC), Canada.

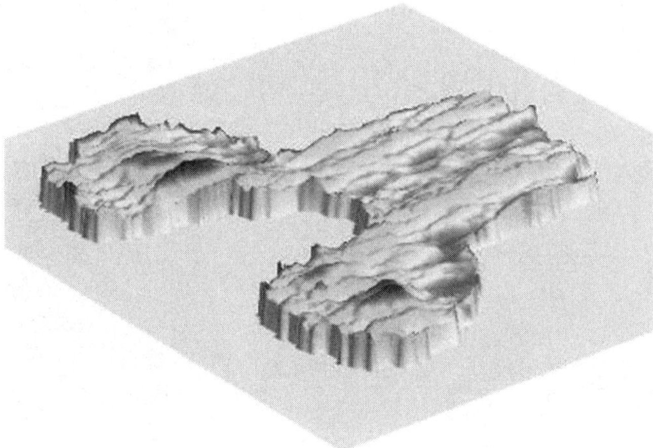

(c)

(d)

FIGURE 6.1 (*Continued*)

expanding on technology-driven opportunities, the use of microarray analysis allowed characterization of the early response of adult human articular cartilage to injury. Dell'Accio et al. (64) described the molecular signaling pathways altering chondrocyte biology after injury. Therapeutic targeting of such pathways may improve current protocols of joint surface defect repair and/or prevent the evolution of such lesions into posttraumatic OA. Drug development based on disease targets has become the new standard in the twenty-first century. Once again, technology advances will be central to achieve high throughout and accelerate drug development.

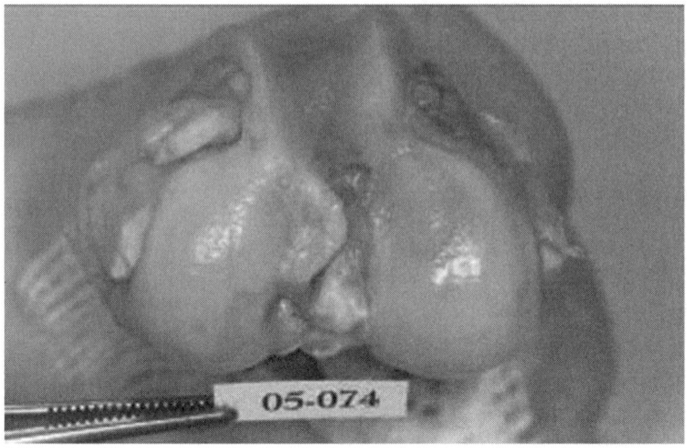

(e)

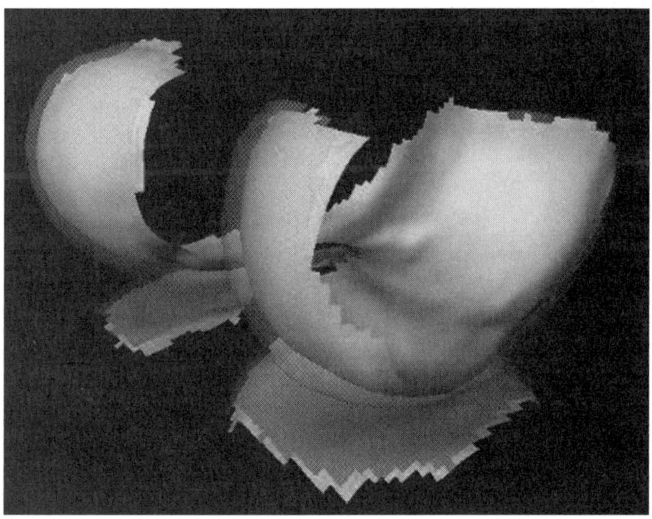

(f)

FIGURE 6.1 (*Continued*)

6.9 THE FUTURE OF BIOMARKERS: DEVELOPMENT OF DRUGS FOR PERSONALIZED MEDICINE

Oncology biomarkers have been correlated with disease risk factors (65), prognosis (66), and biological properties of tumors (67). However, they also provide information regarding the susceptibility of tumors to treatment, thus providing a potential additional step toward personalized medicine (68). Recent developments in pharmacogenetics and pharmacogenomics currently support a shift from generic treatment schemes (i.e., the same treatment for all patients) to therapeutic approaches adapted to individuals. It is now recognized that not only genetics and gender but also many other factors influence drug treatment responses. Personalized medicine is developing rapidly with the use of gene profiling to tailor medical care (e.g., drug and dose level selection) for patients. The latter is emerging with contributions from the pharmaceutical industry under the auspice of the various regulatory authorities. The CDER created a table of valid pharmacogenomic biomarkers (69), which solidifies translational medicine efforts into a formal reference for clinicians. The table provides comprehensive information, including clinical response, risk identification, dose selection guidance, susceptibility, resistance, and polymorphic drug targets. The table illustrates the usefulness of biomarkers as practical tools in the clinic. In the pharmaceutical industry, specific disease targets are now the starting point of research teams that will use various biomarkers to measure the effects of drug candidates. Tyrosine kinase inhibitors such as imatinib (Gleevec®) have occupied the top of the "medical billboard" for the past few years, where this targeted drug development strategy seems to have yielded remarkable public benefit.

An increase in drug labels containing pharmacogenomic information was noted over the past years (70), but the use of biomarkers in personalized medicine has not achieved its potential (71, 72). The benefits of pharmacogenomic biomarkers include efficacy screening such as if found with tumor overexpression of human epidermal growth factor receptor 2 (HER2/neu) required for treatment with trastuzumab (Herceptin®). Improved pharmacokinetic analysis is another important advance of pharmacogenomic biomarkers as illustrated by the impact of the cytochrome *P*450 (CYP) mixed-function oxidase variants on metabolism. CYP 2D6 shows the largest phenotypical variability among the CYPs (largely due to genetic polymorphism). Variants have been identified in patients experiencing reduced as well as increased drug metabolism, which may undermine efficacy or lead to unexpected toxicity in the absence of appropriate phenotype testing.

An important question is whether available pharmacogenomic tools will become widely used in clinical practice. An evaluation group recently concluded that insufficient information was available to recommend genetic evaluation of CYP isoenzymes in the use of selective serotonin reuptake inhibitors (SSRIs) (73). CYP biomarkers included in the FDA table of valid genomic biomarkers are either for informational purposes only or recommended for application; however, none of the current CYP biomarkers are required (i.e., no mandatory testing).

Drug development faces an increase in use of biomarkers that identify subpopulations of patients with specific disease characteristics indicative of either therapeutic efficacy or development of adverse event susceptibility. However, the introduction of these biomarkers as routine clinical tools has progressed very slowly. New differences in drug responses between subpopulations of patients are continually being uncovered. Thus, population dynamics (gender, age, and genetic profile) is playing an increasingly important role in drug development. For a number of anesthetic drugs, gender has been shown to significantly influence pharmacodynamic response, potentially altering the depth of anesthesia. Gender is not the only important genetically determined marker of safety or efficacy.

Genomic variations influencing response to pharmacotherapy of pain are under investigation (74). Candidate genes such as (opioid)-receptors, transporters, and drug-metabolizing enzymes represent major targets of ongoing research aimed to identify associations between genetic profiles and individual drug response (pharmacogenetics). Polymorphisms of CYP enzymes (CYP2D6) also influence analgesic properties of codeine, tramadol, and tricyclic antidepressants. Blood levels of some nonsteroidal anti-inflammatory drugs (NSAIDs) are dependent on CYP2C9 activity, whereas opioid-receptor polymorphisms could support differences observed in opioid-mediated analgesia and side effects (75).

What if we could predict preoperatively how patients might respond to common pain medications, anticoagulants, and antiemetics? In the coming decades, opportunities to develop medications that are specifically designed for patients with unique metabolic characteristics, receptor affinity, or gene expression will arise. Although far from reality at this time, it is possible that someday we will have a preoperative analysis of buccal cells that will guide with opioid selection to achieve appropriate perioperative pain relief, while avoiding side effects such as respiratory depression, nausea, and pruritus. Similarly, tailored interventions based on patient-specific genetic analyses may be possible for hemodynamic management, treatment of perioperative sepsis, and other perioperative issues. For these new biomarkers of safety or efficacy to emerge, there will need to be a structured characterization of patient populations, including correlation with clinical endpoints. The process will entail several challenges, since obtaining patient information with subsequent storage and analysis of personal medical files raises several serious issues such as confidentiality, discrimination, accuracy, and clinical practicality.

Long-term initiatives, including prospective epidemiological clinical studies, will take decades to yield results, but in the mean time, biomarkers have already entered the clinic for specific fields of applications such as oncology. How should the drug development industry react to this situation? The cost of genetic or enzymatic profiling is currently prohibitive and pharmaceutical companies generally seek a traditional "one-size-fits-all" dosing. The cost of genetic and metabolic profiling is high currently but should decrease over the next years, which will make development of "personalized drugs" potentially more attractive. Efforts to identify drugs targeted for specific patient/disease subpopulation may provide safer or confirmed efficacy. Steering drug development with efficacy or safety biomarkers is an appealing approach to increase chances of success in an industry, where very few drug candidates reach bed side. Only the future will tell if research and development efforts based on biomarkers will translate into expected public health benefits.

6.10 BIOMARKERS IN REGULATORY RESEARCH

Biomarkers are embedded in regulatory approval of new drugs. The passage from nonclinical testing to FIH is vastly supported by a panel of biomarkers aimed to demonstrate safety of the drug candidates (Table 6.1). Regulatory guidelines for

TABLE 6.1 Biomarkers in Nonclinical Toxicology and Safety Pharmacology

	Biomarkers	Common Methodology	Comments
General toxicology	Body weight	Usually evaluated once weekly or more often when adverse effects are noted	Used as a nonspecific indicator of toxicity. The growth curve is well characterized in animal species used in toxicology studies, especially in rodents. Comparable age and body weight ranges between control and treated groups is essential to ensure validity of statistical analysis on body weight. Correlated with food consumption and possibly dehydration
	Food consumption	Usually evaluated once daily or weekly. May be evaluated per animal or per cage (all animals from a cage in the same group)	Used as a nonspecific indicator of toxicity. Decreased food consumption is a frequent sign of drug toxicity. Higher sensitivity in rodents due to the larger number of animals in each group
	Electrocardiography (PR, PQ, QRS, QT, and RR)	External electrcardiography (ECG) leads (derivation II). Continuous monitoring of unrestrained animals using jacket in large animals such as dogs and nonhuman primates. Evaluated once before start of dosing and at study completion (e.g., last week of treatment)	Toxicology studies include higher number of animals than safety pharmacology offering the potential for increased statistical power. Toxicology studies also offer the advantage of repeated dose administration compared with safety pharmacology, which is often single dose. Technology advances, including continuous noninvasive ECG monitoring using jackets increase the sensitivity of ECG evaluations in toxicology studies
	Hematology (complete blood count; CBC)	Analysis of plasma: Ethylenediaminetetraacetic acid; EDTA or heparin as anticoagulant. Evaluated once during pretreatment and at study completion	Evaluates toxicity of drug candidates on platelets, erythroid, and myeloid lineages. Interpretation is often correlated with bone marrow histopathological evaluations. Hematology is a standard component of immunotoxicity testing, including lymphocytic, monocytic, and polymorphonuclear cells

(continued)

TABLE 6.1 *(Continued)*

Biomarkers	Common Methodology	Comments
Clinical chemistry	Analysis of serum: serum separator tubes (SST). Evaluated once during pretreatment and at study completion	Biomarkers of hepatic (e.g., alanine aminotransferase for hepatocellular injury), renal, and muscular (or gastrointestinal toxicity). Globulin level is also part of the FDA (but not EMEA) first tier immunotoxicity test
Coagulation	Analysis of plasma: citrate as anticoagulant. Evaluated once during pretreatment and at study completion	Activated partial thromboplastin time (APTT) to assess intrinsic coagulation and prothrombin time (PT) to evaluate extrinsic coagulation
Urinalysis physicochemical (volume, density, color and appearance, pH, glucose, ketones, etc.) and microscopic evaluation (erythrocytes, protein casts, leucocytes, bacteria, etc.)	Urine collection cage to prevent water supply contamination. Evaluated once during pretreatment and at study completion	Used to detect signs of renal toxicity and often correlated with clinical chemistry results. Casts (cylindruria) is an important component of microscopic evaluations. Hyaline, granular or red cell, or epithelial cell casts be indicative of renal parenchyma toxicity
Systemic arterial pressure (systolic, mean, and diastolic pressures, rate)	Indirect sphygmomanometry. Evaluated once during pretreatment and at study completion	Occasionally included in toxicology studies although telemetry is used as a definitive evaluation of systemic arterial pressure. Recognized limitations, including restraining, which induces tachycardia and increased arterial blood pressure. Accuracy of diastolic and mean arterial pressures measured with indirect sphygmomanometry has limited value. Indirect systolic arterial pressure measurement considered valid

	Organ weights	Performed after organs have been trimmed free of fat at necropsy	Used to identify drug-induced atrophy, hypertrophy, edema, or other pathophysiological process that alters organ weight. Organ weight results are correlated with gross necropsy and histopathology observation for interpretation of toxicological findings
Safety Pharmacology	Systemic arterial pressure (systolic, mean, and diastolic pressures)	Direct arterial pressure measurement (fluid-filled or digital catheter). Implantable radiotelemetry transmitters or anesthetized models instrumented with arterial catheters	Investigation of arterial pressure effects (hyper or hypotension). A recent interest for chronobiology (e.g., lost circadian rhythm with some hypertensive drug candidates). Drug candidate with hypertensive effects have high risk of adverse effects in patients given the prevalence of hypertension in the human population. Dog is the most frequent model for cardiovascular safety pharmacology. Rat models often used for screening of arterial pressure after repeated dose administration but unsuitable for evaluation of QT prolongation due to lack of hERG channel. Monkeys required based on pharmacological considerations (e.g., biologics where the target is only expressed in this species or when the pharmacokinetic in the monkey is the most relevant to humans). Pig cardiomyocytes lack the Ito potassium channel (present in humans) but are still used for QT assessments. Pig, with rapid coagulation time compared to other species, is prone to pressure catheter occlusion when using telemetry implant and chronic monitoring

(*continued*)

TABLE 6.1 ***(Continued)***

	Biomarkers	Common Methodology	Comments
	Electrocardiography (ECG)	External, subcutaneous, intracardiac, or pericardiac ECG leads. Implantable radiotelemetry transmitters often in derivation II. Precordial derivations (e.g., V3) also used to improve T-end detection	Subcutaneous ECG leads are usually considered adequate in dogs, pigs, and rats. Monkeys often present important skeletal muscle artifacts when using subcutaneous ECG leads, which favors the use of pericardiac leads
	Respiratory function	Head-out plethysmography. Whole body plethysmography. Head chamber with bias flow for monkeys. Mask with pneumotachometer or thoracic bands in dogs	Rats are often the preferred species for respiratory safety pharmacology for ethical and economical (less testing material required) reasons. Large animal species (such as dogs and monkeys considered for pharmacological reasons (e.g., presence of target)
	Functional observation battery (FOB)	FOB to evaluate the nervous system. Includes a number of physiological parameters such as physical activity, body temperature, grip strength, and mobility	Functional observation battery is most often performed in rats but may occasionally require the use of dogs or monkeys based on pharmacological considerations. General toxicity will often translate into decreased activity level, which is usually interpreted as a nonspecific sign of toxicity
Supplemental safety pharmacology studies	Left ventricular function	Anesthetized animal models or conscious unrestrained telemetered animals with left ventricular catheter	$dP/dT+$, a measure of isovolumetric contraction, is used to assess cardiac contractility. Positive and negative inotropic properties of drugs evaluated

Gastrointestinal (gastric secretion, intestinal transport rate, gastric emptying, etc.)	Gastric cannula for secretion volume, phenol red for gastric emptying, charcoal propulsion for intestinal transport, gastric samples for pH	Gastrointestinal adverse effects often result from pharmacological activity of the test drug and also from chemical or physical properties (e.g., elevated viscosity potentially leading to paralytic ileus in rats)
Electroretinography (a-wave, b-wave, oscillatory potentials)	Anesthetized ERG recording using corneal and subpalpebral electrodes and ground. Includes scotopic (dark) and photopic (background light) evaluations with single response and repeated light flashes (flicker response). Advanced protocols include luminance curve response for calculation of retinal sensitivity (log K)	Used to evaluate functional signs of retinal toxicity as well as reversibility of the effects (e.g., amplitude and latency of a-wave and b-wave or attenuation of oscillatory potentials). ERG baseline results in animals are often less variable than humans due to homogeneity of laboratory animal population
Electroencephalography (spectral analysis, spike trains, sleep scoring, etc.)	External, subcutaneous or dural EEG leads. Emerging technologies enable computerized EEG monitoring and analysis.	EEG investigations required most of the time as follow-up to adverse clinical signs from toxicology studies. Identification of paroxysmal EEG activity which represents an increased risk for seizure. Useful to detect presence of sedation.

nonclinical testing are generally comparable between the European Union, Japan, and the United States as a result of the International Conference on Harmonization of Technical Requirements for Registration of Pharmaceuticals for Human Use (www.ich.org). The organization provides recommendations (ICH guidelines), which reduce or obviate the need to duplicate testing done for development and approval of new drugs among different countries. The ICH guidelines for *in vitro* and *in vivo* preclinical studies allow greater harmonization between countries and a more economical use of human, animal, and material resources and the elimination of unnecessary delay in the global development and availability of new medicines. The selection of biomarkers to include in preclinical studies is influenced by primary goals of these studies, which include identification of (i) an initial safe dose and subsequent dose escalation schemes in humans, (ii) potential target organs for toxicity and reversibility, and (iii) safety parameters for clinical monitoring (76). The process is under strict scrutiny of regulatory agencies as illustrated by the US FDA procedures.

To initiate clinical research in the United States, an Investigational New Drug Application (IND) must be filed with the FDA (77). The design of animal studies for an IND application is a science-based process, where drug characteristics (indication, drug class, administration route, toxicokinetic, mechanism of action, etc.) and regulatory requirements are evaluated to generate a study plan to inform on drug candidate safety. For products intended to treat life-threatening or severely debilitating illnesses, the pharmaceutical company may request an early consultation meeting with the FDA. The pre-IND meeting is intended to reach an agreement on the design of animal studies (including evaluated biomarkers) needed to initiate clinical trials.

As previously mentioned, the study plan endeavors to identify nonclinical biomarkers that could later be used to monitor clinical trials. Animal studies usually include two relevant mammalian species, although a single species may be acceptable in some cases (e.g., mAb for which the target is only expressed in nonhuman primates). The animal species selection is based on the predictive value of biomarkers in animals to the human response. As an example, the rat is not suitable for assessment of the risk of TdP as the ventricular myocardium in this species lacks the slow inward rectifying potassium (I_{Kr}) current (encoded by the hERG gene), which is the major current responsible for ventricular repolarization in humans.

The relevance of biomarkers in animal safety studies also depends on the nature of the drug candidate as illustrated with biologics. Also referred to as *biotechnology-derived* pharmaceuticals, biologics are products that originate from characterized cells using various expression systems (mammalian, insect, bacterial, or yeast), including cytokines, plasminogen activators, recombinant plasma factors, growth factors, fusion proteins, enzymes, receptors, hormones, and monoclonal antibodies (mAbs) (76). Biologics are usually larger molecules thought to be less prone to untoward adverse effects than small (NCE) molecules (78). As a result, *in vitro* hERG assays recommended by the ICH guideline S7B are not considered required for large molecules such as mAbs (>140,000 Da) (79).

6.11 CONCERTED EFFORTS FOR DEVELOPMENT OF BIOMARKERS

Significant financial and human resources are invested in research conducted by the pharmaceutical industry for discovery, lead optimization, and regulatory submission. Research funded by granting agencies such as the National Institute of Health (NIH) will usually lead to publication of results. However, there are comparably fewer publications from the pharmaceutical industry on the development of biomarkers. More importantly, results from clinical trials that may be used to validate biomarkers are often not published. The US FDA approved 90 new drugs between 1998 and 2000. More than half (515/909, 57%) of the 909 drug trials supporting these approvals remained unpublished five years later (80). In 2007, the FDA ruled that the key results from all drug trials must be made publicly available within one year of trial completion or of the drug's approval.

Concerted efforts are required to ensure effective development of biomarkers in drug development. Aligned with this global perspective, the Biomarkers Consortium (www.biomarkersconsortium.org) was created to search for and validate new biomarkers, which may then be applied in research for important indications such as diabetes, cancer, and heart diseases. The Biomarkers Consortium is a public–private biomedical partnership that includes major stakeholders such as FDA, NIH, and the Pharmaceutical Research and Manufacturers of America (www.phrma.org).

6.12 CONCLUSION

Biomarkers will occupy a central role in the development of medicine and will constitute a strategic component of the drug development industry faced with a poor success rate at carrying new therapies from discovery to bed side.

REFERENCES

1. Biomarker Definition Working Group. Biomarkers and surrogate endpoints: preferred definition and conceptual framework. Clin Pharmacol Ther 2001;69:89–95.
2. Temple R. Are surrogate markers adequate to assess cardiovascular disease drugs? J Am Med Assoc 1999;282:790–795.
3. Marshall D, Johnell O, Wedel H. Meta-analysis of how well measures of bone mineral density predict occurrence of osteoporotic fractures. BMJ 1996;312:1254–1259.
4. Cranney A, Jamal SA, Tsang JF, Josse RG, Leslie WD. Low bone mineral density and fracture burden in postmenopausal women. Can Med Assoc J 2007;177:575–580.
5. Antiretroviral Therapy Cohort Collaboration. Life expectancy of individuals on combination antiretroviral therapy in high-income countries: a collaborative analysis of 14 cohort studies. Lancet 2008;372:293–299.
6. Torti C, Quiros-Roldan ME, Cologni G, Nichelatti M, Ceresoli F, Pinti M, Nasi M, Cossarizza A, Lapadula G, Costarelli S, Manca N, Gargiulo F, Magoni M, Carosi G.

Plasma HIV load and proviral DNA decreases after two standard antiretroviral regimens in HIV-positive patients naíve to antiretrovirals. Curr HIV Res 2008;6:43–48.

7. Honda H, Qureshi AR, Heimbürger O, Barany P, Wang K, Pecoits-Filho R, Stenvinkel P, Lindholm B. Serum albumin, C-reactive protein, interleukin 6, and fetuin a as predictors of malnutrition, cardiovascular disease, and mortality in patients with ESRD. Am J Kidney Dis 2006;47:139–148.
8. Oppenheimer GM. Becoming the Framingham study 1947–1950. Am J Public Health 2005;95:602–610.
9. The Joint European Society of Cardiology/American College of Cardiology Committee. Myocardial infarction redefined—a consensus document of the joint European society of cardiology/American college of cardiology committee for the redefinition of myocardial infarction. J Am Coll Cardiol 2000;36:959–969.
10. Baker SG, Kramer BS. A perfect correlate does not a surrogate make. BMC Med Res Methodol 2003;3:16.
11. Berger VW. Does the Prentice criterion validate surrogate endpoints? Stat Med 2004;23:1571–1578.
12. Prentice RL. Surrogate endpoints in clinical trials: definition and operational criteria. Stat Med 1989;8:431–440.
13. Lawrence CL, Bridgeland-Taylor M, Pollard CE, Hammond TG, Valentin JP, Rabbit A. Langendorff heart proarrhythmia model: predictive value for clinical hazard identification and safety margins. Br J Pharmacol 2006;149:845–860.
14. Wallis R. QT-understanding the complexities of defining the translation of animal data to humans. 7th Annual Meeting of the Safety Pharmacology Society; Edinburgh: 2007 Sep 20, 21.
15. [Anonymous]. 2005 May 25. ICH E14: The clinical evaluation of QT/QTc interval prolongation and proarrhythmic potential for non-antiarrhythmic drugs. London. CPMP/ICH/2/04. Available at http://www.ema.europa.eu/docs/en_GB/document_library/Scientific_guideline/2009/09/WC500002879.pdf. Accessed 2011 Jan 7.
16. Eckardt L, Breithardt G, Haverkamp W. Electrophysiologic characterization of the antipsychotic drug sertindole in a rabbit heart model of torsade de pointes: low torsadogenic potential despite QT prolongation. J Pharmacol Exp Ther 2002;300:64–71.
17. Redfern WS, Carlsson L, Davis AS, Lynch WG, MacKenzie I, Palethorpe S, Siegl PK, Strang I, Sullivan AT, Wallis R, Camm AJ, Hammond TG. Relationships between preclinical cardiac electrophysiology, clinical QT interval prolongation and torsade de pointes for a broad range of drugs: evidence for a provisional safety margin in drug development. Cardiovasc Res 2003;58:32–45.
18. Hondeghem LM. QT prolongation is an unreliable predictor of ventricular arrhythmia. Heart Rhythm 2008;5:1210–1212.
19. Lu HR, Vlaminckx E, Hermans AN, Rohrbacher J, van Ammel K, Towart R, Pugsley M, Gallacher DJ. Predicting drug-induced changes in QT interval and arrhythmias: QT-shortening drugs point to gaps in the ICHS7B Guidelines. Br J Pharmacol 2008;154:1427–1438.
20. Fossa AA. Assessing QT prolongation in conscious dogs: validation of a beat-to-beat method. Pharmacol Ther 2008;118:231–238.
21. Nolan ER, Feng MR, Koup JR, Liu J, Turluck D, Zhang Y, Paulissen JB, Olivier NB, Miller T, Bailie MB. A novel predictive pharmacokinetic/pharmacodynamic model of

repolarization prolongation derived from the effects of terfenadine, cisapride and E-4031 in the conscious chronic av node—ablated, His bundle-paced dog. J Pharmacol Toxicol Methods 2006;53:1–10.

22. Ollerstam A, Visser SA, Duker G, Forsberg T, Persson AH, Nilsson LB, Björkman JA, Gabrielsson J, Al-Saffar A. Comparison of the QT interval response during sinus and paced rhythm in conscious and anesthetized beagle dogs. J Pharmacol Toxicol Methods 2007;56:131–144.
23. Pugsley MK, Hancox JC, Curtis MJ. Perception of validity of clinical and preclinical methods for assessment of torsades de pointes liability. Pharmacol Ther 2008;119:115–117.
24. Gintant GA. Preclinical Torsades-de-Pointes screens: advantages and limitations of surrogate and direct approaches in evaluating proarrhythmic risk. Pharmacol Ther 2008;119:199–209.
25. Meuwissen M, Chamuleau SA, Siebes M, de Winter RJ, Koch KT, Dijksman LM, van den Berg AJ, Tijssen JG, Spaan JA, Piek JJ. The prognostic value of combined intracoronary pressure and blood flow velocity measurements after deferral of percutaneous coronary intervention. Catheter Cardiovasc Interv 2008;71:291–297.
26. Lev EI, Kornowski R, Vaknin-Assa H, Porter A, Teplitsky I, Ben-Dor I, Brosh D, Fuchs S, Battler A, Assali A. Comparison of the predictive value of four different risk scores for outcomes of patients with ST-elevation acute myocardial infarction undergoing primary percutaneous coronary intervention. Am J Cardiol 2008;102:6–11.
27. Rees M, Tekkis PP, Welsh FK, O'Rourke T, John TG. Evaluation of long-term survival after hepatic resection for metastatic colorectal cancer: a multifactorial model of 929 patients. Ann Surg 2008;247:125–135.
28. Mitry E, Douillard JY, Van Cutsem E, Cunningham D, Magherini E, Mery-Mignard D, Awad L, Rougier P. Predictive factors of survival in patients with advanced colorectal cancer: an individual data analysis of 602 patients included in irinotecan phase III trials. Ann Oncol 2004;15:1013–1017.
29. Hansson O, Zetterberg H, Buchhave P, Londos E, Blennow K, Minthon L. Association between CSF biomarkers and incipient Alzheimer's disease in patients with mild cognitive impairment: a follow-up study. Lancet Neurol 2006;5:228–234.
30. Pollard CE, Valentin JP, Hammond TG. Strategies to reduce the risk of drug-induced QT interval prolongation: a pharmaceutical company perspective. Br J Pharmacol 2008;154:1538–1543.
31. Gormley M, Dampier W, Ertel A, Karacali B, Tozeren A. Prediction potential of candidate biomarker sets identified and validated on gene expression data from multiple datasets. BMC Bioinformatics 2007;8:415.
32. Hwang T, Sicotte H, Tian Z, Wu B, Kocher JP, Wigle DA, Kumar V, Kuang R. Robust and efficient identification of biomarkers by classifying features on graphs. Bioinformatics 2008;15(24):2023–2029.
33. Sawyers CL. The cancer biomarker problem. Nature 2008;452:548–552.
34. Hunter DJ, Khoury MJ, Drazen JM. Letting the genome out of the bottle—will we get our wish? N Engl J Med 2008;358:105–107.
35. Thompson IM, Pauler DK, Goodman PJ, Tangen CM, Lucia MS, Parnes HL, Minasian LM, Ford LG, Lippman SM, Crawford ED, Crowley JJ, Coltman CA Jr. Prevalence of prostate cancer among men with a prostate-specific antigen level $<$ or $=4.0$ ng per milliliter. N Engl J Med 2004;350:2239–2246.

36. Lin K, Lipsitz R, Miller T, Janakiraman S, U.S. Preventive Services Task Force. Benefits and harms of prostate-specific antigen screening for prostate cancer: an evidence update for the U.S. Preventive services task force. Ann Intern Med 2008;149:192–199.
37. Carter HB. Prostate cancers in men with low PSA levels—must we find them? N Engl J Med 2004;350:2292–2294.
38. Randall DC, Jones DL. Eliminating unnecessary lactate dehydrogenase testing. A utilization review study and national survey. Arch Intern Med 1997;157:1441–1444.
39. Ishii J, Ozaki Y, Lu J, Kitagawa F, Kuno T, Nakano T, Nakamura Y, Naruse H, Mori Y, Matsui S, Oshima H, Nomura M, Ezaki K, Hishida H. Prognostic value of serum concentration of heart-type fatty acid-binding protein relative to cardiac troponin T on admission in the early hours of acute coronary syndrome. Clin Chem 2005;51:1397–1404.
40. McCann CJ, Glover BM, Menown IB, Moore MJ, McEneny J, Owens CG, Smith B, Sharpe PC, Young IS, Adgey JA. Novel biomarkers in early diagnosis of acute myocardial infarction compared with cardiac troponin T. Eur Heart J 2008;29:2843–2850.
41. Katz R. Biomarkers and surrogate markers: an FDA perspective. NeuroRx 2004;1:189–195.
42. Hughes MD. Evaluating surrogate endpoints. Control Clin Trials 2002;23:703–707.
43. Goodsaid F, Frueh F. Biomarker qualification pilot process at the US food and drug administration. AAPS J 2007;9:E105–E108.
44. Petrick N, Brown D, Suleiman O, Myers K. Imaging as a tumor biomarker in oncology drug trials for lung cancer: the FDA perspective. Clinical Pharmacol Ther 2008;84:523–525.
45. Nissen SE, Wolski K. Effect of rosiglitazone on the risk of myocardial infarction and death from cardiovascular causes. N Engl J Med 2007;356:2457–2471.
46. Bigger JT Jr, Fleiss JL, Kleiger R, Miller JP, Rolnitzky LM. The relationships among ventricular arrhythmias, left ventricular dysfunction, and mortality in the 2 years after myocardial infarction. Circulation 1984;69:250–258.
47. The CAST Investigators. Preliminary report: effect of encainide and flecainide on mortality in a randomized trial of arrhythmia suppression after myocardial infarction. N Engl J Med 1989;321:406–412.
48. Pratt CM, Moyé LA. The cardiac arrhythmia suppression trial: casting suppression in a different light. Circulation 1995;91:245–247.
49. Ledford H. Drug markers questioned. Nature 2008;452(7187):510–511.
50. Hughes B. 2007 FDA drug approvals: a year of flux. Nat Rev Drug Discov 2008;7:107–109.
51. Zhang Y, Jordan JM. Epidemiology of osteoarthritis. Rheum Dis Clin North Am 2008;34:515–529.
52. [Anonymous] National Home and Hospice Care Survey, 2002, NCHS, CDC.
53. Sebba A. Tocilizumab: the first interleukin-6-receptor inhibitor. Am J Health Syst Pharm 2008;65:1413–1418.
54. Mancarella L, Bobbio-Pallavicini F, Ceccarelli F, Falappone PC, Ferrante A, Malesci D, Massara A, Nacci F, Secchi ME, Manganelli S, Salaffi F, Bambara ML, Bombardieri S, Cutolo M, Ferri C, Galeazzi M, Gerli R, Giacomelli R, Grassi W, Lapadula

G, Cerinic MM, Montecucco C, Trotta F, Triolo G, Valentini G, Valesini G, Ferraccioli GF, GISEA group. Good clinical response, remission, and predictors of remission in rheumatoid arthritis patients treated with tumor necrosis factor-alpha blockers: the GISEA study. J Rheumatol 2007;34:1670–1673.

55. Sun BH, Wu CW, Kalunian KC. New developments in osteoarthritis. Rheum Dis Clin North Am 2007;33:135–148.
56. Brandt KD. Studies in animal models of osteoarthritis as predictors of a structure-modifying effect of diacerhein in humans with osteoarthritis. Biorheology 2006;43:589–594.
57. Yu LP Jr, Smith GN Jr, Brandt KD, Myers SL, O'Connor BL, Brandt DA. Reduction of the severity of canine osteoarthritis by prophylactic treatment with oral doxycycline. Arthritis Rheum 1992;35:1150–1159.
58. Moreau M, Boileau C, Martel-Pelletier J, Brunet J, Laufer S, Pelletier JP. Licofelone reduces progression of structural changes in a canine model of osteoarthritis under curative conditions: effect on protease expression and activity. J Rheumatol 2006;33:1176–1183.
59. Dougados M, Nguyen M, Berdah L, Maziéres B, Vignon E, Lequesne M, ECHODIAH investigators study group. Evaluation of the structure-modifying effects of diacerein in hip osteoarthritis: ECHODIAH, a three-year, placebo-controlled trial. Evaluation of the chondromodulating effect of diacerein in OA of the hip. Arthritis Rheum 2001;44:2539–2547.
60. Brandt KD, Mazzuca SA, Katz BP, Lane KA, Buckwalter KA, Yocum DE, Wolfe F, Schnitzer TJ, Moreland LW, Manzi S, Bradley JD, Sharma L, Oddis CV, Hugenberg ST, Heck LW. Effects of doxycycline on progression of osteoarthritis: results of a randomized, placebo-controlled, double-blind trial. Arthritis Rheum 2005;52:2015–2025.
61. Raynauld JP, Martel-Pelletier J, Bias P, Laufer S, Haraoui B, Choquette D, Beaulieu AD, Abram F, Dorais M, Vignon E, Pelletier JP, Canadian Licofelone Study Group. Protective effects of licofelone, a 5-lipoxygenase and cyclo-oxygenase inhibitor, versus naproxen on cartilage loss in knee osteoarthritis: a first multicentre clinical trial using quantitative MRI. Ann Rheum Dis 2009;68:938–947.
62. Hunter DJ, Zhang Y, Niu J, Goggins J, Amin S, LaValley MP, Guermazi A, Genant H, Gale D, Felson DT. Increase in bone marrow lesions associated with cartilage loss: a longitudinal magnetic resonance imaging study of knee osteoarthritis. Arthritis Rheum 2006;54:1529–1535.
63. Torres L, Dunlop DD, Peterfy C, Guermazi A, Prasad P, Hayes KW, Song J, Cahue S, Chang A, Marshall M, Sharma L. The relationship between specific tissue lesions and pain severity in persons with knee osteoarthritis. Osteoarthritis Cartilage 2006;14:1033–1040.
64. Dell'Accio F, De Bari C, Eltawil NM, Vanhummelen P, Pitzalis C. Identification of the molecular response of articular cartilage to injury, by microarray screening: Wnt-16 expression and signaling after injury and in osteoarthritis. Arthritis Rheum 2008;58:1410–1421.
65. Ankerst DP, Groskopf J, Day JR, Blase A, Rittenhouse H, Pollock BH, Tangen C, Parekh D, Leach RJ, Thompson I. Predicting prostate cancer risk through incorporation of prostate cancer gene 3. J Urol 2008;180:1303–1308.
66. Xie Y, Minna JD. Predicting the future for people with lung cancer. Nat Med 2008;14:812–813.

67. Weigelt B, Horlings HM, Kreike B, Hayes MM, Hauptmann M, Wessels LF, de Jong D, van de Vijver MJ, Van't Veer LJ, Peterse JL. Refinement of breast cancer classification by molecular characterization of histological special types. J Pathol 2008;216:141–150.
68. Van't Veer LJ, Bernards R. Enabling personalized cancer medicine through analysis of gene-expression patterns. Nature 2008;452:564–570.
69. [Anonymous]. 2008. FDA, center for drug evaluation and research, table of valid genomic biomarkers in the context of approved drug labels. Available at http://www.fda.gov/Drugs/ScienceResearch/ResearchAreas/Pharmacogenetics/ucm083378.htm. Accessed 2012 Jan 8.
70. Frueh FW, Amur S, Mummaneni P, Epstein RS, Aubert RE, Deluca TM, Verbrugge RR, Buckart GL, Lesko LJ. Pharmacogenomic biomarker information in drug labels approved by the United States food and drug administration: prevalence of related drug use. Pharmacotherapy 2008;28:992–998.
71. Lesko LJ. Personalized medicine: elusive dream or imminent reality? Clinical Pharmacol Ther 2007;81:807–816.
72. Davies SM. Pharmacogenetics, pharmacogenomics and personalized medicine: are we there yet? Hematology (Am Soc Hematol Educ Program) 2006;1:111–117.
73. Evaluation of Genomic Applications in Practice and Prevention (EGAPP) Working Group. Recommendations from the EGAPP working group: testing for cytochrome P450 polymorphisms in adults with nonpsychotic depression treated with selective serotonin reuptake inhibitors. Genet Med 2007;9:819–825.
74. Stamer UM, Stüber F. The pharmacogenetics of analgesia. Expert Opin Pharmacother 2007;8:2235–2245.
75. Chou WY, Wang CH, Liu PH, Liu CC, Tseng CC, Jawan B. Human opioid receptor A118G polymorphism affects intravenous patient-controlled analgesia morphine consumption after total abdominal hysterectomy. Anesthesiology 2006;105:334–337.
76. International Conference for Harmonisation. 1997. S6 Preclinical safety evaluation of biotechnology-derived pharmaceuticals. Available at http://www.ich.org/products/guidelines/safety/safety-single/article/preclinical-safety-evaluation-of-biotechnology-derived-pharmaceuticals.html. Accessed 2012 Jan 8.
77. [Anonymous]. 2008. U.S. food and drug administration, title 21, chapter I, subchapter D, Part 312, investigational new drug application. Available at http://www.accessdata.fda.gov/scripts/cdrh/cfdocs/cfcfr/CFRSearch.cfm?CFRPart=312&showFR=1. Accessed 2012 Jan 8.
78. Zhou H. Biologics in the pipeline: large molecules with high hopes or bigger risks? J Clin Pharmacol 2007;47:550–552.
79. Vargas HM, Bass AS, Breidenbach A, Feldman HS, Gintant GA, Harmer AR, Heath B, Hoffmann P, Lagrutta A, Leishman D, McMahon N, Mittelstadt S, Polonchuk L, Pugsley MK, Salata JJ, Valentin JP. Scientific review and recommendations on preclinical cardiovascular safety evaluation of biologics. J Pharmacol Toxicol Methods 2008;58:72–76.
80. Lee K, Bacchetti P, Sim I. Publication of clinical trials supporting successful new drug applications: a literature analysis. PLoS Med 2008;5:e191.

7

THE USE OF MOLECULAR IMAGING FOR RECEPTOR OCCUPANCY DECISION MAKING IN DRUG DEVELOPMENT

RALPH PAUL MAGUIRE

7.1 RECEPTOR OCCUPANCY

Positron emission tomography (PET) is a well-established technique that allows measurement of the concentration of a labeled tracer, in tissue, noninvasively. As a drug development tool, it has enabled the quantitation of drug effect at the site of action and has extended our knowledge of drug distribution from the central blood compartment directly into tissue. In this respect, the spatial resolution of this imaging method is much less important than the sensitivity and the ability to measure quantitatively and noninvasively. The main contribution of PET receptor occupancy to decision making in drug development is in early decision making. In the period 1995–2004, 10 of 35 neurology and psychiatry New Medical Entities (NMEs) approved by the FDA used imaging biomarkers at some stage in development (1) but mainly before phase III. By directly answering the question of drug distribution and organ penetration and determining the relationship between target association and central pharmacokinetic (PK), molecular imaging enables dose optimization or early termination of projects that will ultimately be unsuccessful.

Predictive Approaches in Drug Discovery and Development: Biomarkers and In Vitro/In Vivo Correlations, First Edition. Edited by J. Andrew Williams, Jeffrey R. Koup, Richard Lalonde, and David D. Christ.

7.2 MARKERS OF DISEASE PROGRESSION

Although the main topic of this chapter is receptor occupancy studies with PET, it is worthwhile to mention in passing the application of molecular imaging probes to measure disease progression. In some sense, this is an entirely separate topic. While receptor occupancy measures aim to demonstrate drug–target interaction, disease markers are more closely coupled with outcome or downstream effects of the drug's action. Examples are [F-18]Fluorodeoxyglucose ([F-18]FDG) PET in oncology, used to measure tumor metabolism, or amyloid imaging agents used to measure amyloid load in Alzheimer's disease.

These markers need not always be linked directly to clinical signs of disease progression and could simply be diagnostics of an underlying progressive pathology, which would eventually lead to an event such as cardiac infarct. Markers of this class are more likely to be used in later-stage development, and the decision-making algorithm will be quite different from the application of receptor occupancy.

7.3 DRUG DISTRIBUTION STUDIES

As discussed later, the requirements for a molecular imaging tracer used for receptor occupancy quantitation are not the same as for a successful therapeutic drug. Nevertheless, if a drug is labeled, it may be possible to image drug distribution in the body at low dose. This opens up the opportunity to administer the radiolabeled drug at a very early stage in development to determine organ penetration and drug concentration at the target. However, extrapolation of this data to higher doses makes assumptions of linearity of PK with dose. Since one advantage of this setting is to make measurements at doses below the thresholds for pharmacological effects, it cannot be used to test receptor occupancy of the tracer in a self-blocking study design.

7.4 MOLECULAR IMAGING TECHNOLOGY

The essentials of the PET methodology are the use of a specific tracer ligand and very sensitive spatially resolved instrumentation to quantify the uptake and distribution of the ligand after intravenous injection into a human volunteer or patient. The instrumentation has been under development for the past three decades but has now stabilized and is able to resolve small structures only 2–5 mm apart. The detection efficiency of state-of-the-art scanners is such that, depending on the radioligand, it is possible to make multiple repeat measurements, in a single subject without exposing them to a radiation burden much greater than that from natural radiation sources.

PET radioisotopes, typically carbon-11 or fluorine-18, can be introduced into organic molecules with high receptor affinity to act as a radiolabel. However, not all molecules with high affinity that might make good drugs are suitable PET tracers since other considerations are also important (2, 3): adequate blood–brain barrier penetration and low nonspecific binding, low molecular weight, and no radiolabeled metabolites that can cross the blood–brain barrier and contribute to the uptake signal. In practical application of PET for decision making in drug development, it will also be important to have a ligand that can easily be labeled with C-11 or F-18, which is with a single-step substitution reaction. Although more complicated multistep radiosynthesis is possible, it will be more difficult to rapidly establish standard good manufacturing practice for a complicated synthesis. In contrast to the use of PET ligands to diagnose disease or in basic science, where the PET signal must come from one selective receptor subtype, it is possible to use a PET tracer for receptor occupancy applications, which binds to more than one receptor (4).

There are two scenarios of note here:

1. *Nonselective* PET ligand and a *selective* pharmaceutical under test. In this situation, the signal change observed will be due to occupancy of pharmaceutical only at the receptor, for which the pharmaceutical is selective.
2. *Selective* PET ligand and a *nonselective* pharmaceutical. In this case, the nonselective pharmaceutical may bind to multiple receptors; however, the signal change will only be related to the selective binding site of the PET ligand.

Also, it should be clear that PET ligands do not need to be given in pharmacologically active doses and may not need to have demonstrated safety above microdosing (microgram) levels (5).

7.5 BINDING POTENTIAL

The fundamental measure of receptor–ligand association in molecular imaging is the partition coefficient or concentration ratio, between receptor–ligand complex and free ligand concentration, at equilibrium. A high concentration ratio is expected if the forward reaction rate—of the association—is high. This ratio has been described as the binding potential (BP) and is a characteristic of the tracer.

The reaction of a ligand (L) with a receptor (R) to form a ligand–receptor (LR) complex is an equilibrium reaction:

$$\mathrm{L} + \mathrm{R} \Leftrightarrow \mathrm{LR}$$

The initial concentration of available receptors (B_{avail}) will be decreased as the ligand associates with the receptors, so that the remaining receptor concentration will be ($B_{\mathrm{avail}} - c_{\mathrm{b}}$), where c_{b} is the concentration of receptor–ligand complex. If c_{f} is the free concentration of ligand, the equilibrium constant for this reaction is then given by the ratio of the concentration of the receptor–ligand complex to the product of the concentration of the reactants at equilibrium:

$$K_c = \frac{c_b}{c_f(B_{avail} - c_b)} \tag{7.1}$$

and by defining the dissociation constant K_D as

$$K_c = \frac{1}{K_D} \tag{7.2}$$

Then by rearrangement, the saturation binding equation can be derived.

$$c_b = \frac{B_{avail} c_f}{K_D + c_f} \tag{7.3}$$

Note that this saturation binding equation is general for the ligand–receptor association *and* the drug–receptor association; the K_D is equal to the concentration at which half of the receptors are bound (this can be seen by substituting $c_f = K_D$ in this equation). If the imaging ligand is given at a tracer concentration, such as in the PET experiment, then the free concentration of the ligand is small in relation to the receptor concentration, then this equation can be simplified further to derive the equation for the BP:

$$\mathrm{BP} = \frac{c_b}{c_f} = \frac{B_{avail}}{K_D} \tag{7.4}$$

This equation shows how the BP is proportional to the available receptor density, as long as the K_D is constant.

7.5.1 Calculating Receptor Occupancy from Measurements of BP

The percentage displacement (D)—the signal change under treatment—can be calculated from measurements of the BP at baseline, before treatment BP(0) and during treatment with a drug plasma concentration BP(c).

$$D = \frac{\mathrm{BP}(0) - \mathrm{BP}(c)}{\mathrm{BP}(0)} \tag{7.5}$$

and the displacement versus central PK(c) relationship can be fitted to a saturation binding equation with a maximum displacement (D_{max}) and K_D parameter:

$$D = \frac{D_{max} c}{K_D + c} \tag{7.6}$$

For an ideal tracer, the displacement will be equal to the receptor occupancy. However, it is possible that 100% displacement will not be observed at high concentrations, because, for example, the tracer binds to multiple receptors and the displacing compound binds to only one of those sites. In that case, the D_{max} parameter fitted in this equation will not be 100%; however, the K_D determined from fitting the *displacement* is identical to the half-saturation constant for the receptor, which would be determined by measuring the *receptor occupancy*. The true receptor occupancy at any PK level is estimated from the following equation, which assumes that the maximum receptor occupancy will be 100%.

$$\text{RO} = \frac{c}{K_D + c} \tag{7.7}$$

Note also that it is possible to estimate the receptor occupancy from changes in the area under the curve (AUC) of the PET tracer time–activity curve. Although the BP is not estimated in this case, measurement does not require arterial input measurement or a reference tissue.

It should also be noted from Equation 7.7 that the relationship between receptor occupancy and the plasma drug concentration is nonlinear, so that as the plasma concentration decreases from very high levels, the receptor occupancy will initially only decrease slowly (6). The relationship between receptor occupancy and PK is more linear below K_D, as can be seen by assuming a diminishingly small concentration (c) in the denominator of Equation 7.7.

7.6 PHARMACOKINETIC ASSUMPTIONS

Some of the assumptions of the PET receptor occupancy technique are discussed in relation to trial design later, but it is worthwhile to briefly collate some of the *key* assumptions of the measurement briefly here.

1. The system under measurement is in a constant state during the measurement. For example, the receptor occupancy is not changing during the PET scan. Typically, the scan is of short duration, circa 1–2 h, and this is short in relation to the rate of change in the drug levels in blood and at the receptor, so this assumption is often fulfilled.
2. There is no change in the level of endogenous transmitter at the receptor during the study between baseline and treatment BP measurements and that there is no change in the density of the receptor or the K_D of the PET ligand. There may be exceptions to these assumptions (7), when, for example, the change in available receptors due to neurotransmitter release is under study. However, in general, this chapter discusses the measurement of occupancy of the receptor by an exogenous drug.
3. All PET tracers are given in tracer doses and do not affect the receptor system under study in any way. The PET tracer is not in competition with the drug under study and it is not in competition with endogenous transmitter.

4. There is equilibrium between the free drug concentration at the receptor and in plasma. The aim of the experiment will be to determine the equilibrium receptor occupancy versus PK relationship. Although true equilibrium is almost never reached, it is still important to consider that if the concentrations are changing very rapidly, then the relationship between receptor occupancy and PK will be dependent not only on the instantaneous PK but also on the past PK time course.

7.7 RECEPTOR OCCUPANCY EXPERIMENTAL DESIGNS IN DRUG DEVELOPMENT

Using PET, the relationship between the target occupancy and central PK or dose can be defined. This relationship can be completely described using the receptor saturation equation, which has one parameter, the half-saturation constant K_D. By determining this relationship, the occupancy of the receptor can be determined for any given drug concentration in blood.

Using the concentration of drug in blood and the K_D, the receptor occupancy can be estimated. If the concentration profile after any dosing regime can be projected, then the receptor occupancy can similarly be determined.

The receptor occupancy/PK relationship and K_D is independent of the receptor availability since the receptor occupancy is a proportional occupancy of the available receptors. This leads to the conclusion that the receptor occupancy relationship can be determined in a single trial in one population, for one compound in development.

7.8 HEALTHY VOLUNTEER RECEPTOR OCCUPANCY TRIAL DESIGN

7.8.1 Choice of Doses

Since the aim of the study is to map the receptor occupancy versus PK relationship and determine the half-saturation constant K_D, the experiment should be designed to explore the full range of PK levels that will result in receptor occupancies between less than 30% and saturation. With no further knowledge about the receptor occupancy, it is reasonable to design the experiment to explore the occupancy associated with doses up to and including the safe and tolerable dose. Using K_D data from preclinical studies, it may be possible to estimate a starting dose below this level.

7.8.2 Timing of the Measurement

The measurement will involve a within-subject measurement of receptor occupancy at baseline, without treatment, and at least one follow-up measurement on treatment. From a clinical trial logistics viewpoint, it would be easiest to perform these two measurements on the same day if possible; however, for C-11-labeled compounds, the start of each PET scan will typically be 4 h apart, meaning that only two can be achieved in a working day. Also, the half-life of many successful drugs is of the order of 4–12 h, this means that after dosing the drug, it is necessary to wait for some time before the central PK has changed significantly, remembering that the receptor occupancy/PK relationship will likely need to be mapped over an order of magnitude change in PK.

After oral dosing, it takes a finite time to establish equilibrium between the free concentration at the receptor and the free concentration in plasma. The central PK will increase rapidly and reach a peak level C_{max} at a time T_{max}, and after this time, the blood concentrations will decrease and the brain PK and plasma PK are more likely to be in equilibrium at this stage of the measurement.

7.8.3 Number of Measurements

The number of measurements within one subject need not be more than three baseline and two follow-up measurements. This can be achieved by measurement of the receptor occupancy at different time points on the washout curve after C_{max}. Again, given the relatively slow washout of many drugs, relative to the PET measurement, a total of two measurements on the washout slope are likely, though, more are possible. Radiation dosimetry of the PET measurement is also a consideration.

7.8.4 Numbers of Subjects and Receptor Occupancy Curve Precision

The total number of points on the receptor occupancy/PK relationship will be equal to the number of PET measurements under treatment conditions. Using the above design, this would be then two per subject. By varying the dose in a controlled way, within predefined safety limits, depending on the emerging receptor occupancy data, the whole curve can be mapped with relatively few subjects, certainly less than 15, should yield a curve with sufficient precision for good decision making.

7.9 PATIENT STUDIES

One assumption of the healthy volunteer study is that the measured receptor occupancy in the patient study will be identical to the healthy volunteer value. Although the receptor expression (B_{avail}) may in fact be quite different, there is no theoretical support at this time, which supports the notion that the occupancy would be different.

Nevertheless, there may be drug development constraints, for example, tolerance or treatment time, that mean that it is difficult to give higher doses to healthy volunteers. In this case, a study in patients may be necessary.

If there is firm confidence in the efficacious receptor occupancy that must be achieved, then a possible design is to dose to steady state and confirm the receptor occupancy in the patient population, before moving to efficacy studies.

7.10 RELATIONSHIP WITH PHARMACODYNAMICS

The PET measurement yields information that is beyond central plasma PK, yet the receptor occupancy measure itself is not an outcome measure. PET cannot be thought of as a direct measure of pharmacodynamics (PD). Although higher receptor occupancy could be linked to greater drug effect, there can be adverse effects of too high an occupancy in some receptor systems, while with other receptor types, a low occupancy is adequate and sufficient for efficacy.

7.11 LITERATURE EXAMPLES

The threshold for efficacy cannot be determined by the PET measurement alone. This information has to come from another source. If the target being tested is known and the required receptor occupancy has been demonstrated in the scientific literature, then this can be used as a clear guide to the required threshold for new compounds. With an untested mechanism, it is more difficult. Preclinical models of PD may give some insight into the required occupancy; however, animal models may well be limited in their predictive capacity. Especially in the area of human cognition and psychiatry, animal models may be imprecise in the estimation of the required level of efficacy.

Using estimates of the PK level in plasma and its variation, the receptor occupancy/PK relationship from the PET measurement and an efficacy or safety threshold determined from literature, animal models, or elsewhere, it is possible to decide if adequate occupancy can be achieved at a safe and tolerable dose. If there is insufficient information before efficacy studies to make a decision on progression, then it may still be worthwhile to perform a PET study in order to be able to determine the receptor occupancy levels from plasma levels during later trials. These data can then be used to establish a PD/receptor occupancy relationship and can then be used for future trials with the same target to enable appropriate decision making. Alternatively, in case of a failed efficacy trial, to demonstrate that the required receptor occupancy has been achieved, allowing the hypothesis to be disproved. Perhaps, the most explored receptor occupancy and efficacy relationships are the D_2 receptor in schizophrenia and serotonin transporter (SERT) occupancy in selective serotonin reuptake inhibitor (SSRI) therapy for depression. D_2 antagonist, atypical antipsychotic, medications have been shown to have efficacy at a threshold above 60% and to induce extrapyramidal side effects at occupancies exceeding 90% (8). This allows definition of a therapeutic window for efficacy, although it has been shown that there are exceptions to this receptor occupancy window. It may well be that off-target occupancy at other receptors is responsible for the differences in pharmacology between antipsychotics. Nevertheless, it has been shown that partial agonists may occupy up to 90% (9) at the D2 receptor without inducing extrapyramidal side effects.

Similarly, with SSRIs, the threshold for efficacy appears to be in the same range of occupancies as for the atypical antipsychotics. This leads to the postulate that this threshold could be applicable to antagonists at all g-coupled protein receptor systems.

A review of these systems (10), however, points out that agonist systems do not follow this general rule. Using [C-11]flumazenil to measure benzodiazepine receptor occupancy has shown that an agonist occupancy of only 30% (11) is necessary for efficacy in those systems. Recent studies with the $\alpha 2$ antidepressive agent mirtazipine, using a carbon-11 variant of the drug itself, have shown that doses of 45 mg efficacious in depressed subjects would cause a receptor occupancy of greater than 90% (12) and a similar occupancy (13) has been shown for efficacious doses of the 5-HT2A antagonist M100907 in schizophrenia.

7.12 PET MODELING

Estimation of the BP from tissue concentration data, measured by PET, requires compartmental modeling of the uptake data and ultimately calculation. For most, if not all, PET ligands, a two-tissue compartment model is appropriate, with a measured input time course in plasma and further compartments for nondisplaceable tracer and specifically bound tracer, in tissue. The nondisplaceable tissue tracer compartment comprises free tracer in tissue as well as nonspecifically bound.

The aim of PET data analysis is to estimate the BP of the ligand; however, ideally, this would require arterial measurement of the time course of tracer in arterial blood, determination of the free fraction in plasma, and estimation of the partition coefficient between bound tracer and the plasma concentration. This would allow calculation of the BP_F, the BP, relative to free concentration of tracer in plasma. This and other measures are summarized in a recent nomenclature publication (14).

Using a simpler measurement protocol, without the need for arterial blood sampling, nor the determination of the free plasma concentration of tracer, the partition coefficient between the bound compartment and the nondisplaceable compartment can be estimated—the BP_{ND}. This estimate is related to the BP by the free fraction of the tracer in the nondisplaceable compartment.

$$BP_{ND} = \frac{f_{ND} B_{avail}}{K_D} \tag{7.8}$$

If f_{ND} and K_D are constant, then changes in BP_{ND} will be directly proportional to changes in B_{avail}.

For many PET ligands, estimates of BP can be made using established and standard methods that make use of a reference tissue region. The simplified reference tissue model (SRTM) is widely used. By making an assumption that the reference region has similar characteristics to the nondisplaceable compartment in the target region, the BP_{ND} can be calculated. The Logan plot uses either a brain reference region or arterial plasma data to estimate a total partition coefficient for the tracer in brain. It has also been shown that the AUC of the tracer uptake, after normalization for variation in the input data, can be used to estimate change in the B_{avail} (15).

7.13 DOSE VERSUS PK

As has been stated, the main aim of the PET measurement is to determine the receptor occupancy versus PK relationship. It is also possible to determine a relationship between receptor occupancy and dose, with the same functional form of the relationship as the saturation binding curve for receptor occupancy/PK. The receptor occupancy/dose relationship is likely to be much more variable, simply due to the variability in the PK between subjects for a given dose. In the end, it will be important to understand the variability in the receptor occupancy versus dose relationship in the target population for phase II trials; however, the fundamental PK relationship is between receptor occupancy and PK, as described by Equation 7.7.

It should also be clear that the receptor occupancy/PK relationship is described in terms of the central plasma instantaneous concentration, rather than, for example, the AUC. This underlines the fact that receptor occupancy is a PK parameter. The receptor occupancy/PK relationship is dependent on the equilibrium between drug at the target receptor and drug in plasma. Equally well a relationship has been drug concentration at the receptor and in another compartment, for example, CSF (cerebrospinal fluid) could be determined.

7.14 RECEPTOR OCCUPANCY DETERMINATION VARIABILITY

As has been mentioned, if a receptor occupancy/dose relationship is determined, then the individual variation in PK for a given dose will be the major component of variability. Even when assessing the receptor occupancy/PK relationship, there are other sources of variability (16). The imprecision of determination of activity concentration in the PET scanner is low (17). Test–test data for BP determination in healthy volunteers (18) shows a 5.5% test–retest variability. The variability in the determination of PK also needs to be taken into account, and this will depend on the assay precision but may be of the order of 10%.

7.15 ALTERNATIVE FORMS OF MEASURE OF EXPOSURE

Molecular imaging is the only method to measure central drug receptor interaction *in vivo* in man. However, to answer the question of central penetration into brain, it is possible to sample CSF and evaluate drug concentrations (19). This technique may give insight into whether the drug is able to cross the blood–brain barrier; however, it is not a direct measure of receptor occupancy. In order to compute occupancy from this measure, we need to *assume* a K_D and estimate the occupancy from the drug levels determined. CSF sampling is also not equivalent with determination of free drug concentration at the receptor. There may also be further barriers to drug penetration between the CSF and the intracellular or intrasynaptic compartments (20). Lumbar CSF sampling in man is not equivalent to sampling of ventricular liquor and there may well be a gradient between drug concentration in the brain CSF and drug concentration in spinal CSF (21).

7.16 SPEED OF DEVELOPMENT OF PET LIGANDS

One of the most central requirements for receptor occupancy determination is good molecular imaging probes. In fact, this is one of the biggest bottlenecks to deployment of PET for decision making in drug development. Probe development from novel chemical material may take 1–2 years from initial concept to first clinical trials. This time line is concurrent with the development of a new medicine from candidate screening to first clinical trials. Even if a need for a PET tracer tool in clinical development is identified early in the process, the development of a safe and validated molecular imaging probe needs to be rapid in order to deliver a probe in time to contribute effectively to clinical decision making.

There is a great opportunity for collaboration between industrial and academic partners in the development of imaging agents. As new targets are being discovered, industry–academic scientific teams can work to develop the tools that will be needed both for rapid decision making as well as to evolve the basic science. Critical to this is developing a deeper understanding of the characteristics of chemical material that will yield a safe and useful clinical imaging tool. Not all chemical material or good drugs will make good PET ligands and vice versa, so new decision-making criteria need to be developed and communicated to enable rapid probe development.

REFERENCES

1. Uppoor RS, Mummaneni P, Cooper E, Pien HH, Sorensen AG, Collins J, Mehta MU, Yasuda SU. The use of imaging in the early development of neuropharmacological drugs: a survey of approved NDAs. Clin Pharmacol Ther 2008;84(1):69–74.
2. Laruelle M, Slifstein M, Huang Y. Relationships between Radiotracer Properties and Image Quality in Molecular Imaging of the Brain with Positron Emission Tomography. Molecular Imaging and Biology 2003;5;6:363–375.
3. Schou M, Pike VW, Halldin C. Development of radioligands for imaging of brain norepinephrine transporters In Vivo with positron emission tomography. Current Topics in Medicinal Chemistry 2007;7;18:1806–1816.

4. Shang YL, Gibbs MA, Marek GJ, Stiger T, Burstein AH, Marek K, Seibyl JP, Rogers JF. Displacement of serotonin and dopamine transporters by venlafaxine extended release capsule at steady state: a [123I]2beta-carbomethoxy-3beta-(4-iodophenyl)-tropane single photon emission computed tomography imaging study. J Clin Psychopharmacol 2007;27(1):71–75.
5. EMEA. Note for guidance on non-clinical safety studies for the conduct of human clinical trials and marketing authorization for pharmaceuticals (ICH Topic M3 (R2). London: European Medicines Agency; 2008 Jul. Report nr CPMP/ICH/286/95.
6. Tauscher J, Jones C, Remington G, Zipursky RB, Kapur S. Significant dissociation of brain and plasma kinetics with antipsychotics. Mol Psychiatry 2002;7(3):317–321.
7. Ginovart N, Wilson AA, Hussey D, Houle S, Kapur S. D2-receptor upregulation is dependent upon temporal course of D2-occupancy: a longitudinal [11C]-raclopride PET study in cats. Neuropsychopharmacology 2009; 34(3):662–671.
8. Medori R, Mannaert E, Grunder G. Plasma antipsychotic concentration and receptor occupancy, with special focus on risperidone long-acting injectable. Eur Neuropsychopharmacol 2006;16(4):233–240.
9. Yokoi F, Grunder G, Biziere K, Stephane M, Dogan AS, Dannals RF, et al. Dopamine D2 and D3 receptor occupancy in normal humans treated with the antipsychotic drug aripiprazole (OPC 14597): a study using positron emission tomography and [11C]raclopride. Neuropsychopharmacology 2002; 27(2):248–259.
10. Talbot PS, Laruelle M. The role of *in vivo* molecular imaging with PET and SPECT in the elucidation of psychiatric drug action and new drug development. Eur Neuropsychopharmacol 2002;12(6):503–511.
11. Abadie P, Rioux P, Scatton B, Zarifian E, Barre L, Patat A, Baron JC. Central benzodiazepine receptor occupancy by zolpidem in the human brain as assessed by positron emission tomography. Eur J Pharmacol 1996;295(1):35–44.
12. Smith DF, Stork BS, Wegener G, Jakobsen S, Bender D, Audrain H, Jensen SB, Hansen SB, Rodell A, Rosenberg R. Receptor occupancy of mirtazapine determined by PET in healthy volunteers. Psychopharmacology (Berl) 2007;195(1):131–138.
13. Talvik-Lotfi M, Nyberg S, Nordstrom AL, Ito H, Halldin C, Brunner F, Farde L. High 5HT2A receptor occupancy in M100907-treated schizophrenic patients. Psychopharmacology (Berl) 2000;148(4):400–403.
14. Innis RB, Cunningham VJ, Delforge J, Fujita M, Giedde A, Gunn RN, Holden J, Houle S, Huang SC, Ichise M, Lida H, Ito H, Kimura Y, Koeppe RA, Knudsen GM, Knuuti J, Lammertsma AA, Laruelle M, Logan J, Maguire RP, Mintun MA, Morris ED, Parsey R, Price JC, Slifstein M, Sossi V, Suhara T, Votaw JR, Wong DF, Carson RE. Consensus nomenclature for *in vivo* imaging of reversibly binding radioligands. J Cereb Blood Flow Metab 2007;27(9):1533–1539.
15. Burns HD, Van Laere K, Sanabria-Bohorquez S, Hamill TG, Bormans G, Eng WS, Gibson R, Ryan C, Connolly B, Patel S, Krause S, Vanko A, Van Hecken A, Dupont P, De Lepeleire I, Rothenberg P, Stoch SA, Cote J, Hagmann WK, Jewell JP, Lin LS, Liu P, Goulet MT, Gottesdiener K, Wagner JA, de Hoon J, Mortelmans L, Fong TM, Hargreaves RJ. [18F]MK-9470, a positron emission tomography (PET) tracer for *in vivo* human PET brain imaging of the cannabinoid-1 receptor. Proc Natl Acad Sci USA 2007; 104(23):9800–9805.
16. Lim KS, Kwon JS, Jang IJ, Jeong JM, Lee JS, Kim HW, Kang WJ, Kim JR, Cho JY, Kim E, Koo SY, Shin SG, Yu KS. Modeling of brain D2 receptor occupancy-plasma

concentration relationships with a novel antipsychotic, YKP1358, using serial PET scans in healthy volunteers. Clin Pharmacol Ther 2007;81(2):252–258.

17. Brix G, Zaers J, Ådam LE, Bellemann ME, Ostertag H, Trojan H, Haberkorn U, Doll J, Oberdorfer F, Lorenz WJ. Performance evaluation of a whole-body PET scanner using the NEMA protocol. National electrical manufacturers association J Nucl Med 1997;38(10):1614–1623.
18. Hietala J, Nagren K, Lehikoinen P, Ruotsalainen U, Syvalahti E. Measurement of striatal D2 dopamine receptor density and affinity with [11C]-raclopride *in vivo*: a test-retest analysis. J Cereb Blood Flow Metab 1999; 19(2):210–217.
19. Dembo G, Park SB, Kharasch ED. Central nervous system concentrations of cyclooxygenase-2 inhibitors in humans. Anesthesiology 2005;102(2):409–415.
20. Maurer TS, Debartolo DB, Tess DA, Scott DO. Relationship between exposure and nonspecific binding of thirty-three central nervous system drugs in mice. Drug Metab Dispos 2005;33(1):175–181.
21. Liu XR, Smith BJ, Chen C, Callegari E, Becker SL, Chen X, Cianfrogna J, Doran AC, Doran SD, Gibbs JP, Hosea N, Liu JH, Nelson FR, Szewc MA, Van Deusen J. Evaluation of cerebrospinal fluid concentration and plasma free concentration as a surrogate measurement for brain free concentration. Drug Metab Dispos 2006;34(9):1443–1447.

8

BIOSENSORS FOR CLINICAL BIOMARKERS

SARA TOMBELLI AND MARCO MASCINI

8.1 INTRODUCTION

This chapter reviews recent published works on biosensors for the detection of biomarkers in several medical areas. The definition of biological marker (biomarker) has caused some controversy in the scientific/medical community, but the terminology has been clarified: a biomarker has been defined as "a physical sign or laboratory measurement that occurs in association with a pathological process and that has putative diagnostic and/or prognostic utility" (1) and as "a characteristic that is objectively measured and evaluated as an indicator of normal biological processes, pathogenic processes, or pharmacologic responses to a therapeutic intervention" (2).

Diagnostics through biomarker level detection (i.e., the identification of a particular disease through the measurement of the quantity of specific biomarkers) at hospitals is based either on large-scale automated equipment or enzyme-linked immunosorbent assay (ELISA) techniques based on bioassays that are not suitable for bedside and emergency medicine. Therefore, validated, intelligent, next-generation diagnostic devices and systems based on biosensor technology with new, radically enhanced detection capabilities and integrated sample handling to address the most common diagnostic problems are strongly needed.

The definition of a biosensor has been decided on by IUPAC (3) and recently published by Prof. Turner and Newman (4), who referred to a biosensor as "a compact analytical device incorporating a biological or biologically-derived

Predictive Approaches in Drug Discovery and Development: Biomarkers and In Vitro/In Vivo Correlations, First Edition. Edited by J. Andrew Williams, Jeffrey R. Koup, Richard Lalonde, and David D. Christ.

sensing element either integrated within or intimately associated with a physicochemical transducer."

Biosensors can be classified either on the base of the biological receptor or depending on the transducer. An electrochemical biosensor is a biosensor with an electrochemical transducer, such as ion selective, glass, or gas electrodes for potentiometric measurements and metal or carbon electrodes for amperometric measurements (3). Other types of biosensors are based on piezoelectric (shear and surface acoustic waves) and optical transduction [such as planar waveguide, fiber optic, and surface plasmon resonance (SPR)] (3).

The research in the field of biosensors was initiated by Clark, whose study on the oxygen electrode was published in 1956 (5). Since then, a huge number of biosensors have appeared in the literature and a great number with an application in medical diagnostics. Actually, more than 80% of the commercial devices based on biosensors are utilized in this domain (6) starting with the first commercial apparatus for glucose determination produced by Yellow Spring Instruments (1975) (7).

Apart from the huge space occupied in the market and in the literature by the glucose biosensor recently reviewed by Wang (8), and other enzyme-based catalytic biosensors (9, 10), many examples related to the analysis of clinically relevant analytes by immunosensors have been reported in the past 20 years when this approach first started (11). More recently, in the past decade, DNA-based sensing has appeared for real applications to clinical diagnostics to detect the presence of pathogenic species responsible of infections, to identify genetic polymorphisms, to detect point mutations (12), or for the recognition of disease-related sequences such as the chimeric oncogene breakpoint cluster region gene and a cellular *abl* gene (BCR/ABL), which is the gene present in almost all cases of chronic myelogenous leukemia patients (13).

We have focused on the recent (2007–2009) literature on biosensors for biomarkers for different problematic (cancer, cardiac diseases, and hormones). A separate section centers on DNA biosensors based on hybridization, which can represent a valid alternative to the well-accepted immunosensors.

Particular attention is given to those biosensors that have been tested for the analysis of "real clinical samples," for example, in a diagnostic setting where the biosensor-generated data possibly affects treatment decisions. It is important to note that at the time of writing in 2009, application of biosensors for analysis with "real" clinical samples is still rare. Many papers describing the use of biosensors in this field only have exemplary character. Detailed data as well as validation with established methods for particular parameters are missing in most cases. The analytical potential of biosensors in the medical diagnostics field still has to be strengthened by the demonstration of their applicability to testing of real matrices. The main problem connected with the lack of experimental data on the real samples is the difficult "communication" between technologists and hospitals. A better connection between the researcher in the biosensor field and physicians in hospitals could assure the real understanding of physicians' needs, the choice of correct applications, and the availability of real samples for biosensor optimization and validation.

8.2 BIOSENSORS FOR BIOMARKERS IN ONCOLOGY

Cancer is the one of the major causes of mortality in Western countries and most other countries, and the diagnosis and improvement of cancer treatments represent a major area of unmet need across Europe and all other areas of the world (14).

Currently, the most important cancer diagnostic indicators are morphological and histological characteristics of tumors together with routine clinical assays for single biomarkers, such as prostate-specific antigen (PSA) or carcinoembryonic antigen (CEA) (2), which are part of the list of cancer biomarkers indicated by the Food and Drug Administration as *known valid biomarkers* (15).

Several examples of biosensors for the detection of these biomarkers can be found in the literature (Table 8.1) (16, 17), mostly for the detection of PSA, a biomarker strictly associated with one kind of cancer, prostate cancer.

8.2.1 Biosensors for the Detection of PSA

Prostate-specific antigen has been identified as a biomarker to screen prostate cancer patients and it has been shown that PSA is the most reliable tumor marker to detect prostate cancer at the early stage and to monitor the recurrence of the disease after treatment (36). PSA is found in serum, either free or in complex with various proteinase inhibitors, and a total PSA level of 10 ng/ml or higher is a highly probable indicator for prostate cancer (37). Anyway, a PSA measurement above cutoff value of 4 ng/ml, between 4 and 10 ng/ml, is considered to be suspicious and should be followed by more localized examination such as prostate biopsy (38), which can give a certain positive or negative result and give an indication on the treatment necessary for the patient.

Innovative biosensor strategies could represent alternative strategies for reliable testing of cancer patients to help guide treatment and/or enrollment in clinical trials: several biosensors for PSA detection have been presented in the past few years based on different transduction techniques from electrochemical (39) to piezoelectric (40) and optical (41, 42) methods. Low detection limits have been recently reached by several published electrochemical biosensors employing carbon nanotubes (31, 34) and nanoparticles (32, 33) for signal amplification.

In particular, an immunochromatographic electrochemical biosensor coupled to nanoparticles (32), with a detection limit of 0.02 ng/ml, was tested also in serum samples. This biosensor was validated with a human serum sample and a commercial ELISA PSA kit: the results of the biosensor were consistent with those of ELISA with recoveries for spiked samples of 105–111%.

TABLE 8.1 Recently Published Biosensors for the Detection of Cancer Biomarkers Indicated by the Food and Drug Administration as "Known Valid Biomarkers"

Biomarker	Cancer Type	Normal Levels in Healthy Adults	Biosensor (Transducer/Bioreceptor)	Detection Range	References
α-Fetoprotein[s,rt]	Testicular/liver	≤10 ng/ml	Piezoelectric/antibody	15.3–600 ng/ml, tested in serum	18
			Electrochemical/antibody	1–1000 ng/ml	19
Human chorionic gonadotropin-β[d,p,rt]	Prostate/pancreatic	(Male, nonpregnant female) <5.0 mIU/ml	Electrochemical/antibody	0.5–50 mIU/ml, tested in serum	20
			Electrochemical/antibody	18–450 mIU/ml, tested in serum	21
			Electrochemical/antibody	1–100 mIU/ml, tested in serum	22
CA19-9[rt]	Pancreatic	<40 IU/ml	Piezoelectric/antibody	0.012–0.270 IU/ml	23
			Electrochemical/antibody	0.16–15 IU/ml	24
CA125[s,rt]	Ovarian	0–35 U/ml	Electrochemical/antibody	0.5–55.6 kIU/l	25
			Electrochemical/antibody	0.11–13 U/ml	24
			Fluorescence/antibody	0–400 U/ml	26
CA153[d]	Breast	≤25 U/ml	Electrochemical/antibody	0.084–16 U/ml	24
CEA[rt]	Colon/breast	<2.5 ng/ml	Fluorescence/antibody	0–100 ng/ml, tested in serum and saliva	26
			Electrochemical/antibody	0.2–120 ng/ml	27
			Electrochemical/antibody	0.16–9.2 ng/ml	24

			Piezoelectric/antibody	5 ng/ml	28
			Optical/antibody	0.1–150 ng/ml, tested in serum and nipple aspirate fluid	29
Thyroglobulin[d,p]	Thyroid	2.0–35 ng/ml	Optical/protein-coated surface	1 pg/ml to 1 μg/ml	30
PSA (total)[p,rt]	Prostate	0–4.0 ng/ml	Electrochemical/antibody	Detection limit of 0.25 ng/ml	31
			Electrochemical/antibody	0.05–4 ng/ml, tested in serum	32
			Optical/antibody	Detection limit of 0.01 ng/ml	33
PSA (complex)	—	—	Electrochemical/antibody	1–100 ng/ml	34
NMP22[d]	Bladder	2 ng/ml	Electrochemical/antibody	1.2–200 ng/ml	35
Cytokeratin-7	Used in conjunction with cytokeratin 20 in distinguishing ovarian, pulmonary, and breast carcinomas (CK7+) from colon carcinomas (CK7−)	—	Electrochemical/antibody	10–100 ng/ml	25
HER2/NEU[p]	Breast	<15 ng/ml	Fluorescence/antibody	0–60 ng/ml	26

[d], diagnostic; [p], prognostic; [rt], marker of response to treatment; [s], suggestive.

8.2.2 Biosensors for the Detection of Other Cancer-Related Biomarkers

Among other cancer biomarkers that have been considered as possible target for a single-analyte biosensor, α-fetoprotein (18, 19), CEA (26–29), human chorionic gonadotropin-β (hCG) (20–22, 29), CA19-9 (23), thyroglobulin (30), NMP22 (35), and cytokeratin (25) have been recently taken into account.

Among those which have been validated with real samples, a label-free capacitive immunosensor for the detection of hCG was reported (21) with a detection range between 18 and 450 mIU/ml of hCG and a detection limit of 5.0 mIU/ml, which is much lower than the threshold value of 14.3 mIU/ml in serum considered necessary for a pregnancy diagnosis. The immunosensor was developed by covalently coupling an anti-hCG antibody to an epoxy-silane surface deposited onto glassy carbon electrodes. The principle of detection was based on capacitive measurements by detecting the changes after and before the antigen–antibody interaction. The immunosensor was validated by the analysis of 35 human serum samples in comparison with a radioimmunoassay (RIA): the comparison between the two methods evidenced no systematic differences between them with a correlation coefficient (R) of 0.998.

Apart from the described single-analyte biosensors, multianalyte analysis could be important to overcome the challenges in cancer diagnosis, prognosis, and drug development for cancer treatment, and the development of biosensors with an array format for the simultaneous detection of different tumor markers (24, 26, 43) can be considered of particular significance since most markers are not specific to a particular tumor and the use of panels of tumor markers can increase their diagnostic value (44). One of the multianalyte electrochemical biosensors (24) was created for the simultaneous detection of CA153, CA19-9, CEA, and CA125 using a screen-printed carbon electrode (SPCE) to capture the specific horseradish-peroxidase-labeled antibodies in a competitive assay format (Fig. 8.1).

The detection was achieved by monitoring the mediator-catalyzed enzymatic response to hydrogen peroxide. Clinical serum samples were analyzed with good correlation with a commercial electrochemiluminescence analyzer.

8.3 BIOSENSORS FOR BIOMARKERS IN CARDIOLOGY

The debate on biomarkers for heart failure is still open since this growing public health problem appears to result from a complexity of different factors, from genetic parameters to inflammation and neurohormonal disorders (45).

Laboratory tests are an important part in diagnosing heart infarction (46), and fast and cost-effective diagnostics is needed. At present, three of the most informative markers cardiac troponin I or T (cTnI/T), myoglobin, and natriuretic peptide particularly of the B-type (BNP) are determined by different immunoassay methods, such as ELISA (47), RIA (48), and immunochromatographic tests (49). The first two tests are the most used, but they are time consuming since they require several steps; the last one is a qualitative test. As possible

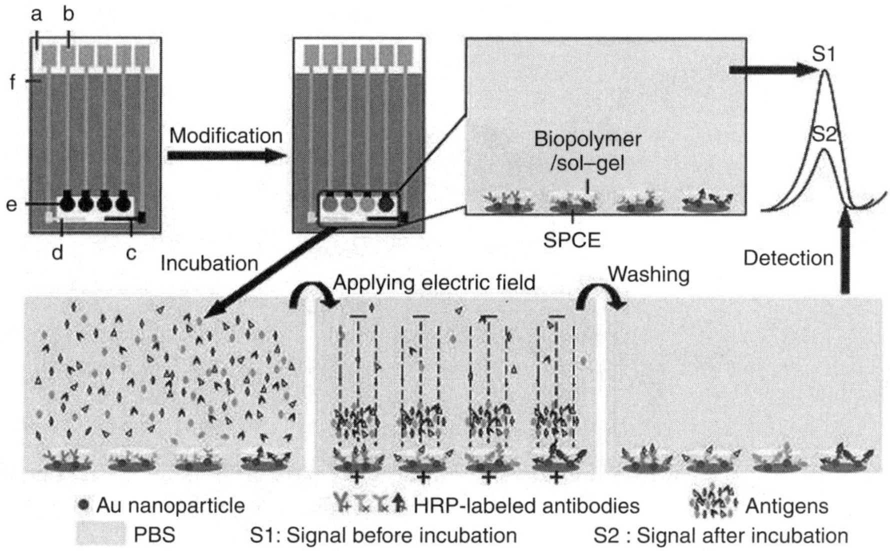

FIGURE 8.1 The multiplexed immunosensor array fabricated on a carbon electrode array containing four graphite working electrodes (e), which was prepared with screen-printed technology on a nylon sheet (a). All working electrodes shared the same Ag/AgCl reference [b (silver ink), d] and graphite auxiliary (c) electrodes. The insulating layer (f) printed around the working area constituted an electrochemical microcell. Anti-CA153, CA125, CA19-9, and CEA antibodies labeled with horseradish peroxidase (HRP) were immobilized onto the biopolymer/sol-gel-modified electrodes. In the presence of gold nanoparticles, the HRP showed enhanced direct electrochemical responses, and the formation of immunocomplexes led to a decrease in the electrochemical signals (S2 vs S1) due to the increasing spatial blocking and impedance caused by the nonconductive immunocomplexes, which blocked the electron transfer between the electrode and HRP-labeled antibodies. *Source:* With permission from Reference 24.

alternative, several biosensors for the detection of TnT and TnI have been published in the past years (50–52) based on electrochemical and optical transduction. Reference ranges for TnI are as follows: $\leq$0.03 ng/ml, no detectable cardiac injury; 0.04–0.49 ng/ml, cardiac muscle injury; and $\geq$0.5 ng/ml, myocardial infarction. For TnT, the reference value is less than 0.1 ng/ml.

Very good sensitivity and selectivity were reached by an optical biosensor coupled to quantum dots (52). The biosensor was based on fluorescence resonance energy transfer (FRET), a signal transduction method that occurs between two fluorescent molecules, a donor and an acceptor. At this regard, protein A was modified with carboxy-functionalized quantum dots, used here as donors, whereas an anti-TnI antibody was modified with a fluorescent dye, the acceptor (Fig. 8.2).

The biosensor achieved a detection limit of 32 nM TnI in buffer and of 55 nM TnI in human plasma. The response time of the biosensor was determined to be less than 1 min, which was much lower than the analysis time required by other TnI diagnostic assays.

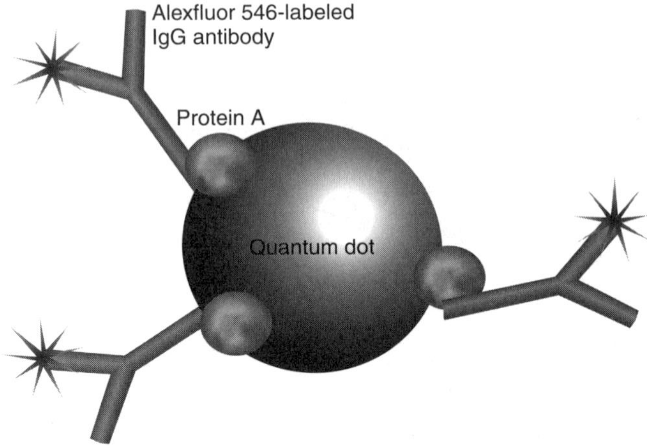

FIGURE 8.2 Schematic of self-assembled optical biosensor architecture. Catskill green quantum dot was utilized as the donor. Its emission peak wavelength (544 nm) overlaps the absorption peak wavelength of the Alexa Fluor 546 (AF-546) fluorescent dye, which was used as the acceptor (546 nm). Carboxyl-functionalized quantum dots were labeled with protein A. Mouse anti-troponin I IgG monoclonal antibodies were labeled with AF-546 fluorescent dye. The biosensor complex was fabricated. By incubating labeled protein A and antibodies, this configuration brings the quantum dot donor and the dye acceptor within an energy-transfer distance. A degree of FRET occurs between the two fluorescent molecules in this state, which is treated as baseline. As the antigen–antibody encounter complex is formed on introduction of the cTnI analyte, a conformational change takes place within the structure of the antibody. This conformational change alters the distance between the quantum dot donor and AF-546 acceptor and a change in energy-transfer efficiency occurs. *Source:* With permission from Reference 52.

Moreover, an SPR immunosensor has been recently presented (50) for the detection of TnT based on the use of immobilized monoclonal antibodies specific for TnT. The biosensor presented a linear response range for TnT between 0.05 and 4.5 ng/ml with a good reproducibility (CV = 4.4%). The amount of TnT was measured in human serum samples and the results were compared with a reference method [Roche Elecsys 2010 immunoassay analyzer based on electrochemiluminescence immunoassay (ECLIA)]. The measurements with the SPR biosensor showed a good agreement with the ECLIA method at 95% confident level.

The importance of biomarkers for inflammation in the early diagnosis of heart failure is subject to intense inquiry since inflammation has a high relevance in heart failure pathogenesis and progression (45, 53). Several clinical studies demonstrated a correlation between inflammatory biomarkers, such as C-reactive protein (CRP), cytokines, interleukin-6, and tumor necrosis factor α, and high risk of the future development of heart failure in asymptomatic older subjects. In particular, CRP, which is used for conventional inflammation diagnosis, can also serve as diagnostic marker for low grade inflammation for risk estimation of

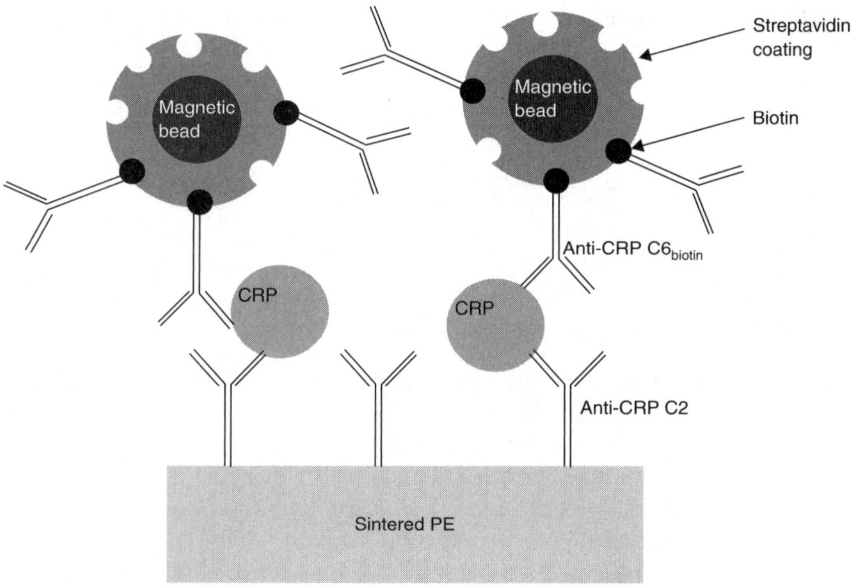

FIGURE 8.3 Capture antibody (anti-CRP, clone 2) was immobilized on polyethylene (PE) sintered filters. Anti-CRP antibody, clone C6, was used as detection antibody, and it was immobilized onto streptavidin-coated magnetic beads. Antigen (CRP) was supplied in different concentrations to the column with the Ab-modified sintered filters (sample size, 0.5 ml). After an additional washing step, pretreated beads with detection antibody $C6_{biotin}$ were allowed to flow through the column. The column was then analyzed in the magnetic reader. *Source:* With permission from Reference 59.

cardiovascular events and it has been recommended as a predictive biochemical marker for risk of coronary disease (2, 54). Normal blood serum concentrations of humans range from 1 to 5 mg/l, and protein levels higher than 5 mg/l are an indication of inflammatory processes (55). In particular, levels of CRP less than 1 mg/l are normal and reflect a low cardiovascular risk. Levels of CRP between 1 and 3 mg/l are indicative of moderate risk, while levels of CRP in excess of 3 mg/l suggest quite elevated vascular risk (56). In routine clinical analysis, CRP levels are determined by ELISA (57), with detection limits down to 0.2 mg/l.

New approaches in medical CRP diagnosis for cardiovascular disease require rapid quantification in native matrices, such as saliva and urine (58), which are not yet available for CRP determination. A novel magnetic biosensor (Fig. 8.3) was presented for the determination of CRP in human serum, saliva, and urine (59). Two CRP antibodies (clone C2 and C6) were used: C2 was used as capturing antibody and it was immobilized onto polyethylene sintered filters in ABICAP® plastic columns. Clone C6 served as secondary antibody and it was biotinylated and attached to streptavidin-coated magnetic beads.

This antibody-magnetic complex interacted with the captured CRP on the primary antibody and it can be quantified by a magnetic reader. A very low detection limit (0.025 mg/l) was achieved with this system, which was also tested in CRP-spiked samples of serum, saliva, and urine. Calibration curves were constructed for CRP in these matrices with good results. In particular, the measurement of CRP-spiked human saliva showed rates of recovery of typically 5% less CRP than spiked. No matrix effect, in comparison with the CRP determination in buffer, was found.

8.4 BIOSENSORS FOR OTHER CLINICAL BIOMARKERS

8.4.1 Biosensors for Hormones Detection

Many of the body's functions are regulated by hormones that are released into the bloodstream from glands or organs. Therefore, hormones can be considered as important biomarkers in several diseases or malfunctions, and some biosensors have been recently presented for their detection (Table 8.2).

Many physiological processes are regulated by hormones released by the pituitary gland, such as thyroid-stimulating hormone (hTSH), growth hormone (hGH), follicle-stimulating hormone (hFSH), and luteinizing hormone (hLH). Among these, hGH has been considered as target for biosensors. hGH is an important diagnostic marker of pregnancy, an important status for inclusion/exclusion in clinical trials, and one of the most important carbohydrate tumor markers. Several immunoassay kits or strategies have been presented, including some electrochemical immunosensors (22, 65, 66). More recently, an optical biosensor based on SPR has been presented with a detection limit of hGH of 6 ng/ml (60). A binding inhibition immunoassay was used to detect hGH both in buffer and untreated serum samples: the biosensor could detect hGH in serum in the range 6 ng/ml–1.3 μg/ml, covering the lower part of the physiological range.

The same SPR biosensor was optimized for the detection of hTSH, hGH, hFSH, and hLH, using the corresponding specific antibodies (61). The biosensor could detect hFSH, hLH, and hGH, with sensitivity adequate to physiological ranges; for hTSH, the detection limit (3 ng/ml = 23 μIU/ml) was not sufficient for the cutoff level for congenital hypothyroidism (23 μIU/ml). hFH and hLH were also detected in human urine without sample pretreatment.

Among other hormones, human parathyroid hormone (PTH) has a clinical significance in parathyroid adenoma, hyperplasia, and cancer (hyperparathyroidism) or in autoimmune disorders (hypoparathyroidism). A giant magnetoresistive biosensor using actuated magnetic particle labels for the detection of PTH has been recently described (62).

The biosensor was based on a one-step sandwich immunoassay in which the tracer anti-PTH antibody is labeled with magnetic particles that are then detected by the device. The system consists of a compact electronic reader and a disposable fluidic cartridge device with an integrated silicon sensor chip (Fig. 8.4). The system comprises on-chip integration of high frequency magnetic field generation for label excitation. The response of the instrument is linearly dependent on the number of beads on the sensor surface: three beads of 300 nm diameter can be

TABLE 8.2 Biosensors for Hormone Detection

Hormone	Clinical Significance	Normal Levels in Healthy Adults (Out of Female Ovulation Period and Menopause)	Biosensor Transducer/Bioreceptor	Detection Range	References
Human growth hormone	Pituitary malformations, damages, or tumors	<5 ng/ml with peaks of secretion in the range 5–45 ng/ml	Optical/antibody	18–542 ng/ml in buffer, 6 ng/ml to 1.3 μg/ml in serum	60
Pituitary hormones	Pituitary malformations, damages, or tumors	hTSH 0.4–4 μIU/ml (20 μIU/ml cutoff value for congenital hypothyroidism), hFSH 2–10 mIU/ml, hLH 2–10 mIU/ml	Optical/antibody	hTSH, 18–367 ng/ml; hFSH, 14–370 ng/ml; hGH, 18–542 ng/ml; hLH, 10–78 ng/ml, tested in serum and urine	61
Parathyroid hormone	Parathyroid cancer, autoimmune disorders	10–65 pg/ml	Magnetic/antibody	Detection limit of 7.6 pg/ml	62
Progesterone	Pregnancy and fertility control	<1 ng/ml (female, preovulation; male) to 2 ng/ml (female, midcycle); up to 90 ng/ml during pregnancy	Optical/antibody	Detection limit 4.9 ng/l	63
Cortisol	Physiological stresses	Morning 3–10 ng/ml; evening 0.6–2.5 ng/ml	Optical/antibody	1.5–10 ng/ml in saliva	64

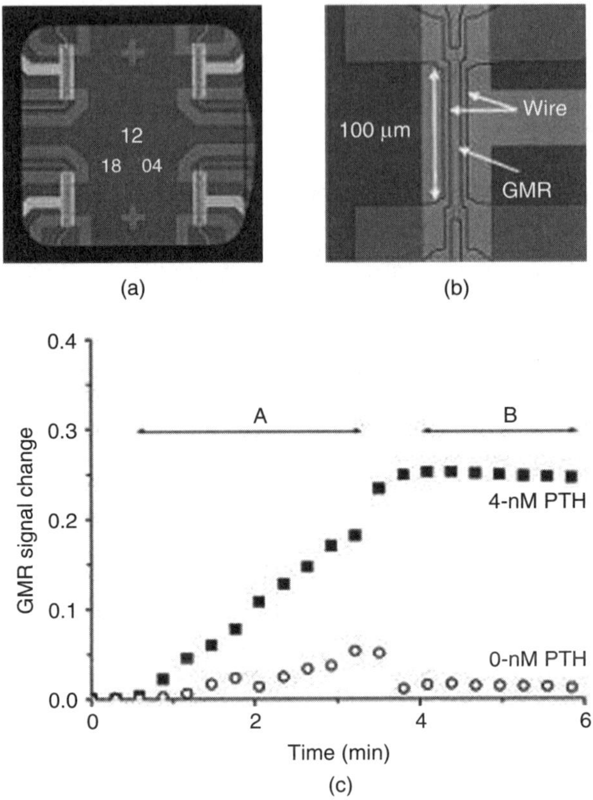

FIGURE 8.4 (a) Light microscopic image of 2-mm^2 silicon sensor chip containing four giant magnetoresistive (GMR) sensors coated with Au. (b) A detailed microscopic image of one sensor, showing the GMR strip between two excitation wires. (c) The relative change in GMR signal as a function of time after the addition of 300-nm streptavidin-coated magnetic particles to a biotinylated sensor surface. In section A of the curve, a magnetic force is applied to bring particles toward the sensor surface. In section B of the curve, a magnetic force is applied away from the surface to remove unbound and weakly bound particles (magnetic wash step). *Source:* With permission from Reference 62.

detected on a sensor surface of 1500 μm^2 for a measurement time of 1 s (67). The biosensor was able to detect PTH, with a detection limit of 0.8 pM in a total assay time of 15 min.

Sex steroids are thought to participate in regulation of immune response and may play a part in modulation of some inflammatory and autoimmune disorders (68). Among these hormones, the measurement of progesterone is important in women since it is involved in the female menstrual cycle, pregnancy, and embryogenesis. Reference concentration ranges of progesterone in plasma are from less than 1 ng/ml (female, preovulation; male) to 2 ng/ml (female, midcycle), with increases up to 90 ng/ml during pregnancy.

Immunoassays are commonly used for the determination of progesterone in serum or saliva. In particular, analysis laboratories use commercially available RIA kits, which are able to detect progesterone with sufficient precision in the clinically relevant concentration range (69). Given the inherent problems, different nonradioactive methods have been developed for measuring progesterone, and among the high number of immunoassay techniques, ELISA combined with a colorimetric measurement is the most widely used method for measuring hormone concentrations (70). The use of immunosensors is another interesting alternative approach. Both electrochemical and optical biosensors have been reported, using screen printing technology coupled to cyclic voltammetry (71) and chronoamperometry (72). For optical sensing, SPR has been used (63, 73). Screen-printed electrodes have been used in the immunosensor development for progesterone detection as solid phase for a competitive immunoassay (74). The EC_{50} value (the analyte concentration necessary to displace 50% of the enzyme label) was calculated as 2 ng/ml, whereas the limit of detection (LOD), calculated by evaluation of the mean of the blank solution (containing the tracer only) response minus two times the standard deviation, was estimated as 32 pg/ml. This detection limit compared well with that (4.9 pg/ml) of other approaches previously published (63).

A recent paper has appeared, focusing on the determination of another hormone, cortisol, by an SPR immunosensor (64). Cortisol is a steroid hormone required for metabolic activities and cardiovascular function. It is considered as an indicator of stress or disease state of a patient with normal serum levels between 20 and 140 ng/ml. Cortisol can also be found in saliva in a concentration range of 1–8 ng/ml for healthy subjects: the most important advantage of detecting cortisol in saliva and not in blood/serum is the good correlation between salivary cortisol and levels of "free" cortisol in serum, which is more biologically active than cortisol bound to transport proteins or albumin. The system recently presented is based on a competition immunoassay with a six-channel portable SPR biosensor employing cortisol-specific monoclonal antibody. In addition, an in-line filter composed of a hollow fiber membrane (20,000 molecular weight cutoff) served to separate small molecules from large molecular weight saliva components such as mucins (Fig. 8.5). The detection limit of the biosensor was 0.36 ng/ml in buffer and 1 ng/ml in saliva.

More recently, an immunosensor based on SPR has been published for the detection of cortisol (75), with a detection limit of 13 pg/ml in buffer and 49 pg/ml in saliva, which is almost reaching the low detection limit obtained by liquid chromatography–mass spectrometry (LC/MS), 5 pg/ml (76).

8.4.2 Biosensors for Other Biomarkers

Several recent biosensors for different biomarkers have been presented, such as an optical biosensor for the detection of Alzheimer's disease (77) or a fluorescence-based biochip for the detection of $CD4^+$ lymphocytes for HIV monitoring (78). Another interesting biosensor has been recently published for the detection of tuberculosis (79).

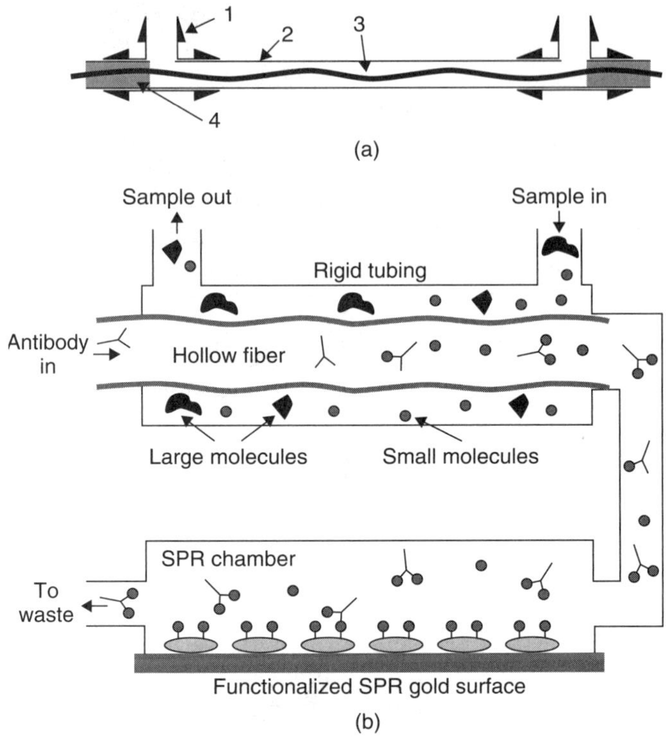

FIGURE 8.5 External, in-line filtering flow cell. (a) Scheme of the in-line filtering flow cell: (1) 1.6-mm Tee connector, (2) 0.75 mm inner diameter tube, (3) hydrophilic hollow fiber, and (4) ends sealed with urethane adhesive. (b) Diluted salivary samples were allowed to flow through the tubing external to the hollow fiber. The antibody solution was flowed countercurrent through the hydrophilic hollow fiber and then through the SPR biosensor system. The in-line filtering flow cell allowed diffusion of small-molecular-weight analytes between the sample stream and the sensor flow stream. *Source:* With permission from Reference 64.

Tuberculosis is a bacterial infection caused by *Mycobacterium tuberculosis*, and the number of active tuberculosis patients is increasing in countries with deficient health care. Classical methods for tuberculosis detection are staining and sputum culture together with DNA-based methods such as polymerase chain reaction (PCR) or restriction length polymorphism. These methods are, however, time-consuming sample drawing and often not enough sensitive for consistently valuable utility (80).

As alternative method, a biosensor for the detection of the tuberculosis-specific antibodies, has been described based on three types of label-free optical devices, a grating coupler in the reflection mode, an interferometric biosensor, and a reflectometric interference spectroscopy device (Fig. 8.6) (79).

All these devices are employing glass surfaces on which the 38-kDa antigen is immobilized. Serum samples were analyzed without any pretreatment,

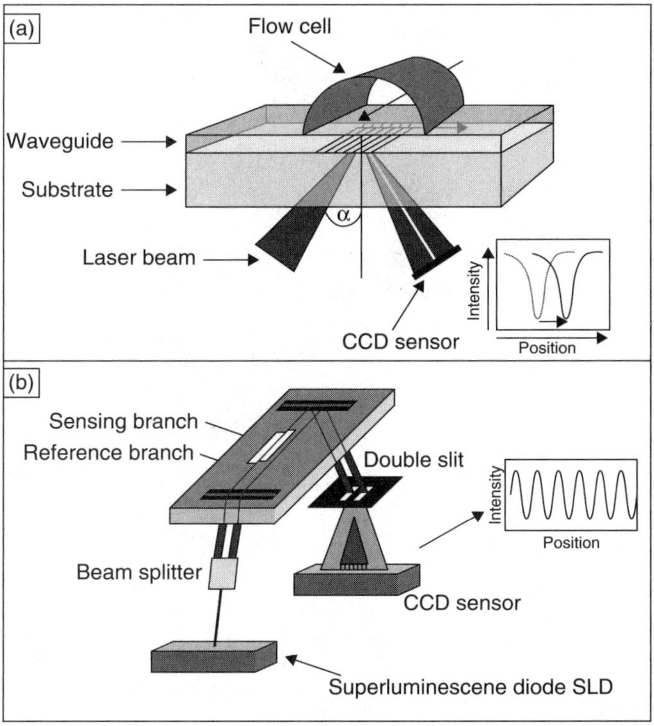

FIGURE 8.6 Configurations of the three used optical label-free devices. (a) Grating coupler device: a convergent HeNe laser beam is irradiated onto the grating, and the reflected intensity is detected by a charge-coupled device (CCD) line sensor. Under the specific coupling angle, a part of the light is coupled into the waveguide and the electromagnetic field of the light wave reaches into the medium, which has a lower refractive index. This results in an exponentially decaying field, the evanescent wave. The position of the reduced intensity in the reflected light is observed and evaluated. Every binding effect at the sensor surface results in a change in the effective refractive index, leading to a shift in the coupling angle. (b) Interferometric biosensor: the light of a superluminescent diode is coupled into the waveguide chip by a grating and is passed through two separate branches under total internal reflection and there is formation of an evanescent wave. At the end of the sensing and the reference branch, the light is outcoupled via a second grating and passes a double slit. The resulting interference pattern is recorded by a CCD sensor. (c) Reflectometric interference spectroscopy (RIfS) system: light is reflected at each interface of a thin transparent film, and as a result of the difference in the optical paths, a shift in the interference pattern can be observed. The outer sensor surface directed toward the flow system is coated with a thin SiO_2 layer. *Source:* With permission from Reference 79.

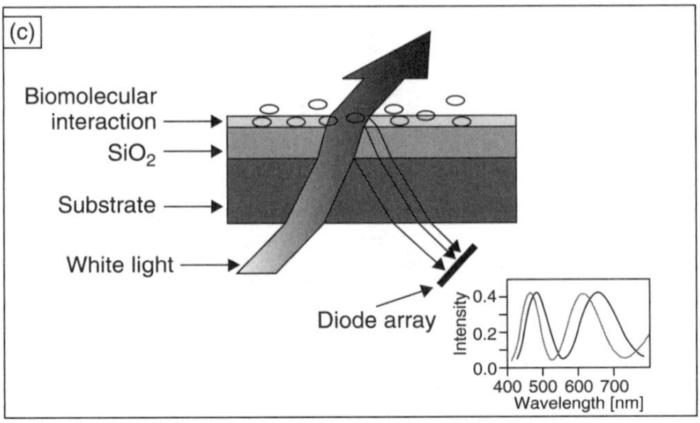

FIGURE 8.6 *(Continued)*

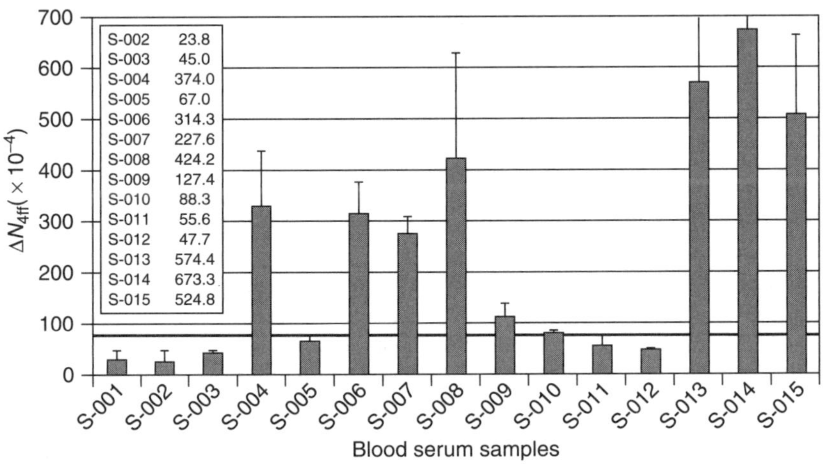

FIGURE 8.7 Summary of all grating coupler measurements with clinical serum samples. Each sample was analyzed on three different grating coupler sensor chips. The samples from S-001 to S-003 were tested by ELISA as tuberculosis negative and S-004 to S-015 as positive. The determined cutoff value was $\Delta N_{\text{eff}} = 75.2 \times 10^{-6}$. *Source:* With permission from Reference 79.

and heparin-stabilized whole blood was diluted 1 : 2 in buffer and spiked with the monoclonal antibody specific for the 38-kDa antigen. The results were in agreement with a tuberculosis-specific ELISA (Fig. 8.7).

8.5 NUCLEIC-ACID-BASED BIOSENSORS FOR BIOMARKERS

The genetic characteristics of patients, such as genotypes or presence of clinically significant polymorphisms, can be used for the evaluation as well as for the identification of a risk of disease for the patient (2).

In recent years, the monitoring methods of clinical disease diagnosis and prognosis for genotypes or polymorphisms include those of chromosome analysis, fluorescence in situ hybridization (FISH), flow cytometry (FCM), real-time quantitative reverse transcription PCR (RT-PCR), or restriction fragment length polymorphism (RFLP) (13).

But there were some limitations in these methods, such as long time consumption, poor precision, and high expense. Compared to other methods, DNA-based biosensors have prominent advantages of being simple, portable, rapid, and inexpensive, and they can be considered as a promising solution for the rapid and inexpensive diagnosis of genetic diseases.

A nucleic acid (NA) biosensor is defined as an analytical device incorporating an oligonucleotide, even a modified one, with a known sequence of bases, or a complex structure of NA (like DNA from calf thymus) either integrated within or intimately associated with a signal transducer (81). Nucleic acid biosensors can be used to detect DNA/RNA fragments or either biological or chemical species. Most NA biosensors are based on the highly specific hybridization of complementary strands of DNA or RNA molecules; this kind of biosensor is also called a *genosensor*. The probe, immobilized onto the transducer surface, acts as the biorecognition molecule and recognizes the target DNA, while the transducer is the component that converts the biorecognition event into a measurable signal. Assembly of numerous (up to a few thousand) DNA biosensors onto the same detection platform results in DNA microarrays (or DNA chips), devices that are increasingly used for large-scale transcriptional profiling and single nucleotide polymorphisms (SNPs) discovery.

In NA biosensors, the detection of the hybridization event has been carried out through different detection technologies, from label-free methods, such as piezoelectric and SPR transduction, to other methods often requiring labels, such as electrochemical techniques. Several reviews have recently appeared in the literature (82, 83), elucidating all the critical aspects related to the transduction step.

Limiting the discussion to applications of this kind of biosensors to the field objects of this chapter, several DNA-based biosensors have been recently developed for the detection of virus and bacteria-related sequences, such as hepatitis B virus (83, 84), human papilloma virus (HPV) (85, 86), and *M. tuberculosis* (87, 88), or for the recognition of disease-related sequences, such as leukemia (13, 89) and cystic fibrosis or breast cancer (90).

In a recent paper (91), an electrical detection technology has been introduced, which is easy to handle, easy to integrate into automated diagnosis systems, and can also be applied in a conventional manner, for example, as a sensor to detect PCR amplicons directly in the PCR vessel. The work deals with an electrical biosensor detection principle for applications in medical and clinical diagnosis based on the electrically detected displacement assay (EDDA). The electrodes used in the work are shaped as a "dipstick" and can be directly placed into PCR tubes containing the DNA amplicons (Fig. 8.8).

The microelectrode array consists of 32 gold electrodes onto which different DNA probes can be immobilized. The detection principle is based on a first

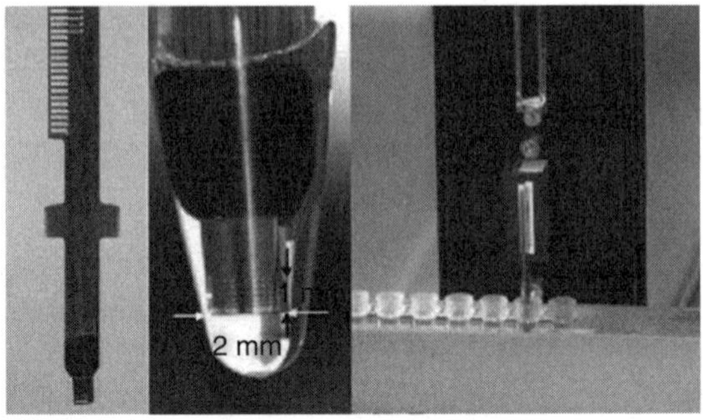

FIGURE 8.8 (a) Photograph of the dipstick-type microelectrode array (MEA) consisting of 32 gold microelectrodes on Si (bottom part of the stick) that is connected to a standard sliding contact on a printed circuit board (macroscopic gold pads on the upper left part of the stick; an identical number of contacts is at the back of the stick) via wire bonding (mounted under the black epoxy glue, next to the Au-on-Si-MEA). (b) An enlargement of the Au-on-Si-MEA, when it is immersed in a standard 100-μl PCR tube. (c) The dipstick connected to the potentiostat. *Source:* With permission from Reference 91.

step of hybridization between the immobilized capture probe and a signaling probe labeled with an electroactive moiety (ferrocene). The signaling probe is hybridized to the capture probe at the end carrying the label, whereas at the other end, it presents an extra sequence that is free for the hybridization with the target PCR amplicon. The amplicon can hybridize to this remote end of the signaling probe, which is released in solution, leading to the decrease in electrochemical signal. The biosensor array was applied to the detection of specific sequences of HPV (HPV-6) both with standard solutions and PCR amplicons.

8.6 CONCLUSIONS

At present, diagnostic work at hospitals is based either on large-scale automated equipment, in most cases based on conventional analytical techniques, or ELISA techniques based on bioassays that are not suitable for bedside and emergency medicine.

To overcome these restrictions, the development of new medical instruments is of high priority for health care, representing an important advancement, to be used in general practice and to provide new automated, self-controlled equipment for patient home care, in case of either postoperative checks or for permanent surveillance in case of chronic diseases or enrollment/exclusion in clinical trials along with other clinical trial-related applications.

Innovative biosensor-based strategies could allow biomarkers testing reliably in a decentralized setting due to their attractive characteristics such as reduced size, cost, and required time of analysis.

This chapter has presented some recent biosensors based either on antibodies or NAs as biomolecular receptors for the detection of different biomarkers. The analytical characteristics of the biosensors together with their innovative aspects have been briefly described.

8.7 GLOSSARY

Biomarker: "A physical sign or laboratory measurement that occurs in association with a pathological process and that has putative diagnostic and/or prognostic utility" and "a characteristic that is objectively measured and evaluated as an indicator of normal biological processes, pathogenic processes, or pharmacologic responses to a therapeutic intervention."

Biosensor: A compact analytical device incorporating a biological or biologically derived sensing element either integrated within or intimately associated with a physicochemical transducer.

Electrochemical Biosensor: A biosensor with an electrochemical transducer, such as ion selective, glass, or gas electrodes for potentiometric measurements and metal or carbon electrodes for amperometric measurements.

Immunochromatography: Test based on a strip of paper coated with an immobilized antibody specific for an antigen.

Optical Biosensor: Biosensor based on planar waveguide, fiber optic, or surface plasmon resonance for the transduction.

Piezoelectric Biosensor: Biosensor based on shear and surface acoustic waves for the transduction.

Selectivity: Selectivity refers to the extent to which a method can determine particular analytes in mixtures or matrices without interferences from other components.

Sensitivity: The sensitivity is the ability to distinguish two different concentrations and is determined by the slope of the calibration curve. Often defined as *the minimum detectable concentration of an analyte*.

REFERENCES

1. Lesko LJ, Atkinson AJ. Use of biomarkers and surrogate endpoints in drug development and regulatory decisionmaking: criteria, validation, strategies. Annu Rev Pharmacol Toxicol 2001;41:347–366.
2. Frank R, Hargreaves R. Clinical biomarkers in drug discovery and development. Nat Rev 2003;2:566–580.
3. Thevenot DR, Toth K, Durst RA, Wilson GS. Electrochemical biosensors: recommended definitions and classification. Biosens Bioelectron 2001;16:121–131.
4. Newman JD, Tigwell LJ, Turner APF, Warner PJ. Biosensors: a clearer view. Cranfield University Publication; Cranfiled, UK; 2004.
5. Clark LC. Monitor and control of blood and tissue oxygen tensions. Trans Am Soc Artif Intern Organs 1956;2:41–48.

6. Dzyadevych SV, Arkhypova VN, Soldatkin A, El'skaya AV, Martelet C, Jaffrezic-Reboualt, N. Amperometric enzyme biosensors: past, present and future. ITBM-RBM 2008;29:171–180.
7. YSI Incorporated. Available at http://www.ysi.com.
8. Wang J. Electrochemical glucose biosensors. Chem Rev 2008;108:814–825.
9. Arya SK, Datta M, Malhotra BD. Recent advances in cholesterol biosensor. Biosens Bioelectron 2008;23:1083–1100.
10. Singh M, Verma N, Garg AK, Redhu N. Urea biosensors. Sens Actuators B 2008;134:345–351.
11. Lin J, Ju H. Electrochemical and chemiluminescent immunosensors for tumor markers. Biosens Bioelectron 2005;20:1461–1470.
12. Dell'Atti D, Tombelli S, Minunni M, Mascini M. Detection of clinically relevant point mutations by a novel piezoelectric biosensor. Biosens Bioelectron 2006; 21:1876–1879.
13. Wang K, Chen J, Chen J, Liu A, Li G, Luo H, Lin X, Chen Y. A sandwich-type electrochemical biosensor for detection of BCR/ABL fusion gene using locked nucleic acids on gold electrode. Electroanalysis 2009;21:1159–1166.
14. www.cancerworld.org
15. Ludwig JA, Weinstein JN. Biomarkers in cancer staging, prognosis and treatment selection. Nat Rev 2005;5:845–856.
16. Tothill IE. Biosensors for cancer makers diagnosis. Semin Cell Dev Biol 2009;20: 55–62.
17. Chen H, Jiang C, Yu C, Zhang S, Liu B, Kong J. Protein chips and nanomaterials for application in tumor marker immunoassays. Biosens Bioelectron 2009; 26:3399–3411.
18. Ding Y, Liu J, Wang H, Shen G, Yu R. A piezoelectric immunosensor for the detection of α-fetoprotein using an interface of gold/hydroxyapatite hybrid nanomaterial. Biomaterials 2007; 2147–2154.
19. Maeng JH, Lee BC, Ko YJ, Cho W, Ahn Y, Cho NG, Lee SH, Hwang SY. A novel microfluidic biosensor based on an electrical detection system for alpha-fetoprotein. Biosens Bioelectron 2008;23:1319–1325.
20. Tan F, Yan F, Ju H. Sensitive reagentless electrochemical immunosensor based on an ormosil sol-gel membrane for human chorionic gonadotropin. Biosens Bioelectron 2007;22:2945–2951.
21. Liao JY. Detection of human chorionic gonadotropin hormone using a label-free epoxysilae-modified capacitive immunosensor. Appl Microbiol Biotechnol 2007;74:1385–1391.
22. Chai R, Yuan R, Chai Y, Ou C, Cao S, Li X. Amperometric immunosensors based on layer-by-layer assembly of gold nanoparticles and methylene blue on thiourea modified glassy carbon electrode for determination of human chorionic gonadotropin. Talanta 2008;74:1330–1336.
23. Ding Y, Liu J, Jin X, Lu H, Shen G, Yu R. Poly-L-lysine/hydroxyapatite/carbon nanotube hybrid nanocomposit applied for piezoelectric immunoassay of carbohydrate antigen 19-9. Analyst 2008;133:184–190.
24. Wu J, Yan Y, Yan F, Ju H. Electric field-driven strategy for multiplexed detection of protein biomarkers using a disposable reagentless electrochemical immunosensor array. Anal Chem 2008;80:6072–6077.

25. Patil SJ, Zajac A, Zhukov T, Bhansali S. Ultrasensitive electrochemical detection of cytokeratin-7, using Au nanowires based biosensor. Sens Actuators B 2008;129:859–865.
26. Jokerst JV, Raamanatan A, Christodoulides N, Floriano PN, Pollard AA, Simmons GW, Wong J, Gage C, Furmaga WB, Redding SW, McDevitt JT. Determination of cancer biomarkers in serum and saliva using quantum dot bioconjugate labels. Biosens Bioelectron 2009;24:3622–3629.
27. He X, Yuan R, Chai Y, Shi Y. A sensitive amperometric immunosensor for carcinoembryonic antigen detection with porous nanogold film and nano-Au/chitosan composite as immobilisation matrix. J Biochem Biophys Methods 2008;70:823–829.
28. Lee Y, Lee S, Seo H, Jeon S, Moon W. Lab-free detection of a bio marker with piezoelectric micro cantilever based on mass micro balancing. Jala 2008;13:259–264.
29. Mukundan H, Kubicek JZ, Holt A, Shively JE, Marinez JS, Grace K, Grace WK, Swanson BI. Planar optical waveguide-based biosensor for the quantitative detection of tumor markers. Sens Actuators B 2009;138:453–460.
30. Choi S, Chae J. A microfluidic biosensor based on competitive protein adsorption for thyroglobulin detection. Biosens Bioelectron 2009;25:118–123.
31. Okuno J, Maehashi K, Kerman K, Takamura Y, Matsumoto K, Tamia E. Label-free immunosensor for prostate-specific antigen based on single-wallet carbon nanotube array-modified microelectrodes. Biosens Bioelectron 2007;22:2377–2381.
32. Lin YY, Wang J, Liu G, Wu H, Wai CM, Lin Y. A nanoparticle label/immunocromatographic electrochemical biosensor for rapid and sensitive detection of prostate-specific antigen. Biosens Bioelectron 2008;23:1659–1665.
33. Choi JW, Kang DY, Jang YH, Kim HH, Min J, Oh BK. Ultra-sensitive surface plasmon resonance based immunosensor for prostate-specific antigen using gold nanoparticle-antibody complex. Colloid Surf A: Physichem Eng Aspects 2008; 313–314:655–659.
34. Kim JP, Lee BY, Lee J, Hong J, Sim SJ. Enhancement of sensitivity and specificity by surface modification of carbon nanotubes in diagnosis of prostate cancer based on carbon nanotube field effect transistors. Biosens Bioelectron 2009;24:3372–3378.
35. Ning G, Lu-Yan W, Wei-Min X, Tian-Hua L, Qian-Li J. Electrochemical immunobiosensor for the rapid determination of nuclear matrix protein 22 (NMP22) antigen in urine samples by Co(III) phthlocyanine/Fe3O4/Au collide coimmobilized electrode. Chin J Anal Chem 2007;35:1553–1558.
36. Loeb S, Catalona WJ. Prostate-specific antigen in clinical practice. Cancer Lett 2007;249:30–39.
37. Benson MC, Whang IS, Pantuck A, Ring K, Kaplan SA, Olsson CA, Corner WH. Prostate specific antigen density: a means of distinguishing benign prostatic hypertrophy and prostate cancer. J Urol 1992;147:815–816.
38. Stenman UH, Leinonen J, Zhang WM, Finne P. Prostate-specific antigen. Semin Cancer Biol 1999;9:83–93.
39. Fernandez-sanchez C, McNeil CJ, Rawson K, Nilsson O. Disposable noncompetitive immunosensor for free and total prostate-specific antigen based on capacitance measurement. Anal Chem 2004;76:5649–5656.
40. Wee KW, Kan GY, Park J, Kang JY, Yoon DS, Park JH, Kim TS. Novel electrical detection of label-free disease marker proteins using piezoresistive self-sensing micro-cantilevers. Biosens Bioelectron 2005;20:1932–1938.

41. Huang L, Reekmans G, Sarens D, Friedt JM, Frederix F, Francis L, Muyldermans S, Campitelli A, Hoof CV. Prostate-specific antigen immunosensing based on mixed self-assembled monolayers, camel antibodies and colloidal gold enhanced sandwich assay. Biosens Bioelectron 2005;21:483–490.

42. Cao C, Kim JP, Kim BW, Chae H, Yoon HC, Pugniere S, Sim SJ. A strategy for sensitivity and specificity enhancements in prostate-specific antigen-alpha-1-anti-chumotrypsin detection on surface plasmon resonance. Biosens Bioelectron 2006;21:2106–2113.

43. Wu J, Yan F, Tang J, Zhai C, Ju H. Disposable multianalyte electrochemical immunosensor array for automated simultaneous determination of tumor markers. Clin Chem 2007;53:1495–1502.

44. Wu J, Fu Z, Yan F, Ju. Biomedical and clinical applications of immunoassays and immunosensors for tumor markers. Trends Anal Chem 2007;26:679–688.

45. Braunwald E. Biomarkers in heart failure. N Engl J Med 2008;358:2148–2159.

46. World Health Organization. Report of the joint International Society and Federation of Cardiology/World Health Organization task force on standardization of clinical nomenclature: Nomenclature and criteria for diagnosis of ischemic heart disease. Circulation 1979;59:607–609.

47. Katus HA, Remppis A, Looser S, Hallermeier S, Scheffold T, Kubler W. Enzyme linked immuno assay of cardiac troponin T for the detection of acute myocardial infarction in patients. Mol Cell Cardiol 1989;21:1349–1353.

48. Cummins B, Auckland ML, Cummins P. Cardiac-. specific Troponin-I radioimmunoassay in the diagnosis of. acute myocardial infarction. Am Heart J 1987; 113:1333–1344.

49. Penttila A, Koukkunen H, Kemppainen A, Halinen M, Rantanen T, Pyorala K, Penttila I. Myoglobin, creatine kinase MB, troponin T, and troponin I-rapid bedside assays in patients with acute chest pain. J Clin Lab Res 1999;29:93–101.

50. Fireman Dutra R, Mendes RK, Lins da Silva V, Kubota LT. Surface plasmon resonance immunosensor for human cardiac troponin T based on self-assembled monolayer. J Pharm Biomed Anal 2007;43:1744–1750.

51. Ko S, Kim B, Jo SS, Oh SY, Park JK. Electrochemical detection of cardiac troponin I using a microchip with the surface-functionalised poly(dimethylsioloxane) channel. Biosens Bioelectron 2007;23:51–59.

52. Cody Stringer R, Hoehn D, Grant SA. Quantum dot-based biosensor for detection of human cardiac troponin I using a liquid-core waveguide. IEEE Sens J 2008;8:295–300.

53. Rackley CE. New clinical markers predictive of cardiovascular disease: the role of inflammatory mediators. Cardiol Rev 2004;12:151–157.

54. Ridker PM, Buring JE, Shih J, Matias M, Hennekens CH. Prospective study of C-reactive protein and the risk of future cardiovascular events among apparently healthy women. Circulation 1998;98:731–733.

55. Black S, Kushner I, Samols D. C-reactive Protein. J Biol Chem 2004;47: 48487–48490.

56. Ridker PM. C-reactive protein: a simple test to help predict risk of heart attack and stroke. Circulation 2003;108:81–85.

57. Dominici R, Luraschi P, Franzini C. Measurement of C-reactive protein: two high sensitivity methods compared. J Clin Lab Anal 2004;18:280–284.

58. Christodoulides N, Mohanty S, Miller CS, Langub MC, Floriano PN, Dharshan P, Ali MF, Bernard B, Romanovicz D, Anslyn E, Fox PC, McDevitt JT. Application of microchip assay system for the measurement of C-reactive protein in human saliva. Lab Chip 2005;5:261–269.

59. Meyer MHF, Hartmann M, Krause HJ, Blankenstein G, Mueller-Chrus B, Oster J, Miethe P, Keusgen M. CRP determination based on a novel magnetic biosensor. Biosens Bioelectron 2007;22:973–979.

60. Trevino J, Calle A, Rodriguez-Frade JM, Mellado M, Lechuga LM. Determination of human growth hormone in human serum samples. Talanta 2009;78:1011–1016.

61. Trevino J, Calle A, Rodriguez-Frade JM, Mellado M, Lechuga LM. Surface plasmon resonance immunoassay analysis of pituitary hormones in urine and serum samples. Clin Chim Acta 2009;403:56–62.

62. Dittmer WU, de Kievit P, Prins MWJ, Vissers JLM, Mersch MEC, Martens MFWC. Sensitive and rapid immunoassay for parathyroid hormone using magnetic particle labels and magnetic actuation. J Imm Methods 2008;338:40–46.

63. Yuan J, Oliver R, Li J, Lee J, Aguilar M, Wu Y. Sensitivity enhancement of SPR assay of progesterone based on mixed self-assembled monolayers using nanogold particles. Biosens Bioelectron 2007;23:144–148.

64. Stevens RC, Soelber SD, Near S, Furono CE. Detection of cortisol in saliva with a flow-filtered portable surface plasmon resonance biosensor system. Anal Chem 2008;80:6747–6751.

65. Lim TK, Matsunaga T. Construction of electrochemical flow immunoassay system using capillary columns and ferrocene conjugated immunoglobulin G for detection of human chorionic gonadotropin. Biosens Bioelectron 2001;16:1063–1069.

66. Tan W, Sabet L, Li Y, Yu T, Klokkevold PR, Wong DT, Ho CM. Optical protein sensor for detecting markers in saliva. Biosens Bioelectron 2008;24:266–271.

67. De Boer BM, Kahlman JAHM, Jansen TPGH, Duric H, Veen J. An integrated and sensitive detection platform for magneto-resistive biosensors. Biosens Bioelectron 2007;9–10:2366–2370.

68. Grossman CJ. Regulation of the immune system by sex steroids. Endoc Rev 1984;5:435–455.

69. Reinsberg J, Jost E, Van Der Ven H. Performance of the fully automated progesterone assays on the Abbott. Clin Biochem 1997;30:469–471.

70. Basu A, Shrivastav TG, Maitra SK. A direct antigen heterologous enzyme immunoassay for measuring progesterone in serum without using displacer. Steroids 2006; 71:222–230.

71. Xu YF, Velasco-Garcia M, Mottram TT. Quantitative analysis of the response of an electrochemical biosensor for progesterone in milk. Biosens Bioelectron 2005;20:2061–2070.

72. Hart JP, Pemberton RM, Luxton R, Wedge R. Studies towards a disposable screen-printed amperometric biosensor for progesterone. Biosens Bioelectron 1997; 12:1113–1121.

73. Gillis EH, Gosling JP, Sreenan JR, Kane M. Development and validation of a biosensor-based immunoassay for progesterone in bovine milk. J Immunol Methods 2002;267:131–138.

74. Minunni M, Tombelli S, Centi S, Mascini M. Measurement of trace medical analytes. Volume 2, The handbook of biosensors and biochips. UK: John Wiley & Sons; 2007. pp. 1055–1067.

75. Mitchell JS, Lowe TE, Ingram JR. Rapid ultrasensitive measurement of salivary cortisol using nano-linker chemistry coupled with surface plasmon resonance detection. Analyst 2009;134:380–386.

76. Kataoka H, Matsuura E, Mitani K. Determination of cortisol in human saliva by automated in-tube solid-phase microextraction coupled with liquid chromatography—mass spectrometry. J Pharm Biomed Anal 2007;44:160–165.

77. Hegnerová K, Bocková M, Vaisocherová H, Krištofiková Z, Øíèný J, Øípová D, Homola J. Surface plasmon resonance biosensors for detection of Alzheimer disease biomarker. Sens Actuators B 2009;139:69–73.

78. Thorslund S, Larsson R, Nikolajeff F, Bergquist J, Sanchez J. Bioactivated PDMS microchannel evaluated as sensor for human CD4+ cells—The concept of a point-of-care method for HIV monitoring. Sens Actuators B 2007;123:847–855.

79. Nagel T, Ehrentreich-Förster E, Singh M, Schmitt K, Brandenburg A, Berka A, Bier FF. Direct detection of tuberculosis infection in blood serum using three optical label-free approaches. Sens Actuators B 2008;129:934–940.

80. Garg SK, Tiwari RP, Tiwari D, Singh R, Malhotra D, Ramnani VK, Prasad GBKS, Chandra R, Fraziano M, Colizzi V, Bisen PS. Diagnosis of tuberculosis: available technologies, limitations, and possibilities. J Clin Lab Anal 2003;17:155–163.

81. Palchetti I, Mascini M. Nucleic acid biosensors for environmental pollution monitoring. Analyst 2008;133:846–854.

82. Lucarelli F, Tombelli S, Minunni M, Marrazza G, Mascini M. Electrochemical and piezoelectric DNA biosensors for hybridisation detection. Anal Chim Acta 2008;609:139–159.

83. Ding C, Zhao C, Zhang M, Zhang S. Hybridization biosensor using 2,9-dimethyl-1,10-phenantroline cobalt as electrochemical indicator for detection of hepatitis B virus DNA. Bioelectrochemistry 2008;72:28–33.

84. Yao C, Zhu T, Tang J, Wu R, Chen Q, Chen M, Zhang B, Huang J, Fum W. Hybridization assay of hepatitis B virus by QCM peptide nucleic acid biosensor. Biosens Bioelectron 2008;23:879–885.

85. Xu L, Yu H, Akhras MS, Han SJ, Osterfeld S, White RL, Pourmand N, Wang SX. Giant magnetoresistive biochip for DNA detection and HPV genotyping. Biosens Bioelectron 2008;24:99–103.

86. Dell'Atti D, Zavaglia M, Tombelli S, Bertacca G, Cavazzana AO, Bevilacqua G, Minunni M, Mascini M. Development of combined DNA-based piezoelectric biosensors for the simultaneous detection and genotyping of high risk Human Papilloma Virus strains. Clin Chim Acta 2007;383:140–146.

87. Prabhakar N, Arora K, Arya SK, Solanki PR, Iwamoto M, Singh H, Malhotra BD. Nucleic acid sensor for M. tuberculosis detection based on surface plasmon resonance. Analyst 2008;133:1587–1592.

88. Kara P, Cavusoglu C, Cavdara S, Ozsoz M. Direct electrochemical genosensing for multiple point mutation detection of Mycobacterium tuberculosis during the development of rifampin resistance. Biosens Bioelectron 2009;24:1796–1800.

89. Chen J, Zhang J, Huang L, Lin X, Chen G. Hybridization biosensor using 2-nitroacridone as electrochemical indicator for detection of short DNA species of Chronic Myelogenous Leukemia. Biosens Bioelectron 2008;24:349–355.
90. Castaneda MT, Merkoci A, Numera M, Alegret S. Electrochemical genosensors for biomedical applications based on gold nanoparticles. Biosens Bioelectron 2007;22:1961–1967.
91. Liepold P, Kratzmüller T, Persike N, Bandilla M, Hinz M, Wieder H, Hillebrandt H, Ferrer E, Hartwich G. Electrically detected displacement assay (EDDA): a practical approach to nucleic acid testing in clinical or medical diagnosis. Anal Bioanal Chem 2008;391:1759–1772.

PART III

REGULATORY PERSPECTIVES

9

REGULATORY PERSPECTIVES ON BIOMARKER DEVELOPMENT

RAJANIKANTH MADABUSHI, LAWRENCE LESKO, AND JANET WOODCOCK

9.1 INTRODUCTION TO FOOD AND DRUG ADMINISTRATION'S CRITICAL PATH INITIATIVE AS IT PERTAINS TO BIOMARKERS

With the Critical Path Initiative, the Food and Drug Administration (FDA) has taken the lead not only to identify and prioritize the most pressing drug development problems of today but also to be part of the current and future solutions (1). "The Critical Path" refers to the steps needed to move a candidate drug through preclinical and clinical development to regulatory approval. Three crucial scientific/technical dimensions must be successfully negotiated on the journey to a new medicinal product. They are (i) ensuring product safety, (ii) demonstrating medical utility, and (iii) industrialization (manufacturing). The Critical Path Initiative calls for utilization of recent advances in scientific knowledge and technology to generate new tools that forge more efficient and informed pathways in all three dimensions. The greatest challenge lies in predicting a potential product's performance as early as possible with the greatest degree of certainty. The FDA Critical Path Opportunities Report of 2006 identified target research areas that are likely to increase efficiency, predictability, and productivity in the development of new medicinal products. The first issue identified on the Critical Path Opportunities List is as follows: "The process and criteria for qualifying biomarkers should be mapped. Clarity on the conceptual framework and evidentiary standards for qualifying biomarkers for various purposes would establish the path for developing predictive biomarkers" (2).

Predictive Approaches in Drug Discovery and Development: Biomarkers and In Vitro/In Vivo Correlations, First Edition. Edited by J. Andrew Williams, Jeffrey R. Koup, Richard Lalonde, and David D. Christ.

Biomarkers are measurable characteristics that reflect physiological, pharmacological, or disease processes in animals or humans. As the definition suggests, there are different types of biomarkers including biochemical, physiological, anatomical, histological, and physical markers. Different types of biomarkers are used at each stage of drug development. Biochemical markers are frequently employed in the basic research of preclinical receptor pharmacology leading to target identification and are also used in drug discovery and lead optimization. These early phases are well integrated with modern scientific advances, and are not the subject of the Critical Path Initiative. However, in preclinical development, biomarkers are the link between *in vitro* and *in vivo* experimentation related to drug effects and this relationship is used to predict the safety of the product in humans. In clinical development, biomarkers may be used for showing proof-of-mechanism, setting dose and regimen selection based on effectiveness and/or safety, and to select or deselect patients for clinical trials. Rarely when a biomarker has been established as a surrogate endpoint, it is used to establish evidence of clinical effectiveness. Finally, in clinical practice, biomarkers provide potential diagnostic, predictive, and prognostic value. The role of biomarkers through the critical path for development of new medicinal products can be visualized in Figure 9.1.

Given the tremendous impact of biomarkers on drug discovery and development, FDA's Critical Path Opportunities Report of 2006 goes on further to state that a new generation of predictive biomarkers could dramatically improve the efficiency of product development, help identify safety problems before a product

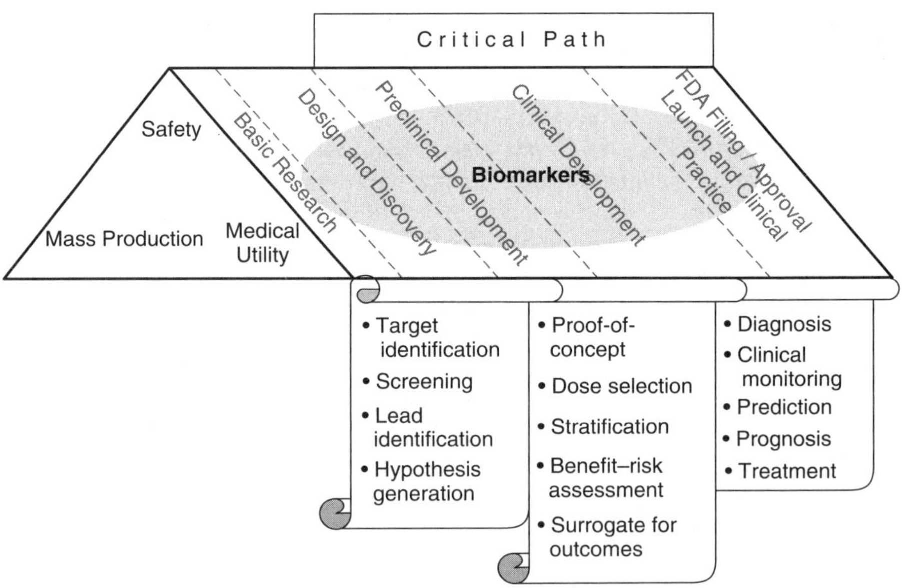

FIGURE 9.1 Impact of biomarkers on the critical path for the development of new medicinal products.

is on the market (and even potentially predict safety problems before it is tested in humans), and facilitate the development of new types of clinical trial designs that will produce better and more informative data faster (2).

9.2 CURRENT UTILITY OF BIOMARKERS IN DRUG DEVELOPMENT

Biomarkers derived from new technologies are rapidly incorporated into drug discovery programs. In contrast, most biomarkers used in preclinical and clinical development (i.e., drug development programs) are traditional in nature, although this is rapidly changing. The biomarkers used most commonly and traditionally in the clinic (leaving ordinary safety tests aside) are bioanalytical tests for drug exposure. The extent of regulatory involvement in the use of a biomarker will depend on the implications of its use in the development program. Typically, biomarkers used to evaluate safety, or those used later in the development program in decision making about patients, have the highest level of regulatory scrutiny.

The role of biomarkers in various stages of drug development is summarized below:

1. *Early Decision Making.* Early in development, biomarkers find utility in establishing the mechanism of action of the drug or defining the targets on which the new molecules exert their effect, for example, inhibition of ADP-dependent platelet aggregation for thienopyridines or the inhibition of angiotensin converting enzyme (ACE) and changes in the levels of angiotensin-I and angiotensin-II, which are used for demonstrating the mechanism of action for ACE inhibitors.
2. *Bridging Animal-to-Human Safety Findings.* Traditional markers of organ system injury are used extensively in toxicological studies and in human trials to evaluate drug safety. The lack of sensitive, well-understood biomarkers to signal early, reversible stages of specific drug-induced toxicities, especially in humans, is a major problem in drug development.
3. *Evaluation of Mechanism.* Target binding studies or assays of specific pharmacodynamic (PD) actions are used to confirm desired activity (i.e., proof-of-concept) and bridge animal and human results.
4. *Dose and Regimen Finding.* Markers of PD effects that are relevant to clinical endpoints may be used to establish doses and regimens in phase 2 exposure–response and clinical efficacy studies. Drug concentration in plasma itself is an important biomarker that is used extensively throughout the clinical development process.
5. *Markers of Drug Clearance.* Biomarkers of drug transport and metabolism may be used to explore the basis for intrasubject variability in drug exposure or response. Markers of diminished organ function (e.g., renal, hepatic) are used to assess the need for dose modifications in patients with organ compromise.

6. *Selection of Patient Population.* Various biomarkers have long been used to define specific patient populations for entry into clinical trials. Diagnostic biomarkers are used to establish the presence of a defined disease or disease subset. Likewise, prognostic biomarkers are used to select subgroups within a disease population with a smaller range of expected outcomes, or who may be thought to respond better to the experimental intervention. Recently, targeted therapies, aimed at the subset of patients exhibiting a specific target, have become more common as in the area of oncology. The use of new biomarkers to select patients to more likely have a responsive disease, or to deselect those with presumed less-responsive disease, will become more common as clinical trials become more "enriched" with likely responders.
7. *Assessing Response to Therapy.* Biomarkers are often used to determine the degree of therapeutic response (e.g., imaging in cancer, assessment of microbiologic cure in infectious disease) and to titrate treatment (e.g., diabetes).
8. *Surrogate Endpoint.* Biomarkers that are proximal to a clinical endpoint but occur earlier in time allow for more rapid assessment of efficacy. Sometimes biomarkers that have been demonstrated to correlate with the clinical outcome are used to substitute for the clinical response in efficacy trials, for example, blood pressure as a substitute for myocardial infarction/stroke-early mortality or LDL (low density lipoprotein) cholesterol lowering as a surrogate for myocardial infarction and early mortality.
9. *Use of Biomarkers in the Drug Label.* Perusal of almost any prescription drug label will reveal a plethora of instructions for use related to biomarkers, including diagnosis, dosing, monitoring of therapy, and safety evaluation. Most of these markers are clinically well established; however, recently, newer and qualified genomic-based markers are beginning to appear in drug labeling.

9.3 THE PRIMARY UTILITY OF BIOMARKERS IN DRUG DEVELOPMENT IS NOT AS SURROGATE ENDPOINTS

A biomarker that is intended to substitute for a clinical endpoint in a new drug registration trial is known as a *surrogate endpoint*. A surrogate endpoint is expected to predict clinical outcome (benefit or lack of benefit or harm) based on mechanistic, epidemiologic, therapeutic, pathophysiologic, or other scientific evidence (3). All surrogate endpoints can be considered derived from biomarkers; however, the universe of biomarkers is much broader than those that might be considered for surrogacy (4). Further, it is also advocated by many that, to become a surrogate, it needs to be correlated with outcome in clinical trials of more than one drug with the same mechanism of action targeted for the same indication (5, 6). Therefore, the path for making the leap from biomarker to surrogate is not easy. The relationship between the biomarker, pathophysiology of the disease,

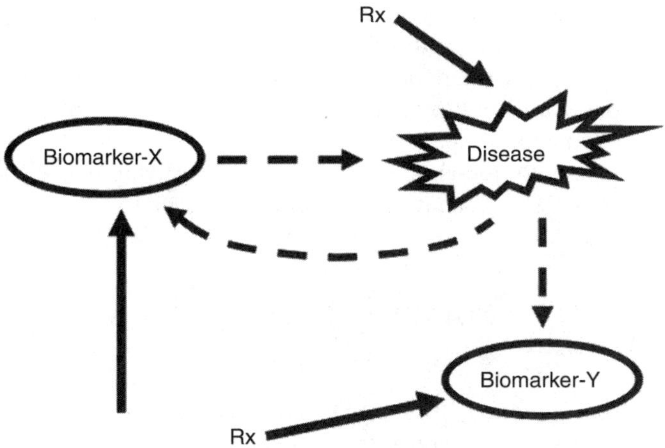

FIGURE 9.2 Interplay of the relationships between disease, biomarker, and pharmacology of the drug.

the mechanism of action of the therapy, and the clinical outcome is complicated. Figure 9.2 shows one example of a simplified schematic of the interplay between the above mentioned components.

An examination of the biomarker-surrogate literature raises a number of serious limitations and concerns including the following:

1. The biomarker may be correlated with progression of the disease but might not be involved with the causal pathway of the disease process. A treatment might show changes in a relevant biomarker; however, these changes may or may not lead to a change in the clinical outcome. This raises the risk of false positives. As a corollary, false negatives may also be possible if there are no biomarker changes, but a treatment does induce a change in clinical outcome (see below).
2. There could be several causal pathways for the disease, and a treatment could be affecting a biomarker of a single pathophysiologic process that might not be clinically important.
3. The biomarker might not be in the pathway of the effect of the treatment on the disease or the biomarker might be insensitive to the treatment effect, that is, produce "false negatives." Either case will falsely lead to the conclusion of lack of treatment effect based solely on the lack of biomarker change.
4. The treatment might have multiple mechanisms of action and the biomarker may reflect only one of the pathways. This could lead to any of the above mentioned conclusions depending on the sensitivity of the change in biomarker and the clinical importance of the pathway. Furthermore, the treatment may exert a mechanism that leads to an adverse event and may shift the perspective of risk/benefit when using a biomarker as opposed to a clinical endpoint.

Biomarker literature extensively explores the utility of biomarkers as surrogates for clinical outcomes. However, this represents a very narrow appreciation of the broad potential for biomarkers in improving drug development. Surrogacy is only one of the many applications of biomarkers as presented in Section 9.3.2. Systematic development of appropriate biomarkers throughout the discovery and development cycle hold tremendous potential for improving the science and predictability of drug development as providing safer, more effective high value therapeutics to the public.

9.4 EXAMPLES OF BIOMARKER USES IN REGULATORY DECISION MAKING

For a drug to be approved by the FDA, it must first demonstrate substantial evidence of effectiveness derived from adequate and well-controlled clinical investigations (7). However, the regulations do not specify whether the evidence of effectiveness must come from an evaluation of actual clinical benefit or from an evaluation of a clinically meaningful surrogate endpoint, which substitutes for such a clinical benefit.

The FDA has a legal basis for using surrogate endpoints in ordinary and accelerated drug approvals leading to market access of new drugs or drug products. Subpart H of the Code of Federal Regulations (21 CFR 314 subpart H) and subpart E of biological license application (BLA) regulations (21 CFR 601 subpart E) allow accelerated approval of certain new drugs and biological products based on biomarkers. This is best exemplified by the proposal that accelerated approval for antiretroviral drugs could be on studies that show a drug's contribution toward shorter-term reductions in HIV (human immunodeficiency virus) RNA (e.g., 24 weeks), while traditional approvals could be based on trials that show a drug's contribution toward durability of HIV RNA suppression (e.g., for at least 48 weeks) (8). Another example is the approval of imatinib (Gleevec®) for the treatment of chronic myelogenous leukemia based on cytogenetic and hematologic response (9).

Further, the FDA Modernization Act of 1997 states that confirmatory evidence, when combined with evidence from one adequate and well-controlled study, can support effectiveness as required for ordinary drug approvals. The quantity of evidence needed to support effectiveness, other than two adequate and well-controlled clinical trials, is discussed in Section II of the FDA Guidance for Industry, entitled "Providing Clinical Evidence of Effectiveness for Human Drug and Biological Products" (10). This guidance states that one adequate and well-controlled clinical efficacy study can sometimes be supported by evidence from a well-controlled study or studies using a pharmacologic effect, as a biomarker, that is not an established surrogate endpoint. Acceptance of this evidence of efficacy is based on (i) the quantity of evidence showing that there is a strong theoretical or mechanistic link between the pharmacologic effect and clinical outcome; and (ii) the quantity of data showing that there is a strong link between the pharmacologic effect and clinical outcome based on prior experience with the pharmacological class, and a clear understanding of the pathophysiology and mechanism of drug action. Demonstration of dose response also provides confirmatory evidence.

Aside from the use of surrogate endpoints in adequate and well-controlled clinical trials to support the effectiveness necessary for market access, biomarkers that do not meet standards for becoming surrogate endpoints have other value in regulatory decision making and may be used in analyses, for example, drug–drug interactions or subgroup analysis, that complement the results of adequate and well-controlled efficacy trials.

The FDA Guidance for Industry, entitled "Providing Clinical Evidence of Effectiveness for Human Drug and Biological Products," provides an important perspective on the potential usefulness of biomarkers, PK–PD relationships and surrogate endpoints. Some examples of these uses are provided below:

1. The most commonly used biomarkers are plasma drug and/or metabolite concentrations. They act as surrogate endpoints for efficacy and safety in the approval of generic drug products. Further, for products intended for local action such as inhalation of aerosols and measurable blood levels are neither observed nor desired. Absence of blood levels is taken as a measure of safety. For such locally acting drug products, endpoints based on PD biomarkers provide an appealing alternative for bioequivalence testing because of their greater sensitivity to detect formulation differences. For example, forced expiratory volume (FEV) or histamine measurements have been the basis of bioequivalence testing of new bronchodilator formulations to be used for treating asthma (11).
2. Biomarkers, beyond those of plasma drug concentrations, that are more distal in the causal chain leading to the clinical outcome and were investigated in early clinical trials may be suitable for assessing the clinical significance of changes in systemic drug exposure due to intrinsic and extrinsic patient factors such as age, gender, genetics, smoking habit, degree of renal impairment, and drug–drug interactions. Several clinical pharmacology regulatory guidance recommended that sponsors define therapeutic equivalence limits using PK–PD relationships for studies of drug–drug interactions, and the effects of renal or hepatic impairment to determine label claims and the need to adjust doses (12, 13). For example, HMG-CoA reductase inhibition and bleeding times may be used as biomarkers to assess drug–drug interactions with cholesterol-lowering statins and anticoagulants, respectively, rather than to rely only on changes in drug exposure.
3. Biomarkers are useful in providing adequate evidence to bridge from a preexisting database of efficacy to support an efficacy decision in new situations or settings, such as in bridging adult data to pediatrics or efficacy data between different regions of the world. Thus, under certain circumstances, regulations allow the use of well-established exposure–response knowledge from one population for the approval in another. For example, D,L-sotalol hydrochloride is approved to treat life-threatening ventricular fibrillation and tachycardia, and for maintenance of sinus rhythm in patients with symptomatic atrial fibrillation and flutter in adults.

A clinical study assessing the antiarrhythmic and β-blocking effects of sotalol on QTc and heart rate in pediatrics ranging from neonates to 12-year-old children formed the basis of approval for sotalol's use in pediatric patients. This study led to judicious dosing recommendations for pediatric patients based on extensive biomarker data (14). A similar example is the determination of dosing recommendation for argatroban injection in pediatric patients with heparin-induced thrombocytopenia (HIT). Argatroban is approved for prophylaxis or treatment of thrombosis in adult patients with HIT, as well as in adult patients at risk for, or with overt, HIT undergoing percutaneous coronary intervention (PCI). Exposure–response analysis of the plasma concentration and the activated plasma thromboplastin time (aPTT, biomarker) derived from an open-label trial in pediatric patients (from birth to 16 years) and healthy adults formed the basis for deriving the dose and dosing regimen in pediatric patients (15, 16).

4. Further, when certain conditions are met, different dosage forms (e.g., a controlled-release product for an established immediate-release (IR) product) can be approved based on biomarker data. For example, metoprolol IR was first approved for treating angina and hypertension in adults. Later, once-a-day metoprolol extended release (XL) formulation was developed by the sponsor for convenience and compared to the IR formulation administered twice daily. Typically, new molecular entities require two clinical trials in target population for sufficient length of time to obtain market access. However, given the prior experience with metoprolol IR formulation, a single trial conducted in healthy volunteers to compare the time courses of β-1 blockade measured as reduction of exercise induced tachycardia and the concentration β-1 blockade relationship after IR and XL doses formed the basis of metoprolol XL approval (17).

9.5 IMPORTANCE OF TECHNOLOGY, BIOANALYTICAL VALIDATION, AND CLINICAL QUALIFICATION AND THE ASSOCIATED CHALLENGES IN DEVELOPMENT OF NEW BIOMARKERS FOR USE IN DRUG DEVELOPMENT

Before the genomic era, individual biomarkers were associated with disease endpoints (such as cardiovascular disease, CVD) on the basis of anticipated biological pathway (e.g., inflammation) and ease of measurement (e.g., C-reactive protein, fibrinogen, D-dimer). Many of these biomarkers were evaluated individually in large-cohort studies to determine their incremental predictive ability over, for example, that of the Framingham Risk Score (18–20). As a result of the human genome project, the availability of technological and biological resources has introduced potentially new paradigms in drug discovery and development. Major advances in the basic science of drug discovery, and more recently the science of drug development, have led to an enormous increase in the number of new drug targets and novel therapies such as genetic-, genomic-, and protein-based targets to treat human disease (21, 22). Simultaneously, the rapid advancements in the field of *in vivo* imaging such as positive emission tomography (PET) and methods of tissue collection have created an unprecedented opportunity to discover and develop biomarkers. The new *-omic* (genomics, proteomics, and metabolomics) technologies and microarray-based methods hold vast potential to identify more predictive biomarkers, alone or as complements to existing domains of biomarkers. These technologies offer the ability to examine changes in a great many molecules at once in single experiments involving perturbation of biological systems at the molecular level. Currently, biomarker assays span diverse methodologies, ranging from the relatively low risk technology end such as immunohistochemistry (IHC), through electrophoresis, reverse transcription PCR to ELISA, to the higher-risk technology end, including platforms for genomics (gene chip), proteomics (SELDI TOF, surface enhanced laser desorption/ionization time-of-flight mass spectrometry), and multiplex ligand-binding assays (23–26). The advent of new technology, while resulting in new and more biomarkers to aid drug discovery and development, also present with new challenges for what constitutes validation and qualification for an intended use.

Validation is a free standing term, which means many things to many people. As one component, analytical validation is a fairly well-understood term. Bioanalytical method validation involves a systematic evaluation of all the processes required to demonstrate that a particular technique is reliable for its intended purpose (27). This type of validation is common in drug development. It involves assessing the assay and its measurement performance characteristics and determination of the range of conditions under which the assay will give reproducible and accurate data (28). This underlying spirit of the bioanalytical validation ensures one to make reasonably accurate and precise interpretation of the data generated on the analyte. However, it should be noted that this definition of bioanalytical method validation was developed predominantly for small molecules that are exogenous and are intended to be used in toxicology and pharmacokinetic (PK) studies. Although biomarker laboratory analyses can have many similarities to those used in toxicology and ADME (absorption, distribution, metabolism, and excretion) studies, the variety of novel biomarkers and the nature of their applications often preclude the use of previously established bioanalytical validation guidelines in biomarker research. Biomarkers are often endogenous macromolecular species. Detection techniques for macromolecules are vastly different compared to small molecules and hence the challenges associated with method validation are also different. Various issues associated with analytic validation of biomarkers include the following: equipment evaluation, procedure standardization, quality assurance protocols, reproducibility and replication, interoperator variability, the production and measurement of reference standards (which, in the case of imaging, would include the use of standardized “phantoms” and reference imaging sets), sensitivity to change after interventions or disease progression, and intersubject biological variance (29). Analytic validation of imaging biomarkers requires special attention to the choice of image reconstruction algorithm, image segmentation algorithm (to isolate classes of lung structures—airways, vessels, parenchyma, etc.—from one another), automated versus observer-dependent image analysis, and, specifically for the lungs, data acquisition during conditions of standardized lung volume (30). Table 9.1 compares and contrasts typical validation paradigms for small molecules and biomarkers.

Many new biomarker assays will most closely resemble, or actually be intended as, *in vitro* diagnostic tests (IVDs). The Office of *In Vitro* Diagnostics, in FDA's Center for Devices and Radiologic Health, has published extensive guidance on the development and analytical validation of IVDs, including guidance on genetic test development and "*In Vitro* Diagnostic Multiplexed Indexed Assays" (IVDMIAs), which are generally biomarker assays utilizing multiple analyte that include many different types of genetic tests (31).

A useful starting point in determining the process of a biomarker assay validation is to consider the assay to be used and the type of data that the assay will generate, as well as the intended purpose of the data for decision making. On the basis of the type of biomarker data generated, biomarker assays can be categorized into following types:

Definitive Quantitative Assays. In these assays, calibration is performed using a reference standard that is well defined and is representative of the endogenous biomarker. Bioanalytical results are expressed in continuous numeric units of definitive reference standards (e.g., human insulin or steroid assay). Such assays represent ideal situation for biomarker measurement.

Relative Quantitative Assays. These assays depend on a response–concentration calibration function. The reference standard is either not fully characterized or not available in purified form, or is not fully representative of the endogenous biomarker. In such assays, precision performance can be validated but accuracy can only be estimated (e.g., cytokine immunoassays). For such assays, more emphasis should be placed on the temporal changes in biomarker concentrations rather than absolute changes.

Quasi-Quantitative Assays. These assays do not use a reference standard or calibration curve. However, the analytical response is continuous and the result is expressed in terms of characteristic of the test sample (e.g., antidrug antibody assays, enzymatic assays, and flow cytometric assays). These assays can only demonstrate precision but not accuracy.

Qualitative Assays. These assays generate categorical data that lack proportionality to the amount of analyte in the sample. Such data may be nominal (e.g., presence or absence of a single nucleotide polymorphism or gene mutation in a sample of DNA) or ordinal (e.g., immunohistochemical assays).

Clinical qualification represents a continuous, iterative, fit-for-purpose evidentiary process of linking a biomarker with biological processes of interest and/or clinical end points in the range of patients (e.g., age, sex, ethnicity) in whom it is intended to benefit or reduce risk (32). The spirit of this process is to

TABLE 9.1 Comparison of Drug, Biomarker, and Diagnostic Assays[a]

Assay		Biomarker Assay for Drug Development	Biomarker Assay for Diagnostic
Intended use	Bioequivalence, PK	Safety, mechanism of action, PD	Distinguish diseased from healthy
Method category	Most assays are definitive quantitative	Most assays are relative or quasiquantitative	—
Nature of analyte	Exogenous in most cases	Endogenous	—
Calibrators/ standards	Well characterized. Standards prepared in study matrix	Typically not well characterized, may change from vendor to vendor, lot to lot. Standards/calibrators are made in matrix different than study samples	—
Validation samples (VS) and quality control (QC)	Made in study matrix. 4–5 VS levels and 3 QC levels	Made in study matrix. 5 VS levels and 3 QC levels. If study matrix is limited (e.g., tissue samples) may use surrogate matrix	QC often in lyophilized form, supplied by the vendors, commonly 2 or 3 levels
Assay sensitivity	LLOQ defined by acceptance criteria	LLOQ and LOD	LOD is often used
Validation of accuracy	True accuracy can be achieved by testing spike recovery	In majority of cases, only relative accuracy can be achieved. Endogenous background needs to be considered if spike recovery is used	Measured result compared to an accepted reference value obtained by an accepted method
Validation of precision	2–6 replicate sample per run, 3–6 runs	2–6 replicate samples per run, 3–6 runs	3 replicate samples per run, one run per day for 5 days. Samples ran in random order
Stability testing	Freeze/thaw, bench top, and long-term measured by spiking biological matrix with drug	Freeze/thaw, bench top, and storage stability with study samples, when available. If not, with spiked samples	Focus on stability of reagents rather than analyte. Long-term analyte stability not routinely tested
Assay acceptance criteria	4–6–20/30 rule	Establish confidence interval or 4–6–X rule	2 SD ranges, Westgard rules, Levy–Jennings chart
Regulatory requirements	GLP compliant	No specific guidelines	Methods are FDA approved, result generation follows CLIA and CLSI guidelines in United States

[a]Ref. 28.

establish the utility of the biomarker for its intended purpose. For example, FDA's exposure–response guidance classifies biomarker types as following, within the specific context of decision making based on exposure–response (33):

- Biomarkers that are more remote from the clinical benefit endpoint but provide a description of some of the major pathways of disease progression or drug action (e.g., degree of binding to a receptor or inhibition of an agonist).
- Biomarkers reflecting drug action but having incomplete or uncertain relationship to clinical outcome (e.g., inhibition of ADP-dependent platelet aggregation, ACE inhibition).
- Biomarkers thought to reflect the pathologic process of disease response to treatment and be at least candidate surrogates (e.g., brain appearance in Alzheimer's disease, brain infarct size, various radiographic/isotopic function tests).
- Biomarkers thought to be valid surrogates for clinical benefit (e.g., blood pressure, cholesterol, viral load).

Many of the biomarkers used in medical product development today have been in use for many years, even decades. These longstanding biomarkers were empirically derived for the most part. The validity of preclinical and clinical biomarkers has been traditionally settled by debate, consensus, and the passage of time. Owing to the lack of a systematic process with drug development for evaluation and qualification, new biomarker development has stalled. The acceptance of biomarker utility is limited by uncertainty on several levels. In many ways, biomarker acceptance is a benefit/risk assessment, for example, the benefit of a true result as compared to the harm of a false result. One of the biggest sources of uncertainty, and confusion, is with respect to the sensitivity and selectivity of the biomarker. This uncertainty generally leads to an inaccurate definition of the context for which the biomarker should be qualified. The next level of uncertainty is the difficulty in establishing the biomarker context. As an example, FDA's guidance on pharmacogenomic data submissions classifies biomarkers as exploratory, probable valid, or known valid and provides some insights into validation or qualification of biomarkers (34). A known valid biomarker is defined as "a biomarker that is measured in an analytical test system with well-established performance characteristics and for which there is widespread agreement in the medical or scientific community about the physiological, toxicological, pharmacological, or clinical significance of the results. A probable valid biomarker is defined as "a biomarker that is measured in an analytical test system with well-established performance characteristics and for which there is a scientific framework or body of evidence that appears to elucidate the physiological, toxicological, pharmacological, or clinical significance of the test results." The process by which an exploratory biomarker can be qualified as a valid biomarker is not clearly established, and is often case by case, although it is a priority for FDA. Along these lines, the FDA has established a biomarker qualification process, and the qualification of seven preclinical nephrotoxic biomarkers was the first success (see below). This pilot structure has been designed around the model of the Interdisciplinary Pharmacogenomic Review Group (IPRG) to allow contributions of expertise from different FDA Centers, such as the Center for Drug Evaluation and Research (CDER), the Center for Biological Evaluation and Research, the Center for Devices and Radiological Health, and the National Center for Toxicological Research, as well as across clinical divisions and from nonclinical toxicology reviewers in CDER. The main responsibility of this review team is assessment of qualification for novel biomarkers, for example, drug safety, using appropriate preclinical, clinical, and statistical considerations.

Given the challenges in qualifying biomarkers, FDA has advocated pursuing their development independent of a specific drug development program. In particular, consortia among industry, academia, and regulators seem well suited to carry out some of these works. It is clear that one sector alone lacks the time, resources, and motivation to independently develop specific biomarkers. Several important consortia have been formed to pursue specific biomarkers. As mentioned earlier, one outstanding example is the "Predictive Safety Testing Consortium" (PSTC) formed under the auspices of the nonprofit C-Path Institute, in Tucson, AZ (35), comprised of 16 pharmaceutical company partners, with FDA European Medicines Agency (EMEA) participation, the consortium has targeted specific drug-induced organ toxicities such as drug-induced renal injury or vasculitis, for improved safety biomarker development. The consortium advanced a set of seven drug-induced renal injury biomarkers to the point of submission to FDA and for acceptance in animal toxicology studies. These biomarkers were accepted as qualified for this use in 2008 (36). The consortium plans to continue clinical qualification of these renal injury biomarkers, and to pursue biomarkers in other organ systems, with the goal of improving the animal-to-human safety transitions for investigational compounds.

Another consortium doing important qualification work is "The Biomarker Consortium" at the Foundation for NIH (FNIH). A partnership among NIH, FDA, the pharmaceutical and biopharmaceutical industries, medical device companies, academia, and many patient groups, The Biomarker Consortium is engaged in both development and qualification of biomarkers in a wide range of diseases. The biomarkers are primarily related to disease processes and treatment response. The participation of the NIH institutes greatly enhances the power of this partnership. As The Biomarker Consortium is attempting some fairly ambitious projects, it is expected that several years of work will be required before significant results are attained. Another consortium addressing biomarkers in drug development is the "Serious Adverse Events" (SAE) Consortium. This nonprofit group has multiple pharmaceutical industry members, with FDA liaison membership and academic and nonprofit participation. This initiative is using genomic technology to assess potential genetic risk factors for serious, drug-induced toxicities such as Stevens–Johnson syndrome and drug-induced liver injury across drugs and drug classes. These toxicities, when rare, are currently difficult or impossible to predict during a drug development program but are a frequent cause of later market withdrawal. Genetic associations, when present, may lead to better understanding of these toxicities, and to biomarkers that identify people at risk. In the long term, it is envisioned that drugs might be tested with a chip that contains many qualified gene variants that predispose individuals to adverse events to make predictive screening more feasible and cost effective.

In instances where a biomarker must be developed in the context of a specific drug development program, the scientific approach for developing the qualification process must be planned as early as possible in the drug discovery and preclinical period of drug development with a blueprint to bring the biomarker into clinical trials and to establish the link between the biomarker and the specific clinical parameter it represents (e.g., patient population response to therapy). FDA is developing guidance for codevelopment of an investigational drug and diagnostic pair, in development programs where diagnostic-directed therapy is contemplated. A concept paper, albeit out-of-date, can be found at http://www.fda.gov/downloads/drugs/scienceresearch/researchareas/pharmacogenetics/ucm116689.pdf.

9.6 TAXONOMY OF EVIDENCE NEEDED TO VALIDATE AND QUALIFY BIOMARKERS AS FIT-FOR-PURPOSE

There is a critical need for rigorous assessment of the process, procedures, and criteria used to evaluate or validate biomarkers in order for them to gain widespread acceptance. The spirit of validation is to demonstrate that a method is reliable for the "intended application." The application may vary by stage of drug development, by the nature of the drug development program, or by type of decision, for example, developmental versus regulatory. Hence the method validation for biomarker assays should be considered as a continuous and evolving process. Typically, the rigor of method validation should increase from the initial validation proposed mainly for exploratory purposes to more advanced validation dependent on the evidentiary status of the biomarker and/or the use of the results. Such a fit-for-purpose strategy allows for efficient biomarker and drug development with appropriate gathering of supportive evidence. Figure 9.3 demonstrates the concept of fit-for-purpose method validation.

The biomarker method validation processes include four activity circles of prevalidation, exploratory method validation, in-study method validation, and advanced method validation. The processes are continuous, iterative, and driven by the intended purpose of the biomarker data. The prevalidation step is the most vital step in the process. It is at this stage the purpose of the biomarker development program is well defined, the nature of the biologic marker to be measured is delineated, and the choice of assay is selected. These three factors influence the establishment of proposed assay acceptance criteria. On the basis of the utility, intended application of the biomarker, decision to be made, and the stage of the drug development, one could choose from two general categories of method validation (28): (i) exploratory validation with crucial components,

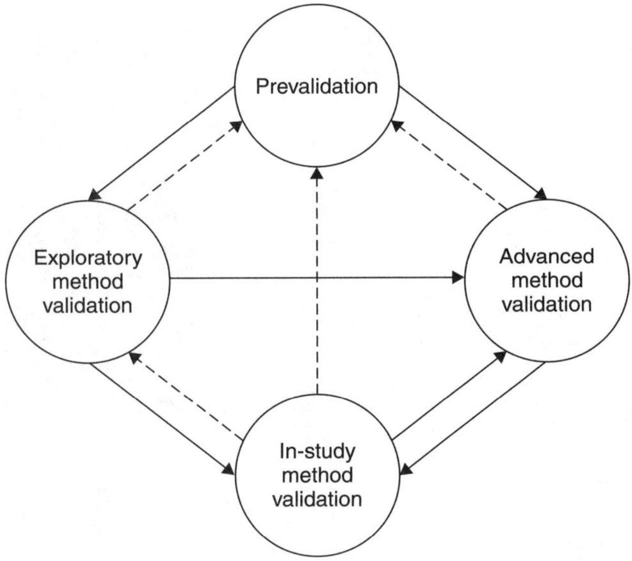

FIGURE 9.3 Conceptual framework of fit-for-purpose method validation.

including accuracy, recovery, precision, relative selectivity, initial target ranges, analyte integrity in matrix, and dilutional linearity; and (ii) advanced validation with the graded addition of other necessary components, including additional specificity, sensitivity, parallelism, expanded reference range, extended stability, method robustness, and document control. For exploratory work, one should not avoid an assay just because tight controls are not in place. An exploratory validation should be reasonable. If a go-no-go decision is made on the compound or program, then biomarker assay may deserve advanced validation. A description of various elements for developing and validating a fit-for-purpose bioanalytical method for biomarkers is presented in Table 9.2 (28).

Various authors have proposed biomarker taxonomies that have subtle similarities and differences. Recently, taxonomy explicitly for biomarkers in drug development has been proposed (37) (Table 9.3). This classification takes into account the intended application, stage of drug development, type of the decision to be made, and the validity of the biomarker assay method.

This schema captures the progression of qualification of biomarker with accruing evidence leading to increasing utility and/or decision making during drug development. However, it should be emphasized that many highly useful biomarkers will never be intended to substitute for clinical evidence of effectiveness and thus would never become surrogate endpoints in the classical sense. The evidentiary standards required for the qualification of biomarkers may be approached from a classical "tolerability of risk" (TOR) theory. This requires defining the value of true result with a biomarker compared to the harm of a false result with the biomarker defined by the stake holders for the specific intended

TABLE 9.2 Fit-for-Purpose Elements of Biomarker Assay Development and Validation (28)

Parameters Assay Elements	Preanalytical and Method Development	Exploratory Method Validation	Advanced Method Validation
Reagents and reference material	Consistent and accessible source—do due diligence	Initial characterization. Stability initiated	Well characterized. Inventoried. Establish stability. Establish change control
Target range	Estimate in biomarker work plan. Define expectation of LLOQ and ULOQ	Acquiring data	Establish from incurred samples
Dynamic range (lower and upper quantitation limits)	Determine preliminary assay range with precision profile over target range	Use three validation runs	Use at least 6 runs (in study validation data can be utilized). Establish LLOQ and ULOQ
Sensitivity	Define minimum detectable range. Define requirements of sensitivity LOD and LLOQ	Estimate sensitivity. Consider LOD versus LLOQ	Establish sensitivity
Curve fitting	Choose appropriate calibration model fitting method and tools	Confirm choice of calibration	Use six validation runs to confirm calibration model
Selectivity and specificity	Reagent specificity from supplier or literature. Assess matrix effects and minimize if possible. Determine minimum required dilution (MRD). Sample and substitute matrix	Investigate likely sources of interference, including the therapeutic agent	Extensive testing of interference and risk recommendation. Assessment of biomarker heterogeneity and isoforms
Parallelism	N/A	Use incurred samples, if available	Investigation in targeted population. Determine maximum tolerable dilution

Dilution linearity	Determine if applicable, as defined in the biomarker plan (test range)	Use spiked samples	Use spiked samples and dilution VS if applicable
Precision and accuracy (analytical)	Establish expectations early on in biomarker work plan. Consider heterogeneity	Use three validation runs	Use of total of at least six runs (in study validation data can be utilized)
Relative accuracy/recovery (biological)	Establish expectations early on in biomarker work plan	Use spiked incurred samples at multiple concentrations. Addition recovery	Use multiple donors
Robustness (reagent and change control)	Determine need. Consider availability of biological matrix	NA	Establish tolerability on crucial elements
Sample handling, collection, processing, and storage	Establish feasible conditions	Establish short-term and bench top stability. Optimize conditions and effects on assay	Establish freeze/thaw and long-term sample stability
Documentation	Biomarker work plan. Drafts procedures. Assess outsourcing options	Appropriate documentation to support the intended use of the data	Appropriate documentation to support the intended use of the data

In-study validation criteria are not listed in this table; they are defined through empirical testing of the pre-, exploratory, and advanced validation. Refer to Section "In-Study Validation and Sample Analysis Acceptance Criteria" for elements of in-study validation.

The recommendation is an example for typical immunoassay to obtain adequate statistical data.

TABLE 9.3 Fit-for-Purpose Based Classification of Biomarkers (37)

Biomarker	Description	Drug Development Use	Example
Exploration	Biomarkers are research and development tools accompanied by *in vitro* and/or preclinical evidence, but there is no consistent information linking the biomarker to clinical outcomes in humans	Hypothesis generation	Gene expression
Demonstration	Biomarkers are associated with adequate preclinical sensitivity and specificity and linked with clinical outcomes but have not been reproducibly demonstrated in clinical studies. This category corresponds to "probable valid biomarkers" in nomenclature suggested in draft guidance from FDA[12]	Decision making, supporting evidence with primary clinical evidence	Adiponectin
Characterization	Biomarkers associated with adequate preclinical sensitivity and specificity and reproducibly linked clinical outcomes in more than one prospective clinical study in humans. This category corresponds to "known valid biomarkers" in nomenclature suggested in guidance by FDA[12]	Decision making, dose finding, secondary/tertiary claims	Fasting plasma glucose
Surrogacy	A holistic evaluation of the available data demonstrates that the biomarker can substitute for a clinical endpoint. The designation of "surrogate end point" requires agreement with regulatory authorities	Registration	Hemoglobin A1C

purpose or decision making. The qualification criteria can be established for a biomarker based on the types of error associated with the predictive ability of the biomarker and the limits of acceptance with regard to the error as defined by the stakeholders. This may not always be quantitative but may be based on the perception of benefit and risk. The type of false results that can be expected are categorized as type 1, falsely detecting a biological signal that is not real, and type 2, failing to detect a real biological signal. This also defines the amount of evidence required to qualify the biomarker for the intended purpose. If the value of the true result is high and the harm of falsehood is low, then the weight of evidence required is lower because there is higher tolerance for uncertainty and ambiguity. Conversely, if the value of the truth is low and the harm of falsehood is high, there is lower tolerance for error and a higher weight of evidence would be required to demonstrate that there is a tight control over performance and that the biomarker yields true results far more frequently than false ones.

From a regulatory perspective, it is very much desirable to decrease/control the type 1 error when the efficacy of a new molecular entity is being assessed. Further, from a safety point of view, one would desire to control the type 2 error rate, as the regulatory agency's primary duty is to maximize public safety. From a drug developer's viewpoint, the monetary harm associated with type 1 error with efficacy and type 2 error in predicting the safety with a biomarker are significant as these will result in attrition at later stages of drug development or withdrawal of drugs from the market. However, from a patient's perspective, any efficacy and safety error would represent harm either at a monetary level (high cost of approved drugs) or lack of better alternatives (due to false negative efficacy). This does not mean that the regulatory agencies are not worried about the type 2 errors associated with the biomarker's predictive ability of efficacy or type 1 error associated with safety; however, regulatory agencies are less able to control for these aspects. Such situations occur in early drug development where molecules and doses are screened for decisions about proceeding into later stages of drug development. Hence it is imperative for all stakeholders to reach consensus as to the overall average value of a true result and the harm of a false result. A proposed framework (Fig. 9.4) for the flow of information for creating evidentiary standards for fit-for-purpose biomarker qualification has been published (38).

This framework also defines a prototypical "evidence map" to be used for determining (i) the type of evidence that is relevant to the utility and (ii) the weight or grade of evidence that is needed based on the overall rating of the value and harm.

For qualifying a biomarker based on the principles of TOR, and especially for a biomarker intended as a diagnostic test, it will be essential to establish the receiver operating characteristics (ROC) of the assay and its positive and negative predictive values. The role of positive and negative predictive values can be illustrated in the utility of HLA-B*5701 (biomarker) screening to reduce the incidence of the hypersensitivity reaction to abacavir. Abacavir is a nucleoside reverse-transcriptase inhibitor with activity against the HIV, available for

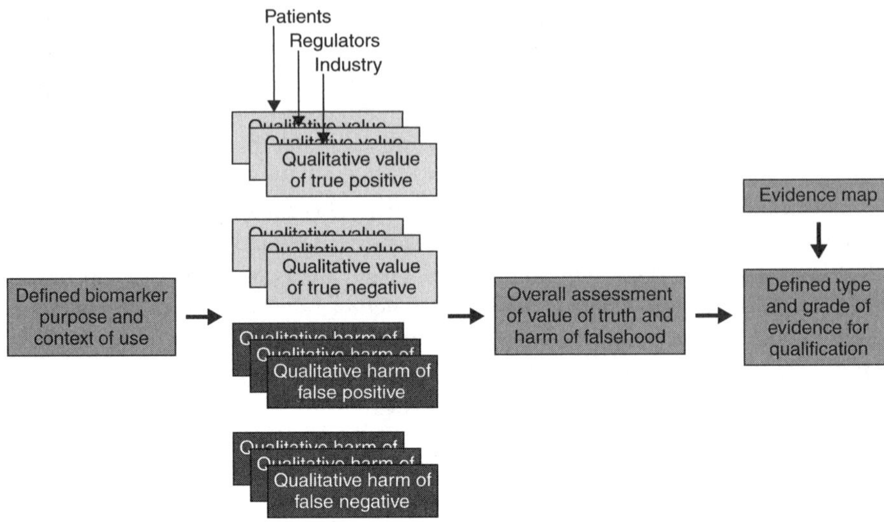

FIGURE 9.4 Frame work for fit-for-purpose biomarker qualification for drug development and regulatory utility (38).

once daily use in combination with other antiretroviral agents. The most important adverse effect of abacavir that limits its use in therapy and mandates a high degree of clinical vigilance is an immunologically mediated hypersensitivity reaction affecting 5–8% of Caucasians (39, 40). This hypersensitivity reaction is difficult to clinically distinguish from concomitant infection, reaction to other drugs, or inflammatory diseases. An association between a diagnosis of hypersensitivity reaction to abacavir and carriage of the major histocompatibility complex class I allele HLA-B*5701 was observed independently by several exploratory research groups. However, a definitive clinical diagnosis of hypersensitivity is often difficult to be made. The HLA-B*5701 screening has a negative predictive value of 100% and a positive predictive value of 47.9%. The poor positive predictive value does not make the HLA-B*5701 testing invalid. Since the hypersensitive reaction rate is low and the prevalence of the HLA-B*5701 (5.6%) is comparable to the event rate, testing for the genetic biomarker could identify high risk individuals and reduce the hypersensitivity reaction rates by approximately 50%. This was established in a prospective study where the incidence of clinically diagnosed hypersensitivity in the prospective-screening group was 3.4% compared to 7.8% in the control group (41).

9.7 FUTURE PERSPECTIVES ON REGULATORY INVOLVEMENT IN BIOMARKER DEVELOPMENT

The next generation of biomarkers will be based on many of the emerging -omic technologies specifically tailored toward understanding of the disease pathology and progression, disease prognosis, and the treatment effects—both beneficial and harmful. They will provide a critical step toward optimal biomarker usage, such as in personalized medicine as diagnostic tests. Advances in pharmacogenomics are likely to result in the development of biomarkers that will play important roles such as entry, stratification, or exclusion criteria for clinical trials during drug development. Genotypes, single nucleotide polymorphisms (SNPs), and haplotypes will increasingly be used to serve as biomarkers predictive of clinical phenotypes. Since the pathophysiology of most disease is multifactorial, future development in pharmacogenomics is likely to focus on microarrays to study the differential expression of as many as 10,000 genes in a single experiment (42, 43). Profiling studies are being designed to correlate RNA expression signatures with key biological events related to prognosis and drug efficacy and safety events. Simultaneous advances in bioinformatic tools for handling and interpreting the molecular patterns are making RNA expression profiling a reality. Metabonomics and proteomics are providing additional means for genoptyping and/or phenotyping patient samples (44). New imaging technologies also hold a vast potential for use of biomarkers for an array of purposes in drug development. The advent of translational research technologies in the form of small animal microCT, micro ultrasound, and microPET detectors is now beginning to allow identical protocol designing and *in vivo* assessments across species, enabling truly comparative pharmacology, direct proof-of-concept studies and improving decision making (45). However, there are significant challenges in adoption of these technologies in the later stages of drug development since a high level of reliability and documented performance are needed. Standardization—for example, of tissue collection techniques, of imaging acquisition protocols, of microarray interpretation—is a significant issue that is being addressed by various factions of the Federal government and in the private sector. For example, the FDA and many other groups have been involved in the Microarry Quality Control (MAQC) Consortium (46) that is developing standards in the area of gene expression assays. The National Cancer Institute is working along with many other stakeholders, to establish tissue collection and handling standards for tumor biopsies. Each new technology will require development of public standards and protocols before its wide adoption and efficient use. Additionally, the bioinformatics processes used to generate and analyze data from data-intensive assays will require rigorous statistical evaluation.

As the standardization and analytical validation of new biomarker assays progresses, the FDA and other worldwide drug regulatory agencies will have an additional task—to establish the standards and processes needed for acceptance of this new information in regulatory decision making. While this has been done in the past on an ad hoc basis, the sheer volume of new biomarkers calls for development of a formal processes for biomarker qualification. The CDER at FDA is in the process of establishing formalized procedures. However, these represent additional workloads for staff scientist in an already underresourced regulatory program. This is one rationale behind the Critical Path Initiative: to involve all stakeholders in the scientific work that must go forward to establish a new generation of biomarkers. However, even given the participation of the broader scientific community, at the end of the day the regulators must make the tough regulatory calls and establish the evidentiary thresholds for biomarker qualification. In order for the promise of these new technologies to be efficiently translated into public benefit, the regulators must undertake this additional scientific activity and, in fact, institutionalize it as part of regulatory processes. To a great extent, the pace of adoption and success in improving drug safety and effectiveness of these new technologies will be dependent on the ability of regulators to actively devise, and engage in, the new scientific process of explicit biomarker qualification.

REFERENCES

1. Food and Drug Administration. The critical path initiative—innovation-stagnation: challenge and opportunity on the critical path to new medical products. 2004.
2. Food and Drug Administration. The critical path opportunities report. 2006.
3. Biomarkers Definitions Working Group. Biomarkers and surrogate endpoints: preferred definitions and conceptual framework. Clin Pharmacol Ther 2001;69(3):89–95.
4. The food and drug modernization Act of 1997. Title 21 code of federal regulations Part 314 Subpart H Section 314.500.
5. Temple R. Are surrogate markers adequate to assess cardiovascular disease drugs? JAMA 1999;282:790–795.
6. Lon E. The use of surrogate endpoints in clinical trials: focus on clinical trials in cardiovascular diseases. Pharmacoepidemiol Drug Saf 2001;10:497–508.
7. Federal food, drug, and Cosmetics Act. Chapter V—Drugs and Devices. Subchapter A—Drugs and Devices. Available at http://www.fda.gov/RegulatoryInformation/Legislation/FederalFoodDrugandCosmeticActFDCAct/FDCActChapterVDrugsandDevices/default.htm. Accessed 2011 Dec.
8. US Food and Drug Administration. 2002. Guidance for industry. Antiretroviral drugs using plasma HIV RNA measurements—clinical considerations for accelerated and traditional approval. Available at http://www.fda.gov/downloads/Drugs/GuidanceComplianceRegulatoryInformation/Guidances/ucm070968.pdf. Accessed 2011 Dec.
9. Cohen MH, Williams G, Johnson JR, Duan J, Gobburu J, Rahman A, Benson K, Leighton J, Kim SK, Wood R, Rothmann M, Chen G, U KM, Staten AM, Pazdur R. Approval summary for imatinib mesylate capsules in the treatment of chronic myelogenous leukemia. Clin Cancer Res 2002;8:935–942.
10. Food Drug Administration. The Center for Drug Evaluation and Research, Center for Biologics Evaluation and Research. 1998. Guidance for industry: providing clinical evidence of effectiveness for human drug and biological products. Available at http://www.fda.gov/downloads/Drugs/GuidanceComplianceRegulatoryInformation/Guidances/ucm078749.pdf. Accessed 2011 Dec.
11. Chen ML, Shah V, Patnaik R, Adams W, Hussain A, Conner D, Mehta M, Malinowski H, Lazor J, Huang SM, Hare D, Lesko L, Sporn D, Williams R. Bioavailability and bioequivalence: an FDA regulatory overview. Pharm Res 2001;18(12):1645–1650.
12. Food Drug Administration. The Center for Drug Evaluation and Research, Center for Biologics Evaluation and Research. 1999. Guidance for Industry: In Vivo Drug Metabolism/Drug Interaction Studies - Study Design, Data Analysis, and Recommendations for Dosing and Labeling. Available at http://www.fda.gov/downloads/Drugs/GuidanceComplianceRegulatoryInformation/Guidances/UCM072119.pdf. Accessed 2011 Dec.

13. Food Drug Administration. The Center for Drug Evaluation and Research, Center for Biologics Evaluation and Research. 2010. Guidance for industry: pharmacokinetics in patients with impaired renal function—study design, data analysis and impact on dosing and labeling. Available at http://www.fda.gov/downloads/Drugs/GuidanceComplianceRegulatoryInformation/Guidances/UCM204959.pdf. Accessed 2011 Dec.
14. BETAPACE (Sotalol hydrochloride) Label. Available at http://www.accessdata.fda.gov/drugsatfda_docs/label/2011/019865s020lbl.pdf. Accessed 2011 Dec.
15. Available at http://www.fda.gov/downloads/Drugs/DevelopmentApprovalProcess/DevelopmentResources/ucm071734.pdf. Accessed 2011 Dec.
16. Argatroban Injection Label. Available at http://www.accessdata.fda.gov/drugsatfda_docs/label/2008/020883s014lbl.pdf. Accessed 2011 Dec.
17. Abrahamsson B, Lücker P, Olofsson B, Regårdh CG, Sandberg A, Wieselgren I, Bergstrand R. The relationship between metoprolol plasma concentration and beta 1-blockade in healthy subjects: a study on conventional metoprolol and metoprolol CR/ZOK formulations. J Clin Pharmacol 1990;30 Suppl2: S46–S54.
18. Wilson, D'Agostino RB, Levy D, Belanger AM, Silbershatz H, Kannel WB. Prediction of coronary heart disease using risk factor categories. Circulation 1998;97:1834–1847.
19. Greenland, Knoll MD, Stamler J, Neaton JD, Dyer AR, Garside DB, Wilson PW. Major risk factors as antecedents of fatal and nonfatal coronary heart disease events. J Am Med Assoc 2001;290:891–897.
20. Ridker, Rifai N, Rose L, Buring JE, Cook NR. Comparison of the C-reactive protein and low-density lipoprotein cholesterol levels in the prediction of the first cardiovascular event. N Engl J Med 2002;347:1557–1565.
21. Insel TR, Lehner TA., A new era in psychiatric genetics? Biol Psychiatry 2007;61: 1017–1018.
22. Shaw, Korecka M, Clark CM, Lee VM, Trojanowski JQ. Biomarkers of neurodegeneration for diagnosis and monitoring therapeutics. Nat Rev Drug Discov 2007;6:295–303.
23. Jones CD, Yeung C, Zehnder JL. Comprehensive validation of a real-time quantitative bcr-abl assay for clinical laboratory use. Am J Clin Pathol 2003;120:42–48.
24. Johann DJ, McGuigan MD, Patel AR, Tomov S, Ross S, Conrads TP, Veenstra TD, Fishman DA, Whiteley GR, Petricoin EF 3rd, Liotta LA. Clinical proteomics and biomarker discovery. Ann N Y Acad Sci 2004;1022:295–305.
25. Scholler N, Crawford M, Sato A, Drescher CW, O'Briant KC, Kiviat N, Anderson GL, Urban N. Bead-based ELISA for validation of ovarian cancer early detection markers. Clin Cancer Res 2006;12:2117–2124.
26. Seligson DB. The tissue micro-array as a translational research tool for biomarker profiling and validation. Biomarkers 2005;10(Suppl 1): S77–S82
27. Shah VP. The history of bioanalytical method validation and regulation: evolution of a guidance document on bioanalytical methods validation. AAPS J 2007;9: E43–E47.
28. Lee JW, Devanarayan V, Barrett YC, Weiner R, Allinson J, Fountain S, Keller S, Weinryb I, Green M, Duan L, Rogers JA, Millham R, O'Brien PJ, Sailstad J, Khan M, Ray C, Wagner JA. Fit-for-purpose method development and validation for successful biomarker measurement. Pharm Res 2006;23(3):312–328.
29. Schuster DP. The opportunities and challenges of developing imaging biomarkers to study lung function and disease. Am J Respir Crit Care Med 2007;176:224–230.

30. Reilly J. Using computed tomographic scanning to advance understanding of chronic obstructive pulmonary disease. Proc Am Thorac Soc 2002;3:450–455.
31. Office of *in vitro* diagnostic device evaluation and safety. Available at http://www.fda.gov/medicaldevices/productsandmedicalprocedures/invitrodiagnostics/default.htm. Accessed 2011 Dec.
32. Wagner JA. Overview of biomarkers and surrogate endpoints in drug development. Dis Markers 2002;18:41–46.
33. U.S. Food Drug Administration. 2003. Guidance for industry. Exposure-response relationships: study design, data analysis, and regulatory applications. Available at http://www.fda.gov/downloads/Drugs/GuidanceComplianceRegulatoryInformation/Guidances/UCM072109.pdf. Accessed 2011 Dec.
34. US Food and Drug Administration. 2006. Guidance for industry: pharmacogenomic data submissions. Available at http://www.fda.gov/downloads/Regulatory Information/Guidances/ucm126957.pdf. Accessed 2011 Dec.
35. Predictive Safety Testing Consortium. Available at http://www.c-path.org/pstc.cfm. Accessed 2011 Dec.
36. Available at http://www.c-path.org/pdf/PSTC_nephro_VXDS_summary_final.pdf. Accessed 2011 Dec.
37. Wagner JA, Williams SA, Webster CJ. Biomarkers and Surrogate end points for Fit-for-purpose development and regulatory evaluation of new drugs. Clin Pharmacol Ther 2007;81(1):104–107.
38. Altar CA, Amakye D, Bounos D, Bloom J, Clack G, Dean R, Devanarayan V, Fu D, Furlong S, Hinman L, Girman C, Lathia C, Lesko L, Madani S, Mayne J, Meyer J, Raunig D, Sager P, Williams SA, Wong P, Zerba K. A prototypical process for creating evidentiary standards for biomarkers and diagnostics. Clin Pharm Ther 2008;83(2):368–371.
39. Hetherington S, McGuirk S, Powell G, Cutrell A, Naderer O, Spreen B, Lafon S, Pearce G, Steel H. Hypersensitivity reactions during therapy with the nucleoside reverse transcriptase inhibitor abacavir. Clin Ther 2001;23:1603–1614.
40. Hernandez JE, Hernandez J, Edwards M, Fleming J, Powell W, Scott T. Clinical risk factors for hypersensitivity reactions to abacavir: retrospective analysis of over 8,000 subjects receiving abacavir in 34 clinical trials. Program and abstracts of the 43rd Interscience Conference on Antimicrobial Agents and Chemotherapy, Chicago, September 14–17, 2003; Washington (DC): American Society for Microbiology; 2003. p. 339. Abstract.
41. Mallal S, Phillips E, Carosi G, Molina JM, Workman C, Tomazic J, Jägel-Guedes E, Rugina S, Kozyrev O, Cid JF, Hay P, Nolan D, Hughes S, Hughes A, Ryan S, Fitch N, Thorborn D, Benbow A. PREDICT-1 Study Team, HLA-B*5701 screening for hypersensitivity to abacavir. N Engl J Med 2008;358:568–579.
42. Debouck C, Goodfellow PN. DNA microarrays in drug discovery and development. Nat Genet 1999;21:48–50.
43. Hilutenen MO, Niemi M, Yla-Herttuala S. Functional genomics and DNA array techniques in atherosclerosis research. Curr Opin Lipidol 1999;10:515–519.
44. McDonald WH, Yates JR. Shotgun proteomics and biomarker discovery. Dis Markers 2002;18:99–105.
45. Cherry SR, Gambhir SS. Use of positron emission tomography in small animal research. ILARJ 2001;42:219–232.
46. MAQC. Available at http://www.fda.gov/ScienceResearch/BioinformaticsTools/MicroarrayQualityControlProject/default.htm. Accessed 2011 Dec.

10

PERSPECTIVES FROM THE EUROPEAN REGULATORY AUTHORITIES*

IAN HUDSON

10.1 INTRODUCTION

European regulators recognize that appropriately validated and qualified biomarkers have the potential to have a fundamental effect on the way the drugs are developed and used. Biomarkers have many potential applications; for example, they may be used by companies in selecting which drug candidates to take forward and which to terminate, they could be used to reduce the size and duration of preclinical studies, and they may be used to monitor pharmacological response or toxicity in preclinical studies. Within the clinical development program, they could be used to bridge between preclinical findings and early clinical studies and monitor for pharmacological effect or toxicity. As the clinical development program progresses, they could potentially be used in a number of ways, for example, deciding on dose escalation in early studies or dose selection to take forward into larger studies. Biomarkers could be used to identify or select specific patient subgroups more, or less, likely to respond to a medicine or have an adverse reaction. Once dosing has started, biomarkers may be used to monitor response or monitor safety/toxicity. They could potentially affect the size, duration, and cost of clinical development programs and could ultimately be used to

*The views in this chapter represent those of the author and should not be attributed to those of the MHRA or other organization the author may be affiliated with.

Predictive Approaches in Drug Discovery and Development: Biomarkers and In Vitro/In Vivo Correlations, First Edition. Edited by J. Andrew Williams, Jeffrey R. Koup, Richard Lalonde, and David D. Christ.

support registration if appropriately validated. In the clinical development process, dose might be altered depending on the presence, absence, or level of a specific biomarker within each specific patient and hence tailoring the treatment to the individual, moving toward personalized medicines.

After the drug development process, biomarkers (once appropriately validated) may then be adopted into routine clinical practice, selecting patients or monitoring for safety or efficacy, or adjusting dose. They may be used as a diagnostic test that could lead to a much greater degree of tailoring of the use of medicines to individual patients and personalized medicines.

Biomarkers could potentially enable some compounds to be developed that otherwise might not have been possible because of the risk of irreversible toxicity that could be prevented by monitoring for early changes in sensitive biomarkers before permanent damage occurs.

It is for these reasons that there needs to be close involvement of regulators in the development and validation of biomarkers to ensure industry and regulators come to a common view as to where and how novel biomarkers may be used in the drug development and licensing process.

10.2 APPROVAL OF BIOMARKERS

The regulatory approval mechanisms for biomarkers may involve different regulators/regulatory agencies depending on how biomarkers will be used.

Those involved in the assessment of medicinal products, the national competent authorities or the EMA (European Medicines Agency), will look at data generated by applicants in so far as relevant for the safe and effective use of the medicinal product and to help define the appropriate populations in which the risk–benefit is positive. They will also ensure appropriate information is included in the Summary of Product Characteristics for the medicinal product. However, the development and approval of a biomarker as a diagnostic or test is handled differently—under the legislative framework of the *in vitro* diagnostics directive (IVD) (Directive 98/99/EC of the European Parliament and of the Council of 27 October 1998 on *In Vitro* Diagnostic Medical Device), and the appropriate regulatory bodies that oversee this area are the notified bodies.

The Directive groups IVD into four categories according to the risks associated with relative danger to public health and/or patient treatment by an IVD failing to perform as intended. The level of control is proportionate to the risk; for lower risk, the manufacturers self-declare conformity with the relevant requirement of the Directive. For highest risk products, the manufacturers system will have to be verified by a notified body. Advice is available from regulatory agency such as the MHRA (www.mhra.gsi.gov.uk/howweregulate/Device/invitrodiagnosticmedicaldevicedirective/index/htm). For a biomarker to be developed or a diagnostic to be used with a medicinal product, developers have to develop the diagnostic in parallel with the medicines and ensure appropriate authorization for both aspects.

10.3 CURRENT STATUS OF USE OF BIOMARKERS IN THE LICENSING OF DRUGS

European regulators are starting to see some applications for drugs, which involve the use of novel biomarkers to target the population for which the drug should be used; indeed, some have already been licensed. In addition, safety or efficacy data based on presence or absence of a biomarker is becoming available for a number of already licensed drugs, prompting the debate as to how much data is required to change the label for an existing drug, how best to amend licences to take into account the new information, and how to ensure this new information is adopted into clinical practice.

10.3.1 Newly Licensed Drugs

A number of new drugs have been licensed in Europe where, during the course of the development, the applicant ascertained and provided valuable pharmacogenomic information that would be relevant to the choice of patients who derive benefit from the drug, and therefore, the license has been restricted on the basis of this information. These mainly relate to the field of oncology. The first example was Herceptin (trastuzumab) for treatment of metastatic breast cancer in patients whose tumors overexpress HER2. Herceptin was first authorized by the EU Commission in August 2000.

Subsequently, a number of other agents have been approved. These include the following [given as a trade (proprietary) name]:

- Erbitux (cetuximab) for the treatment of patients with epidermal growth factor receptor (EGFR)-expressing, KRAS wild-type metastatic colorectal cancer (first authorized by the EU Commission in June 2004).
- Tyverb (lapatinib), in combination with capecitabine, for the treatment of patients with advanced or metastatic breast cancer whose tumors overexpress ErbB2 (HER2) (first authorized by the EU Commission in June 2008).
- TRISENOX (arsenic trioxide) for induction of remission and consolidation in adult patients with relapsed/refractory acute promyelocytic leukemia (APL), characterized by the presence of the t(15;17) translocation and/or the presence of the promyelocytic leukemia/retinoic-acid-receptor-α (PML/RAR-α) gene (first authorized by the EU Commission in March 2002).
- Glivec (imatinib), which has been licensed for a number of indications where relevant biomarker information is included in the indication. These include adult and pediatric patients with newly diagnosed Philadelphia chromosome (bcr-abl) positive (Ph+) chronic myeloid leukemia (CML), Philadelphia chromosome positive acute lymphoblastic leukemia; adult patients with myelodysplastic/myeloproliferative diseases (MDS/MPD) associated with platelet-derived growth factor receptor (PDGFR) gene rearrangements, adult patients with advanced hypereosinophilic syndrome (HES) and/or chronic

eosinophilic leukemia (CEL) with FIP1L1-PDGFRα rearrangement, and the treatment of adult patients with Kit (CD 117) positive unresectable and/or metastatic malignant gastrointestinal stromal tumors (GIST) (first authorized by the EU Commission in November 2001).

- Vectibix (panitumumab) for the treatment of EGFR-expressing metastatic colorectal carcinoma with wild-type KRAS after failure of certain other chemotherapy (first authorized by the EU Commission in December 2007).
- Celsentri (maraviroc) has been licensed for treatment-experienced adult patients infected with only CCR5-tropic HIV-1 virus detectable (first authorized by the EU Commission in September 2007). Of note, here the restrictions based on the biomarker relate to the virus.
- Tarceva has been licensed for the treatment of non-small-cell lung cancer. With this drug, no survival benefit or other clinically relevant effects of the treatment have been demonstrated in patients with EGFR-negative tumors (first authorized by the EU Commission in September 2005).
- Iressa (gefitinib) has been licensed for adult patients with non-small-cell lung cancer with activating mutations of EGFR-tyrosine kinase (first authorized by the EU Commission for this indication in June 2009).
- Sprycel has been licensed for a number of indications including adult patients with Philadelphia chromosome positive CML (first authorized by the EU Commission in November 2006).
- Tasigna (nilotinib) has been licensed for adult patients with Philadelphia chromosome positive CML in the chronic phase (first authorized by the Commission in November 2007).

The CHMP has recently issued a positive opinion recommending the granting of a conditional marketing authorization for Caprelsa (vandetanib), for the treatment of aggressive and symptomatic medullary thyroid cancer (MTC) in patients with unresectable locally advanced or metastatic disease. For patients in whom rearranged during transfection (RET) mutation is not known or is negative, a possible lower benefit should be taken into account before individual treatment decision. The Commission decision is awaited at the time of preparing this manuscript.

10.3.2 Existing Drugs

New data have emerged for a number of licensed drugs in the EU. In some cases, this has been of sufficient significance to warrant changes to the Summary of Product Characteristics. Examples include the following:

- Carbamazepine and the associated SJS (Stevens–Johnson syndrome) in those who carry the HLA-B*1502 allele, this has now been reflected in the Summary of Product Characteristics.
- Ziagen (abacavir). Studies have shown that carriage of the HLA-B*5701 allele is associated with a significantly increased risk of a hypersensitivity reaction to abacavir.

- Warfarin and the impact of genetic polymorphisms is being discussed currently at various levels. Data are being generated to investigate the utility of these genetic polymorphisms in determining treatment algorithms. Whether this will result in a change to the Summary of Product Characteristics remains to be determined.

Exact details of indications and wording of warnings or restrictions to licenses are available in the Summary of Product Characteristics for each drug.

It is important to note that European and North American regulators operate entirely independent processes, and therefore there are likely to be some differences in the terms of the authorizations (drug approvals) in the United States and in the European Union.

10.3.3 Other Drugs

The EMA received an application for lumiracoxib for Symptomatic relief in the treatment of osteoarthritis of the knee and hip in patients who are non-carriers of the DQA1*0102 allele. Lumiracoxib was first authorized in 2003, however was revoked in 2008 due to concerns over hepatotoxicity. The new application included results of a retroxpective pharmacogenetic study which identified a biomarker, the DQA1*0102 allele which was claimed to identify patients at risk from hepatotoxicity. However, this application was withdrawn by the applicant before the review process had concluded. (http://www.ema.europa.eu/docs/en_GB/document_library/Application_withdrawal_assessment_report/2011/11/WC500118339.pdf).

10.4 REGULATORY ACCEPTANCE OF BIOMARKERS

Regulatory acceptance of biomarkers will follow appropriate qualification of a particular biomarker for a specific purpose or application. This process will define where a biomarker is to be used and how the information from measurement of the biomarker will be integrated into the drug development process. Data requirements will be very much dependent on the proposed use for the biomarker, preclinical or clinical, whether exploratory or not, in addition to or to replace other established methods of monitoring, and whether used for patient selection, dose selection, response prediction, or monitoring toxicity. Also to be considered is whether the use of this biomarker will be taken forward into clinical practice.

If used preclinically, issues such as species specificity and mechanism of toxicity/pathology, correlation with disease evolution, reversibility, and ability to extrapolate to man will be important.

Recognizing the need to have a process to consider novel biomarkers and diagnostics and come to a conclusion on regulatory acceptance for a specific purpose in relation to a medicinal product, the CHMP (Committee for Human Medicinal Products), EMA, and national agencies have established processes to enable a dialog between those developing biomarkers and regulators. This dialog ranges from advice on further studies through to formal qualification of a biomarker for a specific purpose.

10.5 EUROPEAN REGULATORS INTERACTIONS WITH INDUSTRY: GENERAL ASPECTS

Recognizing the potential importance of biomarkers, National and EU regulators have established a number of mechanisms to discuss emerging science related to biomarkers and their applicability to drug development and approval. At a European level, these include discussion platforms through the EMA's Innovation Task Force (ITF), informal briefing meetings with the Pharmacogenetics Working Party (PGWP) and other working parties of the CHMP, scientific advice meetings, and scientific review meetings. In addition, the EMA and CHMP have developed a formal process for biomarker qualification (see below).

At a national level, individual member states play a full role in the activities of the EMA and its various working parties and committees. National agencies also offer scientific advice to applicants that can be used to discuss the potential development and utility of biomarkers. In addition, the UK agency, the MHRA, together with the UK's Association of the British Pharmaceutical Industry held a review of safety biomarkers under the umbrella of the Ministerial Industry Strategy Group. The review took place in October 2008 (See MHRA website—Forum on Safety Biomarkers; Summary of Discussions and Conclusions). This meeting concluded that the area of biomarker identification was still developing but that it was an important area of work with enormous potential benefits for drug development and approval and consequently for public health, for example, through improved safety monitoring of patients. However, the review also concluded that at the present time, a considerable amount of work had been done and that there was still no current clear clinical pathway to utility but that there were encouraging areas such as nephrotoxicity. It was also recognized that regulation should not lead the science, but more that regulation should follow science; however, regulatory dialog was important. Much of the science was at an early stage and therefore not ready for widespread applicability. Future advances would require multidisciplinary approaches with academia playing a large role and consortia activity being critical. It was also recognized that regulators needed to work together internationally.

10.6 EMEA/CHMP/NATIONAL AGENCIES INTERACTIONS WITH INDUSTRY: SPECIFIC INTERACTIONS

10.6.1 Pharmacogenetic Briefing Meetings

The PGWP of the CHMP at the EMA has put in place a mechanism for having informal briefing meetings with applicants with a view to sharing emerging science on the potential utility of genomic biomarkers.

It was recognized that knowledge on pharmacogenetics is at an early stage, and therefore, both industry and regulators would benefit from being able to discuss information in an informal setting. The working party gains by getting input from applicants on the circumstances and rationale under which pharmacogenomic data is generated, and may be used. Regulators can then help to minimize the risks of creating inadvertent obstacles to the use of this novel technology. Applicants gain by getting input from the CHMP expert group, in an informal setting, without regulatory impact on products under development, as the experts would not engage in formal preassessment of the information provided for the briefing meeting. Advice from the briefing meetings may be useful in preparing for scientific advice or for an authorization application, or for an application under the formal biomarker qualification process (see below). Briefing meetings are informal procedures and do not result in a document expressing CHMP's opinion on the matter.

The CHMP has produced a guideline for such meetings, outlining when and how to prepare for a dialog with the PGWP of the CHMP. The guideline also includes appropriate definitions of relevant terms (http://www.emea.europa.eu/pdfs/human/pharmacogenetics/2022704en.pdf).

Typical reasons for having meetings might include when applicants wish to explore a new indication for an approved product based on recent developments in pharmacogenetics or when new pharmacogenetic information becomes available during the development of a medicinal product. The working party would receive a submission of a briefing document from the applicant, then would be present at the meeting, and make a presentation highlighting the key issues. The working party would then discuss the issues with the applicant. Subsequently, the applicant would prepare a summary that would be agreed by the PGWP.

10.6.2 Innovation Task Force Meetings

The EMA has also set up an ITF, which is a multidisciplinary group covering scientific, regulatory, and legal areas. It has been set up to provide a forum for early dialog with applicants (http://www.emea.europa.eu/htms/human/mes/itf.htm). This group also holds briefing meetings with applicants.

The scope of the briefing meetings covers regulatory, scientific, and other issues arising from the development of new therapies and technologies and borderline products. Much of the work of the ITF relates to provision of regulatory advice on whether new medicinal products for emerging therapies and borderline products are eligible for EMA procedures. This is done in conjunction with other EMA scientific committees.

Applicant may request advice from the ITF, and a briefing meeting will be arranged to facilitate the informal exchange of information and the provision of guidance early in the development process. Where necessary, this may be done in liaison with EMA scientific committees, working parties, and expert groups.

Briefing meetings are informal meetings and are meant to complement and reinforce existing formal regulatory procedures such as scientific advice procedures.

10.6.3 EMEA Biomarkers Workshops

The EMA has organized two workshops on biomarkers: the first one with academia and health care professionals, in December 2005; and the second in collaboration with the European Federation of Pharmaceutical Industries and Associations (EFPIA) with all stakeholders, in December 2006. These workshops addressed general issues in the context of development and qualification of biomarkers as well as biomarkers in specific areas such as cancer, cardiovascular diseases, and osteoporosis. Following these meetings, reports were published (www.emea.europa.eu).

10.6.4 Scientific Advice Procedures

Both the EMEA/CHMP and many national agencies in the individual member states, for example, MHRA in the United Kingdom, offer the availability of scientific advice to assist applicants on the design of their development programs. This could be used to discuss the use and acceptability of biomarkers in the development process or to discuss new information that has become available for a licensed drug. Indeed, the formal biomarker qualification process falls within the Scientific Advice Working Party (SAWP) of the CHMP at the EMA. European scientific advice follows a formal process prepared by the SAWP, and the advice is subsequently adopted by CHMP.

The scientific advice offered by national agencies tends to be faster and less formal. It can also help applicants clarify issues before a European procedure. It can be seen to be complimentary to European procedures.

10.6.5 CHMP Biomarker Qualification Process

10.6.5.1 Process. The CHMP, recognizing the potential impact of advances in science in this area developed a qualification process in 2008. This was then subjected to a period of public consultation and published in the form of guidance in January 2009. The guidance is entitled "Qualification of Novel Methodologies for Drug Development: Guidance to Applicants" (http://www.emea.europa.eu/pdfs/human/sciadvice/7289408en.pdf). While the scope of this document may suggest a wider application, the focus is likely to be biomarkers developed by consortia, networks, public/private partnerships, learned societies, and pharmaceutical industry, for use in drug development. However, other novel approaches, such as imaging methods, or other drug development tools could be incorporated under this procedure.

The process starts with an approach by an applicant, who submits data, reports, etc., together with information on the proposed use of the biomarker. In their application, the applicant needs to provide information on the need for the proposed biomarker(s) and also the potential impact of the proposed biomarker should it be qualified. Evidence to support the use of the biomarker, in the form of, data/reports/publications is also submitted. A specialist group with appropriate expertise is then assembled and appointed by the CHMP, led by a CHMP or an SAWP member, tailored to the specific issue under review. This group will prepare an assessment report. It is anticipated that there will be an opportunity for further dialog with the applicant, before the start of the procedure and during and before conclusion of the process. After an initial assessment, questions will be prepared and sent to the applicant and a meeting will be held to discuss additional data submissions or further analyses required. The initial assessment report will be reviewed by the SAWP, who will recommend whether the procedure will be eligible for qualification advice or scientific advice.

The overall final output from the process would be one of two options, either a CHMP qualification advice together with a scientific assessment in the form of a public document or a CHMP scientific advice on future protocols and studies that need to be performed for qualification purposes (this would be in the form of confidential documents given to the applicant). Scientific advice would be given on further development when it was concluded that the information available to support qualification was not currently sufficient.

During the procedure, CHMP adopts the scientific advice or discusses the qualification advice on day 130. Clock stops are possible during the procedure. In addition, the process allows for the opportunity for international collaboration (e.g., discussions with the FDA) and an opportunity for engagement with the scientific community during a period of public consultation (for qualification advice, following CHMP review on day 130). Before the public consultation, the applicant will receive the draft advice and can request removal of confidential information. Depending on the outcome of the public consultation, a workshop may be held with the qualification team and the applicant, before the finalization of the CHMP qualification advice. CHMP will adopt the final qualification advice on day 190—the CHMP may conclude that the proposed innovative development method as an acceptable regulatory standard for the claimed use in a defined context for drug development.

On receipt of the CHMP advice or opinion, the applicant may request further clarification as necessary. As further scientific information becomes available, another feature of the process is that a follow-on procedure can be initiated. While there is no appealing procedure, applicants are encouraged to come for follow-up procedures.

As this is a new process in an area of rapidly changing science, it is planned that there will be an annual review of its operation. The advantages of the public consultation and publication at the end of the procedure are that it will enable a broader input into the scientific debate and ensure that the scientific community has access to the information and consequent regulatory view on its utility. It is not possible at the present time to have joint advice between the EMEA and FDA; however, as there is a confidentiality agreement in place between the agencies, the applicant may choose to approach both agencies to have the procedures ongoing simultaneously so that the EMEA and FDA can communicate with each other as the assessment proceeds and meet the applicant together. Each agency will then formulate its view, cognizant of the other's opinion.

It is anticipated that the EMEA/CHMP will organize training sessions periodically on newly qualified approaches for drug development and where appropriate, update or amend relevant guidelines.

10.6.5.2 Applications Submitted Through the EU Biomarker Qualification Process. The first application that underwent evaluation through this process was for seven urinary renal safety biomarkers, including kidney injury molecule 1 (Kim-1) clusterin (CLU), albumin, total protein, β2-microglobulin, cystatin, and trefoil factor 3. In the evaluation of these biomarkers, there was extensive collaboration between the EMEA and FDA. The assessment has been published on the EMEA web site (EMEA/679719/2008 Rev1, http://www.emea.europa.eu/pdfs/human/sciadvice/67971908en.pdf) in a document entitled "Final conclusions on the pilot joint EMEA/FDA VXDS experience of nephrotoxicity biomarkers. This application was submitted by the C-Path Predictive Safety Testing Consortium who submitted, between June 2007 and January 2008, data to support the use of these biomarkers, to add information to serum creatinine and BUN, while six of the seven were also shown to outperform one or both of these clinical chemistry markers." The applicant was required to state the utility of the biomarkers and the qualification claims for this application were as follows: It was claimed that the renal biomarkers correlate to either tubular histomorphologic alterations or to glomerulopathy with functional tubular involvement. The utility was improving the accuracy for drug-induced kidney histomorphologic changes, to be used on a voluntary basis in preclinical studies, and voluntary use of several of the biomarkers (KIM-1, albumin, total protein, B2-microglobulin, and custatin C) as bridging markers for early clinical studies on a case by case basis, when concerns are generated by findings in animal toxicology studies.

The applicants submitted a mixture of published data as well as the results of their own studies. Their own studies were conducted in two strains of rat, with a range of nephrotoxicants. Results were presented in terms of sensitivity and specificity analyses for the various biomarkers. These were all carefully evaluated by an expert team who assessed the application. Part of the evaluation process included a discussion with the FDA, who had also received the application.

The regulatory review concluded that the seven urinary biomarkers were considered acceptable in the context of nonclinical drug development for the detection of acute drug-induced nephrotoxicity, either tubular or glomerular with associated tubular involvement. It was also concluded that they provided additional and complementary information to BUN and serum creatinine to correlate with histopathological alterations considered to be the gold standard, but that additional data on the correlation between the biomarkers and the evolution and reversibility of acute kidney injury were needed. It was also noted that additional information on species specificity was needed.

In the context of clinical (human) use, the review concluded that it was considered worthwhile to further explore the potential for Kim-1, albumin, total protein, β2-microglobulin, urinary clusterin, urinary trefoil factor 3, and urinary cystatin C as clinical biomarkers for acute drug-induced kidney injury; however, based on current knowledge, the general use for monitoring nephrotoxicity in the clinical setting could not be qualified. The review also suggested that the use of biomarkers in clinical trials should be considered on a case by case basis in order to gather further information on their usefulness in monitoring drug induced renal toxicity in man.

Another application that has been submitted to the EMA is from the Health and Environmental Science Institute (HESI), who established the "Development and Application of biomarkers of Toxicity Technical Committee" to develop a program of activities. These included a program looking at novel biomarkers of nephrotoxicity. In May 2008, the working group submitted an extensive package on urinary markers RPA-1, α-GST, μ-GST, and clusterin to the EMEA. The group claim evidence of superior diagnostic performance of site-specific injury of some markers relative to reference markers in the nonclinical setting. Data were generated from three different nephrotoxic compounds and two rodent strains (http://www.hesiglobal.org/files/public/Facthseet/BIOMARKERSNephrotoxfacts seet0409.pdf).

The CHMP opinion was that the HESI submission increase the level of evidence supporting the use of Urinary Clusterin to detect acute drug-induced renal tubule alterations, particularly when regeneration is present, in male rats and can be included along with traditional clinical chemistry markers and histopathology in GLP toxicology studies which are used to support renal safety in clinical trials. In addition the HESI data indicate that urinary RPA-1 is a biomarker that may be used to detect acute drug-induced renal tubular alterations, particularly in the collecting duct, in male rats and can be included along with traditional clinical chemistry markers and histopathology in GLP toxicology studies which are used to support renal safety in clinical trials. The review acknowledged that the HESI data may support the use of urinary α-GST in detecting proximal tubule injury in male rats. However the opposing effects of proximal and collecting duct injury on α-GST levels raise uncertainty about the usefulness of this biomarker for detecting early mild renal injury. Therefore before α-GST is qualified in this context further studies will be needed to evaluate the mechanistic basis and usefulness of this BM (http://www.ema.europa.eu/docs/en_GB/document_library/Regulatory_and_procedural_guideline/2010/11/WC500099359.pdf). The CHMP also made recommendations towards future qualification experiments.

A further completed qualification process related to Qualification opinion of novel methodologies in the predementia stage of Alzheimer's disease: cerebro – spinal fluid related biomarkers for drugs affecting amyloid burden. CHMP concluded that in patients with MCI a positive CSF biomarker signature based on a low Aβ1-42 and a high T-tau can help predict evolution to AD-dementia type and that the CSF biomarker signature based on a low Aβ1-42 and a high T-tau can be useful for the enrichment of clinical trial populations (http://www.ema.europa.eu/docs/en_GB/document_library/Regulatory_and_procedural_guideline/2011/05/WC500106357.pdf).

The fourth completed opinion was a Qualification opinion of low hippocampal volume for use in clinical trials for regulatory purpose – in pre-dementia stage of Alzheimer's disease. CHMP concluded that low hippocampal volume, as measured by MRI and considered as a dichotomized variable (low volume or not), appears to help enriching recruitment into clinical trials aimed at studying drugs potentially slowing the progress/conversion to AD dementia of the included subjects. Low hippocampal volume might be considered a marker of progression to dementia in subjects with cognitive deficit compatible with predementia stage of AD for the purposes of enriching a clinical trial population. (http://www.ema.europa.eu/docs/en_GB/document_library/Regulatory_and_procedural_guideline/2011/12/WC500118737.pdf)

10.7 ICH ACTIVITIES

It would be important to try to achieve a harmonized approach to biomarkers globally. The ICH process would be the main route to do so, and indeed a number of ICH activities are relevant to biomarkers.

These include the development of ICH Topic E15, Definitions for genomic biomarkers, pharmacogenomics, pharmacogenetics, genomic data, and sample coding categories. This has been adopted by the CHMP in November 2007 (note for guidance on definitions for genomic biomarkers, pharmacogenomics, pharmacogenetics, genomic data, and sample coding categories EMEA/CHMP/ICH/437986/2006) and came into force in May 2008.

This guideline aims to facilitate the consistent definition of terminology, in order to facilitate the integration of the discipline of pharmacogenomics and pharmacogenetics into global drug development and approval.

ICH Topic E16 covers pharmacogenomic biomarker qualification format and data standards. A concept paper was drafted, dated April 2008. Step 2 is proposed for autumn 2009 and step 4 for autumn 2010. Step 2 was June 2009 and step 4 version was approved by the steering Committee and recommended for adoption to the three ICH regulatory bodies in August 2010, adopted by CHMP and effective from December 2010.

The scope of this guideline is the context, structure, and format of qualification submissions for clinical and nonclinical genomic biomarkers related to development of drug or biotechnology products including translational medicine approaches, pharmacokinetics, pharmacodynamics, efficacy and safety aspects. The guideline also covers the submission of data relevant to the validation of new analytical approaches to improve the evaluation of current biomarkers. The guideline does not address either the qualification process or the evidentiary standards for a biomarker to be qualified by regulatory authorities.

This guideline follows on from ICH E15, which harmonized terminology, thereby setting the stage for harmonization on the use of such data in drug development and regulatory settings. This should help those planning to use genomic biomarkers in understanding the types of data that will be required to support the qualification of biomarkers and how such data should be collected and reported to regulatory authorities. This guideline should also help to standardize the format of PG data to be submitted to regulatory agencies (http://www.ema.europa.eu/docs/en_GB/document_library/Scientific_guideline/2010/09/WC500097060.pdf).

10.7.1 Other Regulatory Guidance

The EMA, through it's working parties and CHMP has developed and issued other guidance to support applicants. This includes the following:

- Reflection paper on methodological issues with pharmacogenomic biomarkers in relation to clinical development and patient selection – this document is currently still in draft form.
- Reflection paper on co-development of pharmacogenomic biomarkers and assays in the context of drug development – this document is still in draft form.
- Use of pharmacogenetic methodologies in the pharmacokinetic evaluation of medicinal products – this document is still in draft form Reflection paper on pharmacogenomic samples, testing and data handling.
- Guiding principles: Processing joint Food and Drug Administration and European Medicines Agency voluntary genomic data submissions within the framework of the confidentiality arrangement.
- Pharmacogenetics briefing meeting.
- Position paper on terminology in pharmacogenetics.

These documents are all aimed at assisting applicants in understanding regulatory requirements and preparing for regulatory meetings or submissions to agencies. Further details can be obtained from the EMA website: http://www.ema.europa.eu/ema/index.jsp?curl=pages/regulation/general/general_content_000411.jsp&murl=menus/regulations/regulations.jsp&mid=WC0b01ac058002958e.

10.8 FUTURE CHALLENGES

It is recognized that there is an enormous potential for novel biomarkers to revolutionize the drug development process and also hugely impact clinical practice; however, to date, despite considerable research, there is still a long way to go to realize this potential. Research is ongoing but the impact of novel biomarkers is still relatively small. Thus, considerable further efforts will be required in the progress of biomarkers to ensure the potential is realized.

One of the challenges, for some of these novel areas, will be the development of diagnostic tests along side the development of the medicinal product (companion diagnostics), to enable the drug to be used with the diagnostic. The regulatory route for diagnostics is under the auspices of the IVD of the European

Commission, and focuses more on the performance of the test rather than its clinical utility. Assessment of a new medicinal product will take into account the utility of the test and its prognostic value from a safety and efficacy viewpoint. However, this may be more challenging when new information becomes available for existing drugs and new diagnostics are introduced.

Another of the challenges will be how to apply new information to established drugs when it becomes available and what criteria/strength of evidence is required to change the terms of the licence.

A further challenge will be in communicating information on biomarkers and their implications in appropriate product literature to ensure safe prescribing and to ensure new knowledge is taken up into clinical practice.

More needs to be done to extend current initiatives. It is hoped that one or several of the current safety biomarker developments could in due course prove to be clinically useful, either as an add-on to existing monitoring or as a replacement to an existing standard. In addition, further initiatives need to be started.

It will be important that any initiatives are agreed and harmonized across different geographic regions to enable a consistent approach. This would encompass scientific projects (such as the renal safety biomarker project) as well as definitions, consensus statements, etc., and where possible a common agreement on the use of a biomarker.

11

USE OF BIOMARKER IN DRUG DEVELOPMENT—JAPANESE PERSPECTIVES*

YOSHIAKI UYAMA, AKIHIRO ISHIGURO, HARUMASA NAKAMURA, AND SATOSHI TOYOSHIMA

11.1 INTRODUCTION

Drug has a very important role in improving various health disorders and promoting public health. It is a long way, however, from the discovery of a compound, through the approval by regulatory authorities, to drug administration to patients. A drug development process mainly consists of five stages. The first stage is a discovery stage to understand targeted physiology and pharmacology and to search for candidate compounds. In the second stage, properties of the selected compound are assessed in nonclinical studies (e.g., toxicological, pharmacological, and pharmacokinetic (PK) studies using cultured cells or laboratory animals). The third stage is clinical trials conducted to verify the efficacy and safety in human subjects. The fourth stage is regulatory review and approval based on collected study data, and the fifth stage is postmarketing evaluation.

Data from the US Food and Drug Administration (FDA) (1) indicates a continuous tendency that research and development costs for drugs have increased more than twice since early 1990s, whereas numbers of new active ingredients approved by the US FDA have become smaller in recent years. These tendencies have also been reported in Japan by the Office of Pharmaceutical Industry

*The views expressed herein are the result of independent work and do not necessarily represent the views and findings of the Pharmaceuticals and Medical Devices Agency.

Predictive Approaches in Drug Discovery and Development: Biomarkers and In Vitro/In Vivo Correlations, First Edition. Edited by J. Andrew Williams, Jeffrey R. Koup, Richard Lalonde, and David D. Christ.

Research established by the Japanese Pharmaceutical Manufacturers Association (JPMA) (2). Research and development costs that are spent in the discovery and clinical trial stages constitute a large part of the total amount. Thus, increasing efficiency in searching candidate compounds and in performing clinical trials has become an important issue in drug development. Particularly, in phase II clinical trials, approximately 40% of the development projects has been discontinued because of the either reason, showing no efficacy or safety concerns. The success rate is especially low in anticancer drugs and drugs for central nervous disorders (3).

In addition, approved drugs are not always efficacious for all patients. Many approved drugs, in particular, anticancer drugs have no efficacy in more than 50% of the patients (4), while a drug that has efficacy in a patient may also induce an adverse effect and may be forced to be discontinued. Thus, even after a drug has been successfully developed and approved, it may not always provide satisfactory therapeutic results.

Every drug has its risk. Since drug approval is based on its benefit/risk ratio, increasing benefit and reducing risk of a drug will enable more drug development to be successful and will improve the rate of patients who respond well to the drug without safety concerns.

For improving the benefit/risk ratio of a drug, it is necessary to establish a more reliable index for examining efficacy or safety of a drug. Use of validated biomarker in drug development could be one of the leading candidates.

In this chapter, current knowledge regarding differences of drug effects between the Japanese and other populations is discussed, and the current Japanese regulatory activities relating to biomarkers are also summarized.

11.2 DIFFERENCES IN THE STANDARD DOSE OF DRUGS BETWEEN JAPAN AND UNITED STATES/EUROPEAN UNION

E5 guideline (5) of the International Conference on Harmonisation of Technical Requirements for Registration of Pharmaceuticals for Human Use (ICH), entitled "Ethnic Factors in the Acceptability of Foreign Clinical Data" issued in 1998, allows using foreign clinical data in a new drug application of a new region, if data in a bridging study show that a drug behaves similarly in both regions. This has come to be known as the *bridging strategy*. A bridging study will often provide a comparison of dose–response relationships between the two regions.

Since ICH E5 was officially notified in 1998, a number of new drug approvals based on the bridging strategy have gradually increased in Japan (6) and have reached a steady level in 2002–2004 and decreased in more recent years. In 2004, approximately 28% of the total approvals were based on the bridging strategy. In these experiences, there are several drugs, approved between 1999 and 2006, for which a standard dose in Japan was set differently from that in United States/European Union (Table 11.1).

TABLE 11.1 Examples of Approved Drugs, Based on a Bridging Strategy, Having Different Standard Dose Between Japan and the United States/European Union

	Nonproprietary Name	Reasons for Differences	Approval Year in Japan	Target Disease	Standard Dose and Administration[a]		
					Japan	United States	European Union
1	Sildenafil citrate	Safety	1999	Erectile dysfunction	25–50 mg, 1 h before sexual activity	50 mg (min: 25 mg, max: 100 mg), 1 h before sexual activity	50 mg (min: 25 mg, max: 100 mg), 1 h before sexual activity
2	Donepezil hydrochloride	Trial design	1999	Alzheimer's disease	Starting dose: 3 mg OD After 1–2 weeks, 5 mg OD	5 mg OD, may be increased to 10 mg	United Kingdom[b] 5 mg OD, max: 10 mg
3	Alendronate sodium	PK	2001	Osteoporosis	Treatment: 5 mg OD	Treatment: 10 mg OD Prevention: 5 mg OD	United Kingdom[b] Treatment: 10 mg OD Prevention: 5 mg OD
4	Risedronate sodium	PK Efficacy	2002	Osteoporosis	2.5 mg OD	5 mg OD	5 mg OD
5	Eletriptan hydrobromide	Safety	2002	Migraine	20 mg 40 mg for another attack Max: 40 mg/day	Had not been approved	United Kingdom[b] 40 mg 80 mg for another attack. Max: 80 mg

(continued)

TABLE 11.1 (*Continued*)

	Nonproprietary Name	Reasons for Differences	Approval Year in Japan	Target Disease	Standard Dose and Administration		
					Japan	United States	European Union
6	Omeprazole + clarithromycin + amoxicillin	PK	2002	*Helicobacter pylori* eradication on duodenal/gastric ulcer	Omeprazole 20 mg plus clarithromycin 400 mg plus amoxicillin 750 mg, each given BID for 7 days	Omeprazole 20 mg + clarithromycin 500 mg + amoxicillin 1000 mg, each given BID for 10 days	United Kingdom[b] Omeprazole 20 mg BID or 40 mg OD + clarithromycin 500 mg + amoxicillin 1000 mg, each given BID for 7 days
7	Calcium folinate + tegafur/uracil	Trial design	2003	Colon cancer	300 mg/m^2/day tegafur and 672 mg/m^2/day uracil combined with 75 mg/day oral calcium folinate for 28 days; 7-day interval should be taken for another cycle	Not approved	Spain[b] 300 mg/m^2/day tegafur and 672 mg/m^2/day uracil combined with 90 mg/day oral calcium folinate for 28 days; 7-day interval should be taken for another cycle

8	Pramipexole dihydrochloride	Trial design	2003	Parkinson's disease	Starting dose: 0.25 mg/day Second week: 0.5 mg/day Increased by 0.5 mg/day at weekly intervals until reaching maintenance dose (standard dose: 1.5–4.5 mg/day) Max: 4.5 mg/day	Starting dose: 0.375 mg/day Second week: 0.75 mg/day Increased by 0.75 mg/day at weekly intervals until reaching maintenance dose (standard dose: 1.5–4.5 mg/day)	Starting dose: 0.375 mg/day Second week: 0.75 mg/day Increased by 0.75 mg/day at weekly intervals until reaching maintenance dose (standard dose: 0.375–4.5 mg/day) Max: 4.5 mg/day
9	Etanercept	Efficacy	2005	Rheumatoid arthritis	10–25 mg/day, twice a week	25 mg/day, twice a week	25 mg/day, twice a week
10	Rosuvastatin	PK	2005	Hypercholesterolemia	Starting dose: 2.5–5 mg OD, may be increased to 10 mg OD Max: 20 mg/day	Standard dose: 10 mg OD, with a range of 5–40 mg/day Max: 40 mg/day	The Netherlands[b] Standard dose: 10 mg OD, with a range of 5–40 mg/day Max: 40 mg/day

[a]The dosing information is at the time of approval in Japan and may be different from the current status in each region. Information here is based on the information in the summaries of data submitted by applicant.
[b]Information in a country in Europe was only available in the summaries.

There were four main reasons for setting a different standard dose in Japan, as shown in the second row of Table 11.1; that is, efficacy, safety, PK, and trial design. For example, the standard dose of etanercept was set at 10–25 mg/day (twice a week) in Japan, compared to 25 mg/day (twice a week) in the United States/European Union, because of efficacy differences. Statistically significant effects to placebo based on the response rate of ACR20 (primary endpoint) at 12 weeks were observed at 10 as well as 25 mg/day (6.3% for placebo vs 64.0% for 10 mg/day, 65.3% for 25 mg/day, $p < 0.001$). Thus, the minimum effective dose of etanercept was lower in the Japanese population, and 25 mg/day was also included as a standard dose because the effects of 10 mg/day on inhibitions of the progression of structural damage and improving physical function were not clear and may be lower than those at 25 mg/day, which showed no increased risks in the Japanese.

The second reason was regarding safety. Maximal approved dose of eletriptan hydrobromide was 40 mg/day in Japan in contrast to 80 mg/day in the European Union. This resulted from data demonstrating the dose-dependent adverse events in a bridging study. The frequency of major adverse events, with likes of nausea and somnolence, was significantly higher in the Japanese at 80 mg/day than seen in the foreign population (nausea, 10.4% vs 5.7%; somnolence, 16.9% vs 6.6%). In this case, C_{max} and AUC at the single oral dose (10–40 mg) were significantly lower (~30–40%) in the Japanese than those in the foreign population. Safety was also a reason for the different recommended doses for sildenafil citrate.

The third reason was regarding PK differences. The standard dose of rosuvastatin was set at 2.5–5 mg/day in Japan, compared to 10 mg/day in the United States/European Union, because of a PK finding that drug concentration with 5 mg/day in the Japanese (AUC_{0-24}: 31.3 ± 13.6 ng h/ml) corresponded to that at 10 mg/day in the foreign population (AUC_{0-24}: 30.7 ± 56.4 ng h/ml). In addition, 2.5 mg/day was also included as a starting dose because the effect (mean LDL-C percentage change at 6 weeks from baseline) at 2.5 mg/day (−45.0 ± 6.4%) in the bridging study was similar to that at 5 mg/day (−44.5 ± 2.1%) in the comparable foreign study. The main reason for the PK differences was considered to be ethnic differences in clearance and bioavailability. The PK differences were also a reason for the different recommended doses of alendronate sodium, risedronate sodium, and omeprazole + clarithromycin + amoxicillin.

Finally, in some cases, the different dosing recommendations simply reflected different dose settings in the bridging study. Thus, the dose of donepezil hydrochloride was set at 3 mg OD, as a starting dose, followed by 5 mg OD in Japan versus 5 mg OD followed by 10 mg OD in the United States because of the slightly different protocol used in the bridging study. Differences in the trial design were also a reason for cases of pramipexole dihydrochloride and calcium folinate + tegafur/uracil. More details of the discussions on the review for each application are described in the review reports, which are available on the web (7).

The above findings indicate that a drug behaves sometimes differently between the Japanese and foreign populations. Reasons for these differences could be varied, but even in part, some genetic differences among populations, such as allele types of CYP enzyme and single nucleotide polymorphism (SNP) frequencies, could contribute to PK differences and drug efficacy/safety.

11.3 EXAMPLES OF DIFFERENCES IN BIOMARKER RESPONSES AMONG POPULATIONS

It has been suggested that polymorphic types of CYP enzyme is an important factor to decide drug concentrations. For example, many antipsychotic drugs such as imipramine and olanzapine are metabolized by CYP2D6, and these drug concentrations depend on genetic variants of CYP2D6. Major polymorphic CYP alleles have been recognized to be important for treatment with several antidepressants, antipsychotics, antiulcer drugs, anti-HIV drugs, anticoagulants, antidiabetics, and anticancer drugs (8). However, the frequency of CYP2D6 poor metabolizer (PM) because of inheritance of two CYP2D6 null alleles, such as alleles of *3, *4, and *5, is higher in Caucasians (~5–20%) than that in the Japanese (<1%) (8, 9). Instead, the frequency of CYP2D6 intermediate metabolizer (IM) because of inheritance of alleles associated with decreased enzyme activity, such as alleles of *10, is higher in the Japanese (~40%) than that in Caucasians (~5%) (10), suggesting interethnic differences among populations. Moreover, genetic polymorphisms of UGT (uridine diphosphate-glucuronosyltransferase) 1A1 transporter have been reported to have a potential role in irinotecan-induced adverse reaction, for example, neutropenia. The US FDA announced to approve the label changes of irinotecan products, requiring to consider dose reductions for patients with UGT1A1*28/*28, who are thought to have higher adverse drug reaction (ADR) risk (11). UGT1A1*28 is also an important variant in Japan because considerable numbers of Japanese (~13%) have this type of polymorphism. However, as another genetic variant, *6 has been also suggested to have clinical effects on irinotecan therapy in the Japanese (12, 13). Therefore, the package insert (PI, label) of irinotecan in Japan describes both alleles of UGT1A1*6 and *28 (14).

Other types of biomarkers are related to the drug target. Typical example for this biomarker is HER2 in the treatment with trastuzumab, an antibody against HER2 protein. Breast cancer patients who had overexpressed HER2 have been found to be highly responsive to trastuzumab (15). The efficacy of gefitinib for treatment of non-small-cell lung cancer is also known to depend on genetic variants of EGFR (epidermal growth factor receptor), and it has been reported that the EFGR mutation rate is higher in the Japanese than that in Caucasians (16, 17).

Recently, a strong association of carbamazepine-induced Stevens–Johnson syndrome (SJS) and toxic epidermal necrolysis (TEN) with HLA-B*1502 has been reported in the Han Chinese population (18), but such association was not confirmed in the Japanese population (19). The US FDA announced to include the information about HLA-B*1502 in the label of carbamazepine (20). Interestingly, allele frequencies of HLA-B*1502 is approximately 10–15% in the Han Chinese and very rare (<0.3%) in the Japanese population, respectively (21, 22). Recently, a genome-wide association study found an association of carbamazepine-induced

SJS/TEN with a different allele, HLA-A*3101, in the Japanese population (23). These results suggest that the allele type of HLA associated with SJS/TEN is different among populations and may be related with basal allele frequencies in a population.

Finally, ethnic differences in autoimmune-disease-associated SNPs such as solute carrier family members 4 (SLC22A4) in rheumatoid arthritis have been reported (24). Frequency of L503F SNP and haplotypes containing SNP in SLCA4/5 was 39% in Caucasians but extremely rare in the Japanese population, although roles of the differences in disease susceptibility are not still clear.

As described above, if considering the relationships between drug effects and biomarker, more appropriate dose adjustment and target selection in terms of drug efficacy and safety can be possible. This would result in having more favorable benefit/risk ratio of a drug. However, it should be taken into consideration that allele frequencies of a target biomarker may be different among populations and a biomarker that is useful in a population may not be a good marker in other populations. Therefore, it will be very important to examine from an early stage of drug development as to whether ethnic differences in a biomarker have any effects on drug efficacy and safety. To achieve this, one of a useful development strategy will be a simultaneous global drug development to conduct multiregional clinical trials for dose finding including various major kinds of populations (see below section).

11.4 BIOMARKER AND DRUG DEVELOPMENT

At any stage of drug development, from a research stage to a clinical development stage, many types of biomarkers such as diagnostic/prognostic biomarkers may be found, as shown in Figure 11.1.

Usefulness of a biomarker in drug development starts to be examined from early stages such as the basic research stage and is concluded in clinical development stages as to whether a biomarker can be used as an indicator for predicting drug efficacy/safety in patients. To find a biomarker that has real effects on drug efficacy and safety, case-control comparisons such as genome-wide association study will be useful as a powerful tool (25). For utilizing a biomarker in drug development, the biomarker should be carefully qualified/validated because unqualified/validated biomarker may lead to misinterpretation of data. For example, a false-positive signal in safety issues may result in loss of a useful drug and a false-positive signal in efficacy issues may result in approval of a noneffective drug. Similarly, a false-negative signal also may lead to wrong decision making. Therefore, biomarker qualification/validation is very important before widely using it in drug development, but it may be difficult to complete by one body such as an industry. For establishing qualified/validated biomarkers, collaborations between industries, academia, and regulatory agencies will be necessary. It is also pointed out that development of devices for testing biomarkers in conjunction with drug development is

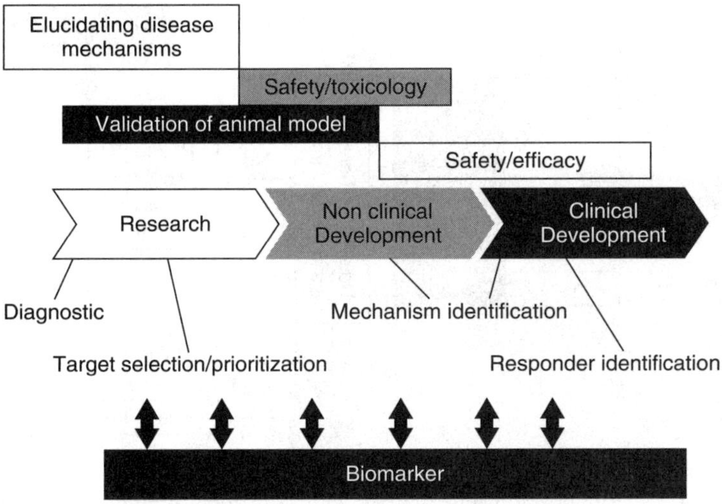

FIGURE 11.1 Biomarker and drug development.

important for utilizing a biomarker in practical therapy for a patient. Therefore, to realize a biomarker-based drug administration, more close collaborations between drug and device industries as well as earlier communication between regulatory agencies and industries in drug developments are indispensable.

As described in the previous section, the simultaneous global drug development to conduct multiregional clinical trials for dose finding including various major kinds of populations may be a useful strategy to examine whether ethnic differences in a biomarker have any effects on drug efficacy and safety. For example, as shown in Figure 11.2, clinical development strategies can be flexibly decided based on the results of a global phase II clinical trial (dose-finding study). If no ethnic differences were obtained in the trial, a phase III confirmatory trial can be conducted as a global clinical trial that includes similar populations entered in the phase II clinical trial. On the other hand, if ethnic differences have effects on drug efficacy or safety, a later phase III confirmatory trial can be conducted separately in each region in a parallel way with modification of some factors such as dose setting and target populations. Thus, in either case, timing of drug development can be synchronized among regions, resulting in an increased possibility of completing clinical drug development and of submitting a new drug application to a regulatory agency in all regions simultaneously or at a similar time. If Japan participates in a global drug development, the above strategy will be effective to examine effects of ethnic factors between the Japanese and other populations. This strategy also helps with appropriate consideration of ethnic factors to resolve a current social problem called *drug lag*, in which the approval time of a new drug in Japan is a few years later than that in the United States/European Union. Therefore, to encourage global drug development including Japan, the Pharmaceuticals and Medical Devices Agency

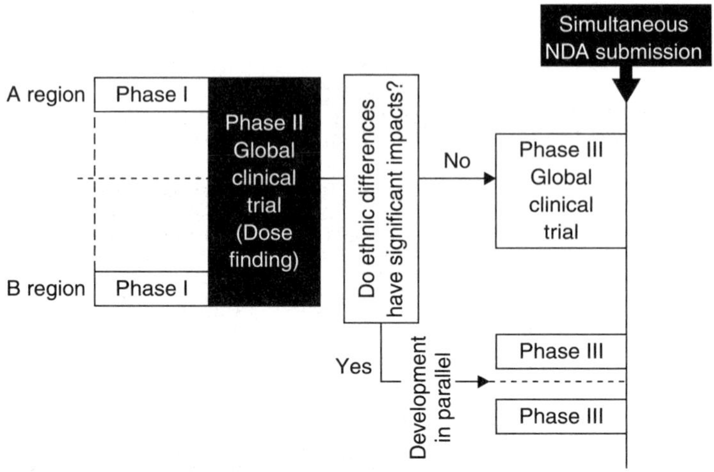

FIGURE 11.2 Drug development strategy with evaluation of ethnic differences. NDA: New Drug Application.

(PMDA) drafted points to consider documents for global clinical trials in March 2007. Final notification entitled "Basic Principles on Global Clinical Trials" was published by the Ministry of Health, Labour and Welfare (MHLW) on September 28, 2007 (Evaluation License Division Notification, No. 0928010) (26). This notification describes how to plan global drug development including Japan and how to design global clinical trials including the Japanese populations as well as other foreign populations.

11.5 EXAMPLES OF BIOMARKER INFORMATION IN PACKAGE INSERTS ON APPROVED DRUGS IN JAPAN

A PI, a label, is a very important source to promote proper use of drugs. To understand how much biomarker information are included in PIs, we searched publicly available PMDA database of PIs for new drugs (27), which were reviewed by the Pharmaceuticals & Food Sanitation Council of MHLW and got approved in Japan from 2002 to 2008. The criteria for selection were to include at least one of the 12 terms (i.e., SNP, genome, genomics, metabolic pathway, genotype, polymorphism, PM, EM (extensive metabolizer), metabolizer, pharmacogenetics, variation). The number of PIs that includes biomarker information has gradually increased in these periods, as shown in Figure 11.3. In 2008, approximately 15% of the PIs of those drugs (46 of 309 drugs) included some biomarker information.

In the 46 drugs, 48 biomarkers were identified and were classified based on the types of information as shown in Figure 11.4. The most popular information was regarding the types of virus/bacterium. For example, efficacy results based on genotypes of hepatitis C virus (HCV) in clinical trials are included in the

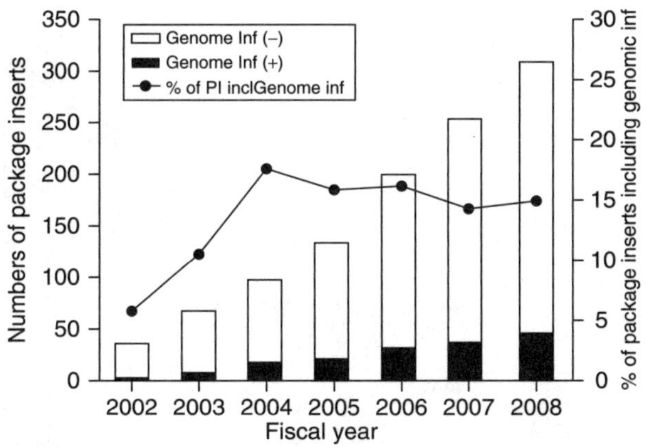

FIGURE 11.3 Trends of biomarker information in package inserts of drugs approved in Japan from 2002 to 2008.

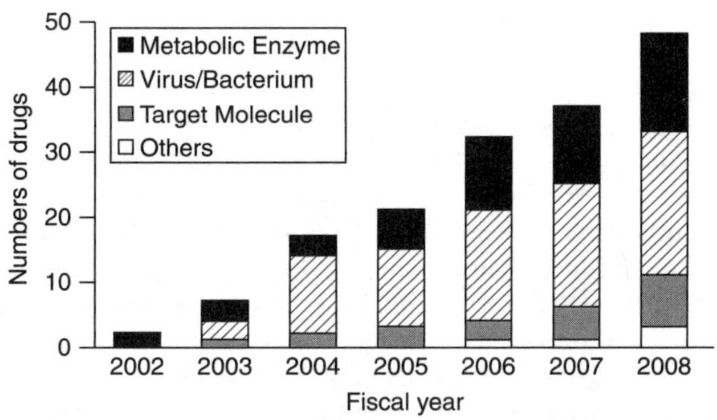

FIGURE 11.4 Types of biomarker information in package inserts of drugs approved in Japan from 2002 to 2008.

PI of ribavirin and PEG-interferon alfa-2a. Similarly, the resistance of drugs to certain HIV genotypes is described in the PIs of anti-HIV drugs such as a combination product of abacavir sulfate and lamivudine. The second popular information regards the types of metabolic enzymes such as cytochrome *P*450 polymorphisms. For example, differences of drug concentrations between PM and EM are described in the PIs of tolterodine tartrate for CYP2D6 and many proton pump inhibitors such as rabeprazole sodium, lansoprazole, and omeprazole for CYP2C19.

Information regarding the target molecule of drugs in the PIs was very limited, but this type of information was included in the PIs of anticancer drugs such as imatinib mesylate for patients with Kit (CD117)-positive metastatic malignant gastrointestinal stromal tumors (GISTs). Other type of information was regarding

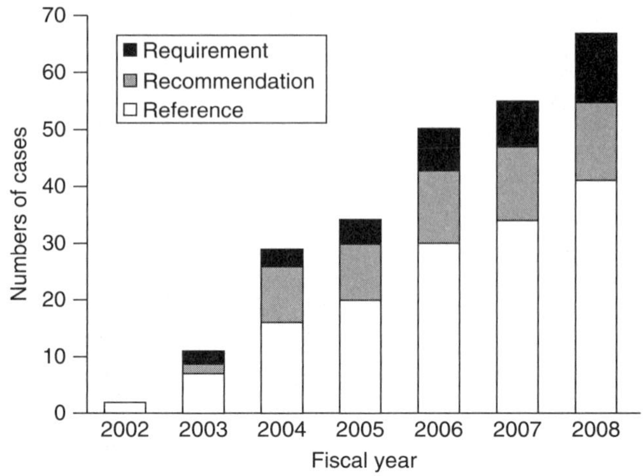

FIGURE 11.5 Status of biomarker information in package inserts of drugs approved in Japan from 2002 to 2008.

genetic test for diagnosis of congenital protein C deficiency in patients who undergo treatment with activated protein C.

Status of biomarker information in the PIs for 48 biomarkers is also summarized in Figure 11.5. Each information was calculated as one case; thus, 67 cases of information were identified because 2 or 3 cases were included in a PI. Many information were included as a reference information and did not intend to require any genetic test before administration of a drug. In 2008, 61% of all cases was categorized as reference and 21% was categorized as recommendation. This type of information was mainly about CYP polymorphism and the genotype of HCV. In 2008, biomarker information in the PIs was a requirement in 18% of all cases. Many of this type of information was diagnostic genetic test such as examining Kit (CD117) for GISTs before imatinib methylate administration and HCV-RNA for chronic HCV infection before PEG-interferon administration.

11.6 REGULATORY ACTIVITIES RELATED TO PHARMACOGENOMICS (PGx)/BIOMARKERS IN JAPAN

11.6.1 Guidance Related to PGx and Biomarkers in Japan

"Guidance on Clinical Pharmacokinetic Studies of Pharmaceuticals" (28) was officially notified in Japan on June 1, 2001, by MHLW (Evaluation License Division Notification, No. 796). In this guidance, it is described that in case genetic polymorphism is likely to affect individual difference in PK, it is recommended for a sponsor to select subjects with or without specific a genetic factor, on the basis of objective criteria such as genotyping tests. Its Q&A section also mentions that ethical issues should be taken into consideration to perform genetic examination and that genetic polymorphisms can be identified either by genotyping or phenotyping, but in the case of genotyping, the genotype that clearly relates to metabolic activity should be utilized. It also says that to investigate genetic polymorphisms of very low frequency in the Japanese population, the results of clinical trials in foreign populations are expected to provide valuable information. In another guidance entitled "Guidance on Methods of Drug Interaction Studies" (Evaluation License Division Notification, No. 813) (29), it is described that in case a polymorphic metabolic enzyme significantly affects metabolism of the investigational drug in individual subjects, it is recommended for a sponsor to study drug interactions, considering phenotypes and/or genotypes of individual subjects. Thus, recommendation has already been provided that PK data, particularly of metabolizing enzyme such as CYPs, should be collected more appropriately by using pharmacogenomics (PGx).

On March 18, 2005, the MHLW published another notification (30) (Evaluation License Division Notification, No. 0318001) to encourage sponsors to submit a list of information to MHLW on being planned, ongoing, and past PGx clinical trials. The purpose of this notification is to collect all available information regarding PGx clinical trials to correctly understand situations of PGx activities in drug development. Submitted information included the purpose of the study, target gene, target disease, race, subject numbers, methods for genetic test, sample storage process, etc. Totally, information of 180 clinical trials from 22 industries was submitted.

In March 30, 2007, at MHLW's request, the Japanese Society of Clinical Pharmacology and Therapeutics published the report entitled "Current Situations & Future Tasks for Utilization of PGx in Drug Evaluation" (31). This report summarizes current knowledge and usefulness of PGx and identifies future tasks that should be taken into consideration by MHLW for promoting drug development using PGx in Japan. In the section describing future tasks, five issues are identified, as shown in Box 11.1.

Box 11.1 FUTURE TASKS IDENTIFIED IN THE REPORT FROM THE JAPANESE SOCIETY OF CLINICAL PHARMACOLOGY AND THERAPEUTICS

- Establishment of general guidance regarding PGx application in clinical trial.
- Clarification of PGx data handling in CTD for a new drug application.
- Considering strategies for promoting drug/device codevelopment for utilizing PGx.
- Establishment of general guidance regarding genomic biomarkers in drug development.
- Establishment of general guidance regarding clinical trial designs using PGx.

The first issue is an establishment of general guidance regarding PGx application in clinical trials. Ethical considerations including how to protect private information in conducting PGx clinical trials based on ICH-GCP (Good Clinical Practice) are included in this issue. The report pointed out to clarify roles of an institutional review board and an ethical committee in a clinical trial site, to consider what kind of information should be provided before and after PGx clinical trials to subjects, and to establish a general process for sample storage for long periods. The report also stated the necessity of international harmonization in this issue for promoting PGx clinical trials including global clinical trials.

The second issue is a clarification of PGx data handling in CTD (Common Technical Document) for a new drug application. The report pointed out to clarify how to submit and evaluate PGx data in CTD for a new drug application.

The third issue is a consideration of strategies for promoting drug/device codevelopment for utilizing PGx. The report pointed out that validated rapid testing methods (devices) with easy use should be necessary to utilize PGx in drug administrations and that the development time of devices for PGx testing should be synchronized with that of drugs as much as possible. Thus, general strategies of how to develop PGx testing devices with drugs should be taken into consideration.

The fourth issue is an establishment of general guidance regarding genomic biomarkers in drug development. The report pointed out that it should be clarified on how regulatory agency can accept biomarker use in drug development. General guidance regarding biomarker use describing points to consider for biomarker validation and its processes will be useful to encourage practical uses of biomarkers in drug development.

The final and fifth issue is an establishment of general guidance regarding clinical trial designs using PGx. Various kinds of designs are possible for PGx clinical trials. The report pointed out that general guidance describing how to design PGx clinical trials to meet with study objects will be useful to facilitate discussion and to promote conducting appropriate PGx clinical trials of which data will be submitted for regulatory review.

On the basis of recommendation in the report, MHLW has been working to move forward PGx issues. For harmonizing international regulatory approaches, these PGx issues should be discussed globally. ICH will be an appropriate forum for that purpose. ICH has established more than 50 harmonized guidelines in terms of quality, safety, and efficacy of drugs. In recent years, ICH established two new guidelines focusing on PGx. First one is the E15 guideline entitled "Definitions for Genomic Biomarkers, Pharmacogenomics, Pharmacogenetics, Genomic Data and Sample Coding Categories" (32), which was implemented in January 2008 in Japan. This guideline is the first guideline for ICH to adopt in an emerging science area, and it is a basis for future ICH guideline relating to PGx. The E15 guideline will help to make common understanding of PGx issues not only in ICH regions but also in non-ICH regions. The second guideline is the E16 guideline entitled "Biomarkers Related to Drug or Biotechnology Product Development: Context, Structure and Format of Qualification Submissions" (33), which was implemented in January 2011 in Japan. This guideline will promote biomarker data submission to regulatory agencies in an internationally harmonized way and international discussion regarding biomarker data among regulatory agencies.

In April, 2008, MHLW published a guideline regarding points to consider for evaluating genetic tests based on DNA chips (34). The purpose of this guideline is to facilitate considerations by industries and accelerate regulatory reviews. It describes what points are important to obtain reliable results examined by genetic tests including DNA chips and related equipment such as software.

On July 27, 2007, the final report (35) of the special committee for "Effective & Safe Drugs Quick to Patients" was published. This committee was established by MHLW for finding a way to accelerate drug development and a review process to eliminate "drug lag." In this report, the committee recommends to use new technology such as new biomarker and microdosing study in drug development. Thus, the report pointed out that those general principles should be published as guidance and be revised appropriately if new scientific knowledge is available.

In September 2008, the guideline entitled "Clinical Trials Using PGx" was published (36). This guideline describes the general principles of clinical trials using PGx, such as a way to collect samples for genetic test; what kinds of points should be described in a protocol and in the feedback of genetic test results to a subject.

In 2009, PMDA started the scientific consultation on PGx/biomarkers. In this process, industries are encouraged to come to PMDA for scientific discussions regarding PGx/biomarkers (37).

As mentioned above, many activities related to PGx have been going on in Japan. International collaborations will be important to make more appropriate harmonized regulatory approaches in this new field.

11.6.2 PMDA Omics Project Team

To manage PGx issues on the regulatory side, PMDA originally established the Pharmacogenomics Discussion Group (PDG) in 2005 and then reorganized it as the PMDA Omics Project (POP) team in 2009 to expand a scope including not only PGx but also other omics such as proteomics and metabonomics. The POP team currently consists of 27 members from various offices in PMDA, such as the Office of Regulatory Science, Office of New Drug, Office of Biologics, Office of Medical Devices, Office of Safety, Office of Compliance, and Office of Review Management (as of July 1, 2011), and have an internal meeting in a regular basis. The mission of POP is summarized in Box 11.2.

Box 11.2 MISSIONS OF POP (PMDA OMICS PROJECT)

- Share data/information and knowledge relating to omics.
- Discuss regulatory issues relating to omics.
- Keep decision consistency among reviewers/offices in PMDA.
- Evaluate omics data that are not directly related to individual drug/diagnostic device developments.

POP has also had informal meetings with industries/academia to understand updated scientific information and a practical situation in omics. The purpose of this informal meeting is to share PMDA's views with industries/academia. Since 2009, as described above, POP is responsible for the scientific consultation on PGx/biomarkers. Therefore, POP has played active roles for promoting omics use in drug developments.

In recent years, POP has also participated in the US FDA/EMEA joint voluntary exploratory data submission (VXDS) meeting as an observer. Strengthening collaboration with other international regulatory agencies such as US FDA and EMEA is important for promoting omics use in drug development appropriately and making harmonized regulatory approaches in omics.

11.7 APPROACHES TO UTILIZE PGx-BASED MEDICINE IN PRACTICAL SITUATIONS

PGx will be more advanced in the next few years as a result of new findings being accumulated every day. In such new fields of emerging science, a stepwise approach as shown in Figure 11.6 would be effective in disseminating PGx-based medicine in clinical practice for responding flexibly to new technologies. This approach is a cyclic process consisting of some activities including (i) communicating PGx information to the public based on the results of drug review by a regulatory agency, (ii) using the information practically in clinical settings and identifying issues to be solved, (iii) performing clinical trials etc. to address the issues and provide data for regulatory review, and (iv) reviewing the obtained new data and taking appropriate measures, such as revision of PGx information in a label/PI. This process will enable provision of appropriate objective PGx information approved by a regulatory agency, which is actually needed in practical clinical settings. Guidance can be established when enough data and experiences are accumulated for standardizing PGx-based medicine in clinical setting (5).

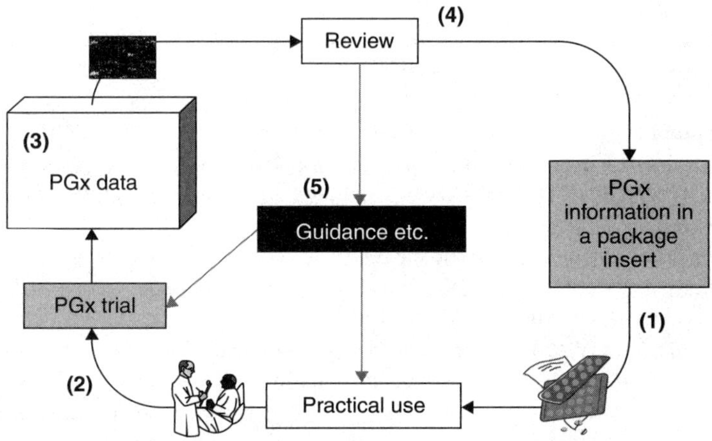

FIGURE 11.6 Approaches for practical utility of PGx-based medicine.

11.8 CONCLUSIONS

Establishing qualified/validated biomarker will be very important to provide drugs having better benefit/risk ratio to patients and to increase the efficiency of drug development. Ethnic differences in a biomarker, however, may have an effect on drug efficacy and safety. Therefore, early detection of such differences in drug development is critical, and early communication with regulatory agencies is important to realize the practical utility of biomarker in drug development. For that purpose, simultaneous global drug development from an early stage, which include various major populations, may be useful strategy to examine the effects of ethnic differences in biomarker on drug efficacy/safety and to avoid the delay of drug development in a region, resulting in an increased possibility to submit a new drug application globally at a similar time. Furthermore, close international collaborations between regulatory agencies will be necessary to establish harmonized regulatory approaches in this emerging science area.

REFERENCES

1. US-FDA. Challenge and Opportunity on the Critical Path to New Medical Products. March 2004.
2. Office of Pharmaceutical Industry Research. Report for the future vision of pharmaceutical industry; Mission and Task toward 2015, 2007.
3. Kola S, Landis J. Can the pharmaceutical industry reduce attrition rates? Nat Rev Drug Discov 2005;3:711–715.
4. Spear BB, Heath-Chiozzi M, Huff J. Clinical application of pharmacogenetics. Trends Mol Med 2001;7:201–204.

5. ICH E5(R1) guideline. Ethnic factors in the acceptability of foreign clinical data, 1998. Available at http://www.ich.org/fileadmin/Public_Web_Site/ICH_Products/Guidelines/Efficacy/E5_R1/Step4/E5_R1__Guideline.pdf. Accessed 2011 Dec 26.
6. Uyama Y, Shibata T, Nagai N, et al. Successful bridging strategy based on ICH E5 guideline for drugs approved in Japan. Clin Pharmacol Ther 2005;78:102–113.
7. Pharmaceuticals and Medical Devices Agency, Home page: information for approved drugs. Available at http://www.info.pmda.go.jp/approvalSrch/PharmacySrchInit? Accessed 2011 Dec 26.
8. Ingelman-Sundberg M, Sim SC, Gomez A, et al. Influence of cytochrome P450 polymorphisms on drug therapies: Pharmacogenetic, pharmacoepigenetic and clinical aspects. Pharmacol Ther 2007;116:496–526.
9. Bradford LD. CYP2D6 allele frequency in European Caucasians, Asians, Africans and their descendants. Pharmacogenomics 2002;3:229–243.
10. Nishida Y, Fukuda T, Yamamoto I, et al. CYP2D6 genotypes in a Japanese population: low frequencies of CYP2D6 gene duplication but high frequency of CYP2D6*10. Pharmacogenetics 2000;10:567–570.
11. FDA. Approval letter for change of irinotecan label. Available at http://www.accessdata.fda.gov/drugsatfda_docs/appletter/2005/020571s024,027,028ltr.pdf. Accessed 2011 Dec 26.
12. Ando Y, Saka H, Ando M, et al. Polymorphisms of UDP-glucuronosyltransferase gene and irinotecan toxicity: a pharmacogenetic analysis. Cancer Res 2000;60:6921–6926.
13. Sai K, Saeki M, Saito Y, et al. UGT1A1 haplotypes associated with reduced glucuronidation and increased serum bilirubin in irinotecan-administered Japanese patients with cancer. Clin Pharmacol Ther 2004;75:501–515.
14. PMDA. The package insert of irinotecan in Japan. Available at http://www.info.pmda.go.jp/go/pack/4240404A1040_1_04/. Accessed 2011 Dec 26.
15. Baselga J, Tripathy D, Mendelsohn J, et al. Phase II study of weekly intravenous recombinant humanized anti-p185HER2 monoclonal antibody in patients with HER2/neu-overexpressing metastatic breast cancer. J Clin Oncol 1996;14:737–744.
16. Lynch TJ, Bell DW, Sordella R, et al. Activating mutations in the epidermal growth factor receptor underlying responsiveness of non-small-cell lung cancer to gefitinib. N Engl J Med 2004;350:2129–2139.
17. Paez JG, Janne PA, Lee JC, et al. EGFR mutations in lung cancer: correlation with clinical response to gefitinib therapy. Science 2004;304:1497–1500.
18. Chung WH, Hung SI, Hong HS, et al. A marker for Stevens-Johnson syndrome, Science 2004;428:486.
19. Kaniwa N, et al. HLA-B locus in Japanese patients with anti-epileptics and allopurinol-related Stevens-Johnson syndrome and toxic epidermal necrolysis. Pharmacogenomics 2008;9:1617–1622.
20. FDA. Alert on Dec 12, 2007, Information for Healthcare Professionals (Carbamazepine (marketed as Carbatrol, Equetro, Tegretol, and generics), 2007. Available at http://www.fda.gov/Drugs/DrugSafety/PostmarketDrugSafetyInformationforPatientsandProviders/ucm124718.htm. Accessed 2011 Dec 26.
21. Tanaka H, Akaza T, Juji T. Report of the Japanese central bone marrow data center. Clin Transpl 1996;139–144.
22. The Allele frequency Net Database, Allele frequencies in Worldwide Populations. Available at http://www.allele frequencies.net/. Accessed 2011 Dec 26.

23. Ozeki T, et al. Genome-wide association study identifies HLA-A*3101 allele as a genetic risk factor for carbamazepine-induced cutaneous adverse drug reactions in Japanese population. Hum Mol Genet 2011;20:1034–1041.
24. Mori M, Yamada R, Kobayashi K, et al. Ethnic differences in allele frequency of autoimmune-disease-associated SNPs. J Hum Genet 2005;50:264–266.
25. The Wellcome Trust Case Control Consortium. Genome-wide association study of 14,000 cases of seven common diseases and 3,000 shared control. Nature 2007; 447:661–678.
26. Evaluation and Licensing Division, Pharmaceutical and Food Safety Bureau, Ministry of Health, Labour and Welfare. Basic principles on global clinical trials, Notification No. 0928010. 2007 Sept 28.
27. Pharmaceutical and Medical Devices Agency. Database for package insert information (in Japanese). Available at http://www.info.pmda.go.jp/psearch/html/menu_tenpu_base.html. Accessed 2011 Dec 26.
28. Evaluation and Licensing Division, Pharmaceutical and Food Safety Bureau, Ministry of Health, Labour and Welfare. Guidance on clinical pharmacokinetic studies of pharmaceuticals. Notification No. 796. 2001 Jun 1.
29. Evaluation and Licensing Division, Pharmaceutical and Food Safety Bureau, Ministry of Health, Labour and Welfare. Guidance on methods of drug interaction study. Notification No. 813. 2001 Jun 4.
30. Evaluation and Licensing Division, Pharmaceutical and Food Safety Bureau, Ministry of Health, Labour and Welfare. Submission of information to regulatory authorities for preparation of guidance on the use of Pharmacogenomics in clinical studies. Notification No. 0318001. 2005 Mar 18.
31. Committee for Genomics in the Japanese Society of Clinical Pharmacology. The report; current situations & future task for utilization of PGx in drug evaluation, 2007.
32. ICH E15 guideline. Definitions for Genomic Biomarkers, Pharmacogenomics, Pharmacogenetics, Genomic Data and Sample Coding Categories. Available at http://www.ich.org/fileadmin/Public_Web_Site/ICH_Products/Guidelines/Efficacy/E15/Step4/E15_Guideline.pdf. Accessed 2011 Dec 26.
33. ICH E16 guideline. Biomarkers related to drug or biotechnology product development: context, structure and format of qualification submissions. Available at http://www.ich.org/fileadmin/Public_Web_Site/ICH_Products/Guidelines/Efficacy/E16/Step4/E16_Step_4.pdf. Accessed 2011 Dec 26.
34. Evaluation and Licensing Division, Pharmaceutical and Food Safety Bureau, Ministry of Health, Labour and Welfare. Guidance; points to consider for evaluating genetic test devices based on DNA Chips, Notification No. 0404002. 2008 April 4.
35. Ministry of Health, Labour and Welfare, The final report of the committee for Effective & Safe Drugs Quick to Patients, Ministry of Health, Labour and Welfare, 2007 July 27.
36. Evaluation and Licensing Division, Pharmaceutical and Food Safety Bureau, Ministry of Health, Labour and Welfare. Guidance; Clinical trials using pharmacogenomics, Notification No. 0930008. 2008 September 30.
37. PMDA. Scientific consultation on pharmacogenomics/biomarkes. Available at http://www.pmda.go.jp/operations/shonin/info/consult/m03_pharma.html. Accessed 2011 Dec 26.

PART IV

PREDICTING *IN VIVO*

12

IN VITRO–IN VIVO CORRELATIONS OF HEPATIC DRUG CLEARANCE

R. Scott Obach

12.1 INTRODUCTION

The search for new pharmaceuticals is a highly challenging area of scientific research, with many pitfalls that can derail the discovery or development of a potential new therapeutic agent. Some of the most common causes of failure are unexpected toxicity and failure of the target mechanism to yield a disease modifying effect. Failures arising from inappropriate absorption, distribution, metabolism, and excretion (ADME) properties are also frequent; however, the advances made in gaining a greater understanding of the underlying biochemical aspects of drug disposition and their importance in the discovery and development process have led to the engagement of ADME scientists earlier in the drug research process to aid in the design of compounds with desirable pharmacokinetic characteristics. Thus, failures of new drugs because of poor ADME characteristics occur less frequently now than 10–15 years ago, when this reason was one of the most common causes. Much of this improvement has been due to the implementation of experiments and methods to predict human pharmacokinetics of new compounds using *in vitro* and animal data.

Failures of new drugs because of ADME properties often result because exposure in humans is lower than desired or needed, forcing the dose to be higher than possible or the frequency of administration so high as to be unreasonable or both. The half-life, which is a main determinant for the frequency of administration needed for efficacy, is driven by the rate of clearance and the extent of distribution of drug from the plasma. As a critical pharmacokinetic

Predictive Approaches in Drug Discovery and Development: Biomarkers and In Vitro/In Vivo Correlations, First Edition. Edited by J. Andrew Williams, Jeffrey R. Koup, Richard Lalonde, and David D. Christ.

property to manipulate through compound design, lowering the clearance is the preferred approach, rather than attempting to alter the volume of distribution. This is because clearance can, in many cases, be changed by simple changes in the structure of a compound, whereas the volume of distribution is often driven by the physicochemical characteristics of a molecule, which require greater changes in the structure that cannot be tolerated while maintaining activity at the pharmacological target. Additionally, since clearance is in many cases governed by metabolism in the liver, lowering clearance can also result in a greater oral bioavailability, due to the reduction in first-pass hepatic extraction. Greater oral bioavailability allows for a decreased dose level.

Drug clearance can also be mediated by the excretion of unchanged drug in urine or bile. Our knowledge of the biochemical and physiological mechanisms of renal and biliary clearance, the proteins that can be involved in these processes, and the genetic variability in these proteins, has only been emerging over the last 10 years or so and is not developed enough to permit using *in vitro* or animal data to make rigorous human clearance predictions. Our knowledge of the metabolic processes for drug clearance is substantially more developed and forms the basis by which predictions of human clearance can be made from *in vitro* data. This is particularly true for the hepatic cytochrome *P*450 (CYP) enzyme families. Knowledge of how to predict clearance when other drug-metabolizing enzymes (e.g., glucuronyltransferases, monoamine oxidases, esterases, and molybdenum cofactor-containing oxidases) are the predominant contributors to the clearance of a given new compound has only been recently emerging for some enzymes.

Much of our basic knowledge on how to predict clearance from *in vitro* data was originally developed using the rat. In 1977, Rane et al. (1) showed that by measuring the enzyme kinetics of metabolism and calculating the intrinsic clearance (CL_{int}), for a set of drugs using rat liver microsomes, hepatic extraction could be estimated. This is probably the first demonstration of an *in vitro–in vivo* correlation (IVIVC) for metabolic clearance. The principles underlying this approach were further developed by others (2–7); however, advances in the field really did not progress until human derived reagents, especially human liver microsomes, became readily available.

12.2 *IN VITRO* SYSTEMS

12.2.1 Liver Microsomes

Liver microsomes offer a convenient, simple, and well-defined source of several important human drug-metabolizing enzymes for determining intrinsic clearance measurements. Many of the most important drug-metabolizing enzymes such as the CYP superfamily, UDP-glucuronyltransferases (UGTs), and flavin monooxygenases (FMOs) are present in liver microsomes. Liver microsomes are actually an artifact of homogenization and subcellular fractionation. They are spherical membrane particles formed when the endoplasmic reticulum is disrupted. Many other important drug-metabolizing enzymes (e.g., sulfotransferases and aldehyde oxidases) are not present in microsomes, and therefore, one must be careful to ensure that the structure of the substrate being studied does not contain substituents susceptible to metabolism by other nonmicrosomal drug-metabolizing enzymes as a potential primary route for metabolic clearance. Nevertheless, because the CYP enzymes play a major role in the metabolism of well over 50% of drugs, the use of liver microsomes supplemented with cofactors necessary for CYP activity is a useful and preferred approach for predicting clearance from *in vitro* data.

Human liver microsomes can be frozen at −80°C for years without losses in activity (8). Thus, large batches can be prepared, characterized for their specific drug-metabolizing enzyme activities, stored in the freezer in small aliquots for use as needed, and individual donors can be pooled to reflect the metabolic activity of the "average" human liver. Enzyme activities will decrease with cycles of freezing and thawing, and it is generally recommended that microsomes be frozen and thawed only once or twice before discarding the remaining unused portion of an aliquot. This stability and tolerance to freezing has enabled commercial vendors to prepare human and animal liver microsomes for sale, which is an important convenience to drug metabolism researchers since the preparation and characterization of microsomes is a rather tedious exercise.

To support CYP activities in liver microsomes, a source of the essential cofactor NADPH must be provided. The catalytic cycle of CYP requires the input of reducing equivalents from NADPH-*P*450 reductase and in some cases cytochrome b_5, and NADPH serves as this source of reducing equivalents. NADPH can be added directly to incubations (at 1.0–1.5 mM) or provided by the addition of an NADPH regeneration system (typically isocitrate dehydrogenase + isocitric acid + NADP or glucose-6-phosphate dehydrogenase + glucose-6-phosphate + NADP). The advantage of using NADPH directly is greater convenience, while the advantage of a regeneration system is a lower cost. Also, using the regeneration system has the added advantage in that the buildup of NADP, which can inhibit the reductase, should not occur. A systematic comparison of the direct addition of NADPH or a regeneration system on *in vitro* intrinsic clearance determinations has not, however, been made. When conducting liver microsomal incubations to support CYP activity, it is important that the NADPH be added last to initiate the reaction, since the combination of CYP and NADPH will result in the generation of reactive oxygen species that will cause a decrease in activity. This consideration is in contrast to the optimal conditions for another microsomal enzyme system, the FMO. This enzyme is labile at ambient temperature when in the ground state form and therefore NADPH is added before the increase in temperature of the microsomes to stabilize the enzyme. This can make it challenging to conduct microsomal incubations for nitrogen-containing compounds, where both CYP and FMO may be involved in metabolism.

Liver microsomes also contain several UGT enzymes that can be involved in the metabolism of new compounds. To date, IVIVC for drugs cleared primarily via glucuronidation has not been widely successful, in part because extrahepatic tissues can have a significant contribution to glucuronidation (9–11). Another potentially confounding factor for UGTs is that on liver homogenization, these enzymes end up on the inside of the spherical microsomes. Optimizing activity *in vitro* requires the use of special conditions such as the addition of detergents or alamethicin (a reagent that can form pores in membranes) to permit entry of the necessary cofactor UDPGA (12). Saccharolactone, an inhibitor of glucuronidases that would hydrolyze any glucuronides formed, is also routinely added to prevent catabolism of the glucuronide metabolites. Obtaining valid UGT enzyme kinetics for use in prediction of CL_{int} requires careful consideration of the *in vitro* conditions (13–15). Recently, it has been demonstrated that the addition of albumin or fatty acid-binding protein to incubations can increase the activity of UGT enzymes (16, 17) and increase the accuracy for predicting CL_{int} (18).

12.2.2 Hepatocytes

Suspensions of hepatocytes offer the key advantages over liver microsomes in that a complete set of hepatic drug-metabolizing enzymes are present, including cytosolic enzymes, and the biochemical pathways for cofactor production are generally active. Uptake transporters, which could also influence the rate of clearance, are also often present and active. However, until recently, the use of hepatocytes for IVIVC has not been common, due to cost, availability, and inconvenience. Advances in cryopreservation techniques have led to an increase in our ability to use hepatocytes for clearance predictions. In the past, hepatocyte experiments were critically dependent on the availability of fresh liver tissue, and the process of isolating the cells had to be initiated immediately on receipt of the tissue. The metabolism experiments also had to be conducted immediately on the availability of viable cells and were done concurrently with experiments to characterize the metabolic activities of the donor liver. At the end of the experiment, there was a chance that the effort was wasted because cell viability was poor or activity was compromised. These difficulties have largely been overcome by cryopreservation. Hepatocytes are still prepared from fresh tissue, but they can be characterized for activities and stored in liquid nitrogen. Cryopreservation also allows for the construction of pooled preparations so that measurements can be made in cells reflective of an average human. The availability of cryopreserved hepatocytes has also led to an improvement in our ability to measure induction of drug-metabolizing enzymes *in vitro*, a topic covered in another chapter.

Conducting drug metabolism experiments using cryopreserved human hepatocyte suspensions, while becoming simpler over the years, is still more complicated than experiments with microsomal incubations and requires additional care. The technique used to thaw the hepatocytes requires great care, and the resulting cell viability must be determined. Incubations need to be conducted under a more defined atmosphere of $95\%O_2/5\%CO_2$, whereas microsomal incubations are typically conducted open to the ambient air. Moreover, sampling the incubation mixture must also be done carefully to ensure the most accurate estimates of CL_{int}.

The experience in predicting *in vivo* clearance using hepatocytes has been mixed, with clearance both under- and accurately predicted (19–22). One key consideration, for example, where the clearance was underpredicted was the observation that sampling of the hepatocyte incubations over time for drug analysis included a mixture of cells and media. It has been found more recently that sampling just the media for loss of the compound may provide a better picture of intrinsic clearance, especially when hepatic uptake processes, in addition to metabolism, also contribute to clearance. Using this approach, CL_{int} determined from the initial rate of decline of the test compound in media was better correlated to *in vivo* clearance (23).

12.3 CONDUCTING *IN VITRO* EXPERIMENTS

12.3.1 *In Vitro* Half-Life Determinations

In an early drug discovery setting, authentic standards of metabolites and assays specific for them or radiolabeled compounds are generally not available. Thus, conventional enzyme kinetic experiments for drug metabolism reactions cannot be conducted because the formation rates of metabolites cannot be determined quantitatively. To overcome this limitation, ADME scientists have adopted an approach, whereby the first-order consumption of the parent compound (k_{deg}) is measured on incubation with liver microsomes to assess the CL_{int} due to CYP-mediated metabolism. Determining the half-life for the disappearance of parent substrate is key to calculating this first-order rate constant. Provided that the substrate concentration is well below K_M, the first-order rate constant of substrate consumption can be converted to an intrinsic clearance value according to the following equations:

$$\textit{In vitro } t_{1/2} = \frac{0.693}{k_{deg}} \tag{12.1}$$

and

$$CL_{int} = \frac{0.693}{\textit{in vitro } t_{1/2}} \times \frac{\text{Incubation volume}}{\text{Amount of microsomal protein}} \tag{12.2}$$

While K_M is generally not known for new compounds, the substrate concentration used is typically low (i.e., $\leq 1.0\ \mu M$), but easily measured analytically, and is assumed to be within the linear range of the enzyme velocity versus substrate curve. The *in vitro* half-life will also depend on the amount of enzyme in the incubation, thus to provide values that can be compared from one compound to the next, the conditions used need to be the same or they must be corrected for the protein concentration. Liver microsomal protein concentration is usually kept below 2 mg/ml to avoid excessive nonspecific binding (discussed more extensively in Section 12.5.2). The disappearance of substrate must also be linear with time (to satisfy initial velocity conditions), so the incubation and sampling times are relatively short, approximately 30–40 min or less.

The experiment is conducted by first mixing the buffer, liver microsomes, and substrate, followed by prewarming the mixture to 37°C for 5 min. An aliquot of the mixture is immediately taken as a time = 0 min, to which subsequent concentration measurements are referenced. The reaction is commenced with addition of NADPH, and aliquots of the incubation mixture are taken at four to six timepoints and the reaction terminated, usually by adding the incubation aliquot to miscible organic solvent or aqueous acid to precipitate the protein. The terminated incubation mixtures are centrifuged to remove the precipitated protein and the supernatant is analyzed for the parent compound remaining by HPLC coupled with a mass spectrometer or other detector. The logarithm of the response factor (concentration) for the parent compound is plotted versus incubation time and the data fit to a first-order decay function. Samples that no longer are on the first-order decay curve are not included in the fitting, as these would generate artificially long half-lives. This approach represents a facile method to determine CL_{int} values for predicting *in vivo* clearance when faced with a diverse set of molecules early in the drug discovery process.

12.3.2 Formal Enzyme Kinetic Experiments

The measurement of enzyme kinetics for drug metabolism reactions, unlike the *in vitro* half-life approach, represents the most precise and definitive method for determining CL_{int}. However, because these experiments require considerably greater effort and the availability of expensive reagents (i.e., metabolite standards or radiolabeled parent compound), they are not amenable to early drug research when hundreds of compounds are being studied for a given therapeutic project. Later in the drug development process, the measurement of enzyme kinetics for the individual metabolic reactions is an essential part of the characterization of new compounds. These measurements are often needed before the experiments designed to determine which enzymes metabolize a new compound ("reaction phenotyping") can be conducted.

In enzyme kinetic experiments, the product(s) formed is measured with a robust analytical method using standard curves made with authentic standards of each metabolite. Initial velocity conditions are required for rigorous enzyme kinetic determinations and linearity with incubation time and protein concentration must be demonstrated for the formation of each metabolite. The longest time for which product formation is linear with time is selected, along with the lowest protein concentration possible that still yields enough metabolite to be quantitated reliably. To determine the enzyme kinetic parameters K_M and V_{max}, the rate of formation of metabolite (v) is measured at multiple substrate concentrations. The data are fitted to a number of typical enzyme kinetic models that commonly describe drug metabolism reactions:

$$v = \frac{V_{max} \cdot [S]}{K_M + [S]} \quad \text{(simple Michaelis–Menten)} \tag{12.3}$$

$$v = \frac{V_{max} \cdot [S]}{K_M + [S] + \frac{[S]^2}{K_i}} \quad \text{(substrate inhibition)} \tag{12.4}$$

$$v = \frac{V_{\max} \cdot [S]^n}{S_{50}^n + [S]^n} \text{ (substrate activation)} \tag{12.5}$$

$$v = \frac{V_{\max} \cdot [S]}{K_M + [S]} + CL_{int(2)} \cdot [S] \text{ (two-enzyme model)} \tag{12.6}$$

Substrate inhibition (Eq. 12.4) and substrate activation (autoactivation; Eq. 12.5) may occur for enzymes with multiple catalytic sites or for enzymes such as CYP3A4, where multiple binding sites within the catalytic "pocket" exist, and manifest as characteristically curved Eadie–Hofstee plots (24, 25). These more complicated enzyme kinetics may not be evident using the half-life approach discussed earlier and would lead to inaccurate predictions of CL. Autoactivation has also been shown for UGT enzymes both *in vitro* and *in vivo* for valproic acid metabolism in sheep (26) and numerous substrates *in vitro* for humans (14, 27).

Intrinsic clearance values can then be calculated for each metabolic reaction by

$$CL_{int} = \frac{V_{\max}}{K_M} \tag{12.7}$$

If a drug is metabolized by multiple pathways, the sum of the CL_{int} terms for each individual reaction yields the total CL_{int} of the parent (since clearance is an additive process).

12.3.3 Screening Approaches with Human Liver Microsomes

Strategically, the determination of metabolic lability is simple enough to be amenable to high throughput approaches. Advances in robotics and HPLC-MS have facilitated the development of 96- and 384-well automated microsomal lability screens, which can be used to assess stability for virtually every new compound synthesized. The intent of gathering these data at this stage is to assign the potential for high, medium, or low CL_{int} rather than precisely predict *in vivo* clearance. As such, the number of sampling times can be abbreviated. Analytical methods to measure substrate consumption utilize rapid chromatography (in many cases loading the analytes on a short reverse-phase HPLC column in high aqueous mobile phase followed by an almost immediate ramping to high organic mobile phase to elute the analyte from the column very rapidly) with mass spectrometric detection. The potentials in the mass spectrometer are tuned to a condition that is applicable to the vast majority of small organic molecules, and the instrument can be programmed to monitor the protonated or deprotonated molecular ions (in positive or negative ion modes, respectively). This approach will not capture data for every compound but a loss of accuracy is accepted to gain rapid and extensive throughput at an early stage of compound profiling.

12.3.4 An Alternative Screening Approaches with Hepatocytes: Utilizing AUC

Hepatocytes are generally not as amenable as microsomes for high throughput screening methods. Determining CL_{int} with hepatocytes can be done by measuring *in vitro* $t_{1/2}$ as described above, or more commonly, using the area under the concentration–time curve (AUC) (since clearance = amount/AUC). Incubations can be done with freshly suspended hepatocytes, under a $95\%O_2/5\%CO_2$ atmosphere at 37°C. Samples are removed at $t = 0$ and subsequent timepoints typically extending out 2–4 h. In most early work, a homogeneous mixture of cells and medium were sampled and added to termination solvent; however, more recently, it has been advocated that only the medium is sampled, as part of hepatocyte CL_{int} is due to active uptake processes. Substrate concentration versus time is plotted and the AUC determined typically by the trapezoidal rule. Intrinsic clearance is calculated as follows:

$$CL_{int} = \frac{[S]_{t=0}}{AUC_{0-\infty}} \cdot \frac{\text{Incubation volume}}{\text{Number of hepatocytes}} \tag{12.8}$$

12.4 APPLYING SCALING FACTORS AND PHYSIOLOGICAL MODELS

To make predictions of clearance in humans and to develop IVIVCs, the *in vitro* CL_{int} data must be scaled to *in vivo* CL_{int} and used in a physiological model of hepatic clearance. This scaling is done by applying simple mathematical factors specific for each system and species. For liver microsomes, several values have been listed for the microsomal content per gram of liver, generally ranging between 30 and 70 mg/g liver (in noninduced liver), and although this value has been disputed (28), a value of 45 mg/g has been most widely used. Furthermore, whether there are differences between species for this value has not been well studied, although there is no physiological rationale behind why this value would differ substantially across mammalian species. The CL_{int} value must also be scaled to the liver weight in each species. In this case, there are well-established differences across the species, with larger species generally having less liver, when corrected for body weight. Values used in mouse, rat, dog (beagle), monkey (cynomolgus), and human are 90, 45, 30, 32, and 20 gm liver/kg body weight (29). Thus, for human, *in vitro* CL_{int} values corrected to milligram microsomal protein are multiplied by a factor of 900 to yield *in vivo* CL_{int} values in units per kilogram body weight. For hepatocytes, a value of 120,000,000 cells/g liver has generally been used for the value of hepatocellularity (28, 30, 31).

The cardinal assumption in the development of IVIVC using hepatic tissue preparations is that systemic clearance of the drug is equal to its hepatic clearance. As described by Wilkinson and Shand (32), hepatic clearance and extraction are dependent on two independent variables, the biochemistry of CL_{int} and the physiology of hepatic blood flow. Therefore, the *in vivo* CL_{int} value that has been scaled from *in vitro* data must be inserted into a model of hepatic clearance, which accounts for blood flow. The most commonly employed model is the well-stirred model (33) that treats the liver as a homogeneous organ (Eq. 12.9). Clearly, this is an oversimplification, but nevertheless, the application of this model provides useful predictions of hepatic clearance from *in vitro* data:

$$CL_h = \frac{Q_h \cdot f_u \cdot CL_{int}}{Q_h + f_u \cdot CL_{int}} \tag{12.9}$$

In this model, hepatic clearance CL_h is a function of hepatic blood flow (Q_h), which provides an upper limit value, the unbound (free) fraction in blood (f_u), and *in vivo* free CL_{int}. The free fraction in blood is determined by multiplying the free fraction in plasma by the blood-to-plasma partition ratio; however, many investigators make the assumption that the blood-to-plasma ratio is unity (but in some cases, this may not be true). It is important to note that this model uses *free* intrinsic clearance. In many cases, it is assumed that the *in vitro* CL_{int} value used to scale to *in vivo* CL_{int} is an unbound value, that is, there is no binding in the *in vitro* experiment. However, this is not always true, and the effects of protein binding are discussed in greater detail subsequently. Values for Q_h differ across species, with lower values observed in larger species. Commonly used values of Q_h are 90, 70, 35, 44, and 21 ml/min/kg body weight for mouse, rat, dog (beagle), monkey (cynomolgus), and human, respectively. Other models of hepatic clearance, such as the parallel tube and dispersion models, have been described and in some cases used for IVIVC, but the advantages offered by these more complex models seem to be minimal.

In addition to hepatic, systemic clearance, *in vitro* CL_{int} data can be used to predict hepatic extraction (E_h), which can be a large contributing factor to oral bioavailability (F). The fraction that evades first-pass extraction (F_h) is inversely related to hepatic extraction:

$$F_h = 1 - E_h = 1 - \left(\frac{CL_h}{Q_h}\right) \tag{12.10}$$

$$F = F_a \cdot F_g \cdot F_h \tag{12.11}$$

Calculating the fraction evading hepatic extraction can then be used, together with the fraction absorbed (F_a) and the fraction escaping extraction by the gut (F_g), to estimate F.

12.5 SPECIAL ASPECTS OF PREDICTING CLEARANCE *IN VITRO*

12.5.1 Cross-Species Correlations

In the vast majority of cases, CL_{int} is predicted long before clinical pharmacokinetic data are available to validate the approach for a molecule or series of molecules. A powerful complement to using *in vitro* metabolism data to predict human clearance of new compounds is the development of cross-species correlations. By determining CL_{int} and hence CL_h for a compound of interest using liver microsomes or hepatocytes from laboratory animal species and comparing these data to the *in vivo* CL determined after intravenous administration, confidence can be gained in the fidelity of the IVIVC for that compound. Comparison of the predicted CL_h and CL measured across species can reveal whether the predictions are in agreement numerically, whether the rank order of clearances in different species (*in vitro* and *in vivo*) are the same, whether there is a correlation that has a systematic difference between CL predicted from *in vitro* data and *in vivo* CL, or whether there is no correlation at all.

If CL in animals are well predicted from the *in vitro* data then predicting human CL from *in vitro* data can be made with confidence. Implicit in this confidence is the assumption that humans do not clear the compound by a completely different mechanism than in animals. Such a supposition is rarely in error, but there can be instances in which the CYP enzyme system is the main route involved in CL in animals, while in humans, metabolic clearance is mediated by other enzymes. It is rare that a compound cleared by metabolism in animals is cleared as unchanged drug in humans.

In many cases, a cross-species IVIVC can be observed, but all the predictions of CL from *in vitro* data differ systematically from the actual values (in many cases underpredicted, but in some cases overpredicted). There is still much value in constructing the correlation, since it can be assumed that there is some systematic, albeit unknown, error contributing to the difference (34). To correct the human CL value predicted from *in vitro* data for this discrepancy, CL_h is simply multiplied by the averaged factor by which the predictions in animals differ from the actual *in vivo* values.

Lack of a cross-species correlation altogether suggests that the prediction of human CL from *in vitro* data cannot be made with confidence. In this case, other prediction approaches, such as allometric scaling, may be warranted.

12.5.2 Nonspecific Binding to Microsomal Components

Recall that scaling to *in vivo* hepatic clearance using Equation 12.9 uses free (unbound) drug concentrations and that the scaled CL_{int} value inserted into the model represents a free CL_{int} value. It had long been assumed that there was negligible binding of compounds in the *in vitro* incubation, such that CL_{int} measured was in fact a free CL_{int} value. However, this is not always the case, and for some classes of compounds (e.g., lipophilic cationic drugs), the extent of binding in the *in vitro* incubation can be high (i.e., >90%) (35, 36). Binding to *in vitro* incubation matrices is due to binding to phospholipid membranes, rather than to the protein component itself (37), and this phenomenon should not be confused with plasma protein binding; one is not a surrogate for the other.

To minimize the impact of this nonspecific binding, low concentrations of protein should be used. When conducting examinations of linearity of velocity with protein concentration, nonspecific binding will be one of the causes of nonlinearity at high protein concentrations. When determining CL_{int} using the *in vitro*

$t_{1/2}$ approach, protein concentrations will generally need to be higher because enough enzyme activity must be present to catalyze considerable consumption of substrate. To more rigorously account for the effect of this binding, CL_{int} values should be converted to free CL_{int} values by measuring the degree of binding occurring under the incubation conditions used. This can be done experimentally using equilibrium dialysis or ultrafiltration or even calculated using computational approaches (38–42). When this factor is ignored, CL_{int} values can be artificially low, leading to the underprediction of CL_h. Furthermore, when CL_{int} data are used in compound design and optimization, increases in lipophilicity often deceptively appear to decrease CL_{int}, when actually the apparent increase in metabolic stability is merely due to increased nonspecific binding.

12.5.3 Important Considerations for Compounds with Low K_M: Estimation of K_M from Substrate Depletion and Saturation of Metabolism

If the K_M value for a drug is low, there is a chance that hepatic concentrations of the drug *in vivo* will be high enough to saturate clearance, leading to plasma concentrations that are greater than expected. Recall that earlier CL_{int} was described merely as the ratio of V_{max} and K_M; however, this is a simplification and is only true when $[S] \ll K_M$ and the reaction is first order (i.e., the rate is proportional to substrate concentration). The actual relationship is as follows:

$$CL_{int} = \frac{V_{max}}{K_M + [S]} \tag{12.12}$$

Thus, as substrate concentration [S] increases, CL_{int} decreases, leading to partial saturation of clearance. Typically, *in vitro* $t_{1/2}$ experiments are done at low [S] (e.g., 1 μM) in the hope that this concentration will be far enough below K_M so that the estimated CL_{int} will be accurate. However, in some instances, this is not the case and K_M values can be near or above 1 μM, posing a problem for *in vitro* $t_{1/2}$ experiments with compounds appearing to be more stable than they actually are. To overcome this, K_M values can be estimated by measuring substrate depletion $t_{1/2}$ values at multiple, low substrate concentrations (43, 44). While these experiments are too resource intensive to be used routinely in early drug research on many compounds, as particular individual compounds become of greater interest, a check can be made to ensure that CL_{int} was determined at a suitably low substrate concentration. Measurement of *in vitro* $t_{1/2}$ at two low substrate concentrations that are 10-fold different or more, and producing values are not widely disparate (e.g., within a factor of 2), can provide assurance that CL_{int} values are acceptable and can be used to predict CL_h.

If the K_M for the metabolism of a new compound is low (i.e., <1 μM), there is a chance that a supraproportional dose–exposure relationship could be observed *in vivo*. Such a relationship can be a problem for a drug that may have a narrow therapeutic index since dose adjustments can be challenging during therapy. Moreover, substrates with low K_M values for specific enzymes can often be associated with the substrate inhibiting the metabolism of other substrates, resulting in drug interactions. A method to predict the dose range at which supraproportional exposures will be observed from *in vitro* K_M data is presently not available.

12.5.4 Autoinactivation and Nonstationary Pharmacokinetics

For some drugs (e.g., paroxetine), repeated administration results in a decrease in clearance due to mechanism-based inactivation of the enzymes involved in their clearance, and this phenomenon can be modeled using *in vitro* approaches (45). Evidence of such a phenomenon can often be discerned from the initial determinations of reaction velocity linearity. If a marked decrease in reaction velocity is observed at early incubation timepoints, then autoinactivation can be occurring. If the *in vitro* $t_{1/2}$ approach is being utilized to determine CL_{int} and predict CL, autoinactivation can be manifested as a nonlinear log [S] versus time plot. A rapid screening method for predicting the impact that autoinactivation can have on the magnitude of the nonstationarity (i.e., how much accumulation will be observed beyond that expected from single-dose kinetics) is not known.

12.5.5 Applying IVIVC to Clearance Mediated by Non-CYP Enzymes

The methods that have been described in this chapter for predicting CL from *in vitro* data have been well explored and documented for CL that is mediated by CYP-catalyzed metabolism. These approaches have not, however, been as well established for other metabolic CL pathways. Conjugation with glucuronic acid is another major metabolic route of clearance, and for compounds cleared primarily by glucuronidation, attempts to predict CL from *in vitro* CL_{int} data have produced mixed success (9, 10); however, more recently, the addition of albumin to incubations has led to better predictions (18). The addition of albumin is proposed to bind fatty acids, present in the incubation, which can otherwise inhibit some UGT enzymes (16, 17). Extrahepatic expression of some UGT enzymes may be confounding this IVIVC. To date, CL prediction methods for drugs cleared by other enzymes such as monoamine oxidases, aldehyde oxidase, or FMO from *in vitro* CL_{int} data have not been established, although the fundamental enzyme kinetic principles described in this chapter should be applicable. One important limitation is that scaling factors for these enzymes have not been described, and some have considerable extrahepatic expression. For example, efforts to predict *in vivo* CL_{int} for drugs metabolized by aldehyde oxidase led to the conclusion that *in vitro* experiments markedly underpredicted the *in vivo* data, yet a good correlation could be established (46). More research and the availability of high quality reagents are needed in this area.

12.6 CONCLUSIONS

Predicting human CL using *in vitro* approaches and cross-species correlations has become an expectation in drug research when selecting compounds for further development that are more likely to have desirable pharmacokinetic attributes. Acceptable pharmacokinetics, especially after oral dosing, is necessary to help ensure optimal therapy because patient compliance will be more easily achieved with convenient dosing regimens. Pharmacokinetic predictions also include prediction of dose level, which is important in early drug research to understand whether the dose needed will be reasonable and safe and to ensure that cost-of-goods estimates are commercially viable. The routine availability of high quality human and animal hepatic microsomes and cryopreserved hepatocytes, coupled with advances in robotic sample preparation, incubation, and analysis procedures, together with computerized data collection and calculation algorithms have made the prediction of human CL an integral part of new drug discovery. While clearance prediction methods are best established for drugs cleared primarily by CYP-mediated oxidation, more research is needed for other important CL pathways.

REFERENCES

1. Rane A, Wilkinson GR, Shand DG. Prediction of hepatic extraction ratio from *in vitro* measurement of intrinsic clearance. J Pharmacol Exp Ther 1977;200:420–424.
2. Baarnhielm C, Dahlback H, Skanberg I. *In vivo* pharmacokinetics of felodipine predicted from *in vitro* studies in rat, dog and man. Acta Pharmacol Toxicol 1986;59:113–122.
3. Houston JB. Utility of *in vitro* drug metabolism data in predicting *in vivo* metabolic clearance. Biochem Pharmacol 1994;47:1469–1479.
4. Iwatsubo T, Hirota N, Ooie T, Suzuki H, Sugiyama Y. Prediction of *in vivo* drug disposition from *in vitro* data based on physiological pharmacokinetics. Biopharm Drug Dispos 1996;17:273–310.
5. Houston JB, Carlile DJ. Prediction of hepatic clearance from microsomes, hepatocytes, and liver slices. Drug Met Rev 1997;29:891–922.
6. Obach RS, Baxter JG, Liston TE, Silber BM, Jones BC, McIntyre F, Rance DJ, Wastall P. The prediction of human pharmacokinetic parameters from preclinical and *in vitro* metabolism data. J Pharmacol Exp Ther 1997;283:46–58.
7. Ito K, Houston JB. Prediction of human drug clearance from *in vitro* and preclinical data using physiologically based and empirical approaches. Pharm Res 2005;22:103–112.
8. Pearce RE, McIntyre CJ, Madan A, Sanzgiri U, Draper AJ, Bullock PL, Cook DC, Burton LA, Latham J, Nevins C, Parkinson A. Effects of freezing, thawing, and storing human liver microsomes on cytochrome P450 activity. Arch Biochem Biophys 1996;331:145–169.
9. Soars MG, Burchell B, Riley RJ. *In vitro* analysis of human drug glucuronidation and prediction of *in vivo* metabolic clearance. J Pharmacol Exp Ther 2002;301:382–390.
10. Miners JO, Knights KM, Houston JB, Mackenzie PI. *In vitro-in vivo* correlation for drugs and other compounds eliminated by glucuronidation in humans: pitfalls and promises. Biochem Pharmacol 2006;71:1531–1539.

11. Mohutsky MA, Chien JY, Ring BJ, Wrighton SA. Predictions of the *in vivo* clearance of drugs from rate of loss using human liver microsomes for phase I and phase II biotransformations. Pharm Res 2006;23:654–662.
12. Fisher MB, Campanale K, Ackermann BL, Vandenbranden M, Wrighton SA. *In vitro* glucuronidation using human liver microsomes and the pore-forming peptide alamethicin. Drug Metab Dispos 2000;28:560–566.
13. Boase S, Miners JO. *In vitro-in vivo* correlations for drugs eliminated by glucuronidation: investigations with the model substrate zidovudine. Br J Clin Pharmacol 2002;54:493–503.
14. Soars MG, Ring BJ, Wrighton SA. The effect of incubation conditions on the enzyme kinetics of UDP-glucuronosyltransferases. Drug Metab Dispos 2003;31:762–767.
15. Engtrakul JJ, Foti RS, Strelevitz TJ, Fisher MB. Altered AZT (3′-azido-3′-deoxythymidine) glucuronidation kinetics in liver microsomes as an explanation for underprediction of *in vivo* clearance: comparison to hepatocytes and effect of incubation environment. Drug Metab Dispos 2005;33:1621–1627.
16. Rowland A, Knights KM, Mackenzie PI, Miners JO. The "albumin effect" and drug glucuronidation: bovine serum albumin and fatty acid-free human serum albumin enhance the glucuronidation of UDP-glucuronosyltransferase (UGT) 1A9 substrates but not UGT1A1 and UGT1A6 activities. Drug Metab Dispos 2008;35:1056–1062.
17. Rowland A, Knights KM, Mackenzie PI, Miners JO. Characterization of the binding of drugs to human intestinal fatty acid binding protein (IFABP): potential role of IFABP as an alternative to albumin for *in vitro-in vivo* extrapolation of drug kinetic parameters. Drug Metab Dispos 2009;37:1395–1403.
18. Kilford PJ, Stringer R, Sohal B, Houston JB, Galetin A. Prediction of drug clearance by glucuronidation from *in vitro* data: use of combined cytochrome P450 and UDP-glucuronosyltransferase cofactors in alamethecin-activated human liver microsomes. Drug Metab Dispos 2009;37:82–89.
19. McGinnity DF, Soars MG, Urbanowicz RA, Riley RJ. Evaluation of fresh and cryopreserved hepatocytes as *in vitro* drug metabolism tools for the prediction of metabolic clearance. Drug Metab Dispos 2004;32:1247–1253.
20. Riley RJ, McGinnity DF, Austin RP. A unified model for predicting human hepatic metabolic clearance from *in vitro* intrinsic clearance data in hepatocytes and microsomes. Drug Metab Dispos 2005;33:1304–1311.
21. Hallifax D, Rawden HC, Hakooz N, Houston JB. Prediction of metabolic clearance using cryopreserved human hepatocytes: kinetic characteristics for five benzodiazepines. Drug Metab Dispos 2005;33:1852–1858.
22. Brown HS, Griffin M, Houston JB. Evaluation of cryopreserved human hepatocytes as an alternative *in vitro* system to microsomes for the prediction of metabolic clearance. Drug Metab Dispos 2007;35:293–301.
23. Soars MG, Grime K, Sproston JL, Webborn PJH, Riley RJ. Use of hepatocytes to assess the contribution of hepatic uptake to clearance *in vivo*. Drug Metab Dispos 2007;35:859–865.
24. Ueng Y-F, Kuwabara T, Chun Y-J, Guengerich FP. Cooperativity in oxidations catalyzed by cytochromes P450 3A4. Biochemistry 1997;36:370–381.
25. Korzekwa KR, Krishnamachary N, Shou M, Ogai A, Parise RA, et al. Evaluation of atypical cytochrome P450 kinetics with two-substrate models: evidence that multiple substrates can bind to cytochrome P450 active sites. Biochemistry 1998;37:4137–4147.

26. Wong H, Tong V, Riggs KW, Rurak DW, Abbott FS, Kumar S. Kinetics of valproic acid glucuronidation: evidence for *in vivo* autoactivation. Drug Metab Dispos 2007;35:1380–1386.
27. Lewis BC, Mackenzie PI, Elliot DJ, Burchell B, Bhasker CR, Miners JO. Amino terminal domains of human UDP-glucuronosyltransferase (UGT) 2B7 and 2B15 associated with substrate selectivity and autoactivation. Biochem Pharmacol 2007;73:1463–1473.
28. Barter ZE, Bayliss MK, Beaune PH, Boobis AR, Carlile DJ, Edwards RJ, et al. Scaling factors for the extrapolation of *in vivo* metabolic drug clearance from *in vitro* data: reaching a consensus on values of human microsomal protein and hepatocellularity per gram of liver. Curr Drug Metab 2007;8:33–45.
29. Davies B, Morris T. Physiological parameters in laboratory animals and humans. Pharm Res 1993;10:1093–1095.
30. Carlile DJ, Zomorodi K, Houston JB. Scaling factors to relate drug metabolic clearance in hepatic microsomes, isolated hepatocytes, and the intact liver. Studies with induced livers involving diazepam. Drug Metab Dispos 1997;25:903–911.
31. Sohlenius-Sternbeck A-K. Determination of the hepatocellularity number for human, dog, rabbit, rat and mouse livers from protein concentration measurements. Toxicol *in vitro* 2006;20:1582–1586.
32. Wilkinson GR, Shand DG. A physiological approach to hepatic drug clearance. Clin Pharmacol Ther 1975;18:377–390.
33. Pang KS, Rowland M. Hepatic clearance of drugs. I. Theoretical considerations of a "well-stirred" model and a "parallel tube" model. Influence of hepatic blood flow, plasma and blood cell binding, and the hepatocellular enzymatic activity on hepatic drug clearance. J Pharmacokinet Biopharm 1977;5:625–653.
34. Obach RS. The prediction of human clearance from hepatic microsomal metabolism data. Curr Opin Drug Discov Devel 2001;4:36–44.
35. Obach RS. Prediction of human clearance of twenty-nine drugs from hepatic microsomal intrinsic clearance data: an examination of *in vitro* half-life approach and nonspecific binding to microsomes. Drug Metab Dispos 1999;27:1350–1359.
36. McLure JA, Miners JO, Birkett DJ. Nonspecific binding of drugs to human liver microsomes. Br J Clin Pharmacol 2000;49:453–461.
37. Margolis JM, Obach RS. Impact of nonspecific binding to microsomes and phospholipid on the inhibition of cytochrome P4502D6: Implications for relating *in vitro* inhibition data to *in vivo* drug interactions. Drug Metab Dispos 2003;31:606–611.
38. Obach RS. Nonspecific binding to microsomes: impact on scale-up of *in vitro* intrinsic clearance to hepatic clearance as assessed through examination of warfarin, imipramine, and propranolol. Drug Metab Dispos 1997;25:1359–1369.
39. Giuliano C, Jairaj M, Zafiu CM, Laufer R. Direct determination of unbound intrinsic drug clearance in the microsomal stability assay. Drug Metab Dispos 2005;33:1319–1324.
40. Austin RP, Barton P, Mohmed S, Riley RJ. The binding of drugs to hepatocytes and its relationship to physicochemical properties. Drug Metab Dispos 2005;33:419–425.
41. Hallifax D, Houston JB. Binding of drugs to hepatic microsomes: comment andassessment of current prediction methodology with recommendation for improvement. Drug Metab Dispos 2006;34:724–726.

42. Sykes MJ, Sorich MJ, Miners JO. Molecular modeling approaches for the prediction of the nonspecific binding of drugs to hepatic microsomes. J Chem Inf Model 2006;46:2661–2673.
43. Obach RS, Reed-Hagen AE. Measurement of Michaelis constants for cytochrome P450-mediated biotransformation reactions using a substrate depletion approach. Drug Metab Dispos 2002;30:831–837.
44. Nath A, Atkins WM. A theoretical validation of the substrate depletion approach to determining kinetic parameters. Drug Metab Dispos 2006;34:1433–1435.
45. Venkatakrishnan K, Obach RS. *In vitro-in vivo* extrapolation of CYP2D6 inactivation by paroxetine: prediction of nonstationary pharmacokinetics and drug interaction magnitude. Drug Metab Dispos 2005;33:845–852.
46. Zientek M, Jiang Y, Youdim K, Obach RS. *In vitro-in vivo* correlation for intrinsic clearance for drugs metabolized by human aldehyde oxidase. Drug Metab Dispos 2010;38:1322–1327.

13

THE POTENTIAL OF *IN SILICO* AND *IN VITRO* APPROACHES TO PREDICT *IN VIVO* DRUG–DRUG INTERACTIONS AND ADMET/TOX PROPERTIES

KENNETH BACHMANN AND SEAN EKINS

13.1 INTRODUCTION

It is widely known that the metabolic transformations of pharmaceuticals profoundly impact bioavailability, efficacy, chronic toxicity, and the rate and route of excretion. Both the parent molecule and the products of metabolic transformations may also interfere with other coadministered compounds. A system of transporters, channels, receptors, and enzymes acts as gatekeepers to foreign molecules, affecting the absorption, distribution, metabolism, excretion, and toxicology (ADME/Tox) responses of a molecule in humans (1). Whether and how molecules interact with these gatekeepers are some of the early questions that can be answered with appropriate *in vitro* and *in vivo* preclinical studies (2), and these are important questions that affect clinical success. Understanding and predicting these interactions, which are based on molecular physicochemical properties, in a timely and resource-efficient manner are key goals of new drug discovery and development. With relatively recent advances in computational chemistry software and the rapid accumulation of empirical data, many of these physicochemical properties can also be rapidly predicted computationally (3–6).

The focus of this chapter is to describe some of the approaches that have been taken for *in vitro–in vivo* correlations (IVIVCs) with focus on drug–drug

Predictive Approaches in Drug Discovery and Development: Biomarkers and In Vitro/In Vivo Correlations, First Edition. Edited by J. Andrew Williams, Jeffrey R. Koup, Richard Lalonde, and David D. Christ.

interactions (DDIs) at the level of CYP enzymes, transporters, and nuclear receptors. In addition, we will describe *in silico* approaches that have been applied to the prediction of pharmacokinetic (PK) properties, as well as interactions with enzymes and ion channels, ultimately to predict ADME/Tox *in vivo*.

13.2 PREDICTING THE CLINICAL SIGNIFICANCE OF DRUG–DRUG INTERACTIONS FROM *IN VITRO* ENZYME INHIBITION EXPERIMENTS

13.2.1 Reversible Inhibition

The unification of the strategy for *a priori* prediction of the clinical significance of DDIs evolved from a meeting of academic, regulatory, and pharmaceutical industry scientists who met under the auspices of the European Federation of Pharmaceutical Sciences (EUFEPS) and first published a consensus document in the European Journal of Pharmaceutical Sciences in 2001 (7). Similar reports appeared shortly thereafter in other key journals in clinical pharmacology and drug metabolism (8–11). The main focus of this strategy was to use *in vitro* drug metabolism data, particularly data pertaining to the inhibition of CYP enzymes by a potential coadministered drug (i.e., the "perpetrator") in the prediction of the clinical significance of DDIs for this perpetrator. As discussed shortly, these *in vitro* enzyme inhibition data can be used alone or in combination with limited *in vivo* (clinical) data about the perpetrator. The *in vitro* experiments are designed to quantitate the inhibition of a specific CYP enzyme by the perpetrator drug by comparing the intrinsic clearance (CL_{int}) of a prototypical substrate in the absence of the perpetrator drug to $CL_{int,i}$, that is, intrinsic substrate clearance after the *in vitro* addition of the inhibitor or perpetrator at a specified concentration. Alternatively, the IC_{50} or K_i (the dissociation constant for the enzyme–inhibitor complex) for the perpetrator could be determined using a range of concentrations. Accurate estimation of the K_i requires, among other things, the appropriate characterization of the type of enzyme inhibition, for example, competitive, noncompetitive, or uncompetitive. The appropriate *in vitro* experiments require that multiple concentrations of the inhibitor must be used as well as a range of substrate concentrations that embrace the substrate K_m, and from these experiments both the type of inhibition and K_i can be deduced. K_i values can also be estimated from the IC_{50} values for an inhibitor. The IC_{50} value is simply defined as the concentration of inhibitor that decreases the biotransformation of a substrate [S] by 50% and is related to the K_i depending on the type of inhibition (assuming that the binding of one ligand does not affect the binding of the other). For competitive inhibitors

$$IC_{50} = K_i\left(1 + \frac{[S]}{K_m}\right) \quad (13.1)$$

$K_i = (0.5)IC_{50}$ when [S] is equal to K_m. This approximation applies to uncompetitive inhibitors as well, when [S] approximates K_m. For noncompetitive inhibitors (or uncompetitive inhibitors where $[S] \gg K_m$), the IC_{50} and K_i values will be equal. Armed with an experimental K_i for a perpetrator and an *in vitro* measure of CL_{int} for the substrate (victim) acquired in the absence of perpetrator, $CL_{int,i}$ can be calculated from:

$$CL_{int,i} = \frac{CL_{int}}{1 + \frac{[I]}{K_i}} \quad (13.2)$$

where [I] denotes the concentration of perpetrator (inhibitor) used.

The matrix chosen for conducting these *in vitro* experiments may have important consequences on the potential magnitude of the interaction and the predicted clinical relevance. Quantitative outcomes can be affected by the choice of *in vitro* system and also by the substrate selected as the victim drug. Some investigators prefer to use human hepatic microsomes, others prefer heterologous systems that express individual CYP enzymes, and others prefer fresh or cryopreserved human hepatocytes or even human liver slices. Currently, the test system most commonly used for these studies and accepted by worldwide regulatory agencies is a pool of metabolically competent human liver microsomes. Some substrates that have been engineered to be highly selective for specific CYPs are highly fluorescent, thus making them suitable for use in high throughput screens for enzyme inhibition, may not always yield K_i values that are comparable to those found when bonafide clinically used drugs are used as victim drugs in the assays and are not recommended for determining clinical relevance.

Indeed, the EUFEPS Conference Report recommended that the use of recombinant CYP enzyme systems be limited to *qualitative* estimates of K_i because of the variable expression of CYP isozymes across different systems as well as the variable stoichiometries of the obligate reductase and cytochrome *b*5 coenzymes (relative to CYP content). This variability can be partially compensated for by the application of relative activity factors (RAFs), which attempt to normalize activity (12). Alternatively, a normalized rate (NR) can be computed by multiplying the reaction rate of a specific CYP isozyme by the mean specific content of that CYP (nmol/mg protein) as found in human liver microsomes. The activity for all CYPs is then summed to give a total normalized rate (TNR), and the percent TNR can be directly related to percent inhibition for each isozyme (13). One major limitation to this approach is that the specific content of some CYP isozymes (e.g., 3A4) can vary by more than an order of magnitude within different human livers. Another approach that has been used to correct for differences in individual enzymes expressed in heterologous systems versus microsomes isolated from human livers involves the application of intersystem extrapolation factors (ISEFs) that are empirically derived correction factors (14). Failure to account for nonspecific substrate binding to microsomes, within cells, or to plates and other incubation vessels can lead to errant estimates of K_i (15).

How might the *in vitro* estimation of a victim drug in the presence of a perpetrator ($CL_{int,i}$) be used to predict the clinical significance of a putative inhibitory DDI? The underlying principle in predicting the clinical significance of a DDI from *in vitro* data is that the area under the plasma-concentration-time curve (AUC) is directly related to the CL_{int}, and therefore, changes in CL_{int} in the presence of the inhibitor (i) will be reflected in changes in exposure (AUC) (16). Since AUC is inversely related to CL_{int}, the relationship can be expressed as

$$\frac{AUC_i}{AUC} = \frac{CL_{int}}{CL_{int,i}} \tag{13.3}$$

Since $CL_{int,i}$ is determined by K_i and the concentration of the inhibitor [I] as shown in Equation 13.2, substituting for $CL_{int,i}$ produces

$$\frac{AUC_i}{AUC} = 1 + \frac{[I]}{K_i} \tag{13.4}$$

The ratio of AUC_i/AUC denotes the change in exposure to the victim in the presence of perpetrator relative to exposure in the absence of perpetrator. At this point, there has been no measurement of either AUC or AUC_i *in vivo*, and prediction of the *in vivo* AUC ratio depends only on *in vitro* estimates and expected *in vivo* concentrations of the inhibitor [I]. The value of [I] chosen for these predictions is recommended to be the mean C_{max} total (protein bound + free or unbound) plasma concentration expected or determined at steady state for the highest proposed clinical dose (17). Additional ways to calculate and choose other fluid concentrations of the perpetrator are listed in Table 13.1.The relationship between the AUC ratio and the $[I]/K_i$ ratio and the risk that an interaction *in vivo* is "low, medium, or high" are shown in Figure 13.1.

In general, a threshold AUC ratio of ≥ 2 is required before a DDI is considered high risk (Fig. 13.1). Likewise, an AUC ratio of ≤ 1.25 (for inhibitory DDIs) signifies low risk or no risk for a clinically significant DDI. It would appear that by default, AUC ratios that fall between 1.25 and 2 are viewed as posing moderate risk of a clinically significant DDI. Now, based on Equation 13.4 and Figure 13.1, it should also be apparent that the same risk (for a clinically significant DDI) analysis can be cast in terms of the $[I]/K_i$ ratio as well, and

TABLE 13.1 Representative Methods for the Selection or Computation of [I]

Choice of [I]	Computation of [I]
Average steady-state plasma concentration	$[I]_{av} = \frac{FD}{CL \times \tau}$
Peak steady-state plasma concentration	$[I]_{max} = \frac{[I]_{av} k\tau}{(1 - \exp^{-k\tau})}$
Maximum hepatic input concentration	$[I]_{in} = [I]_{av} + \left(\frac{k_a F_a D}{Q}\right)$
Concentration in the hepatocytes	$[I]_{hep} = \frac{[I]_{in\ vitro}}{[I]_{medium}} \times [I]_{blood}$
Unbound concentration	$[I]_i \times f_u$

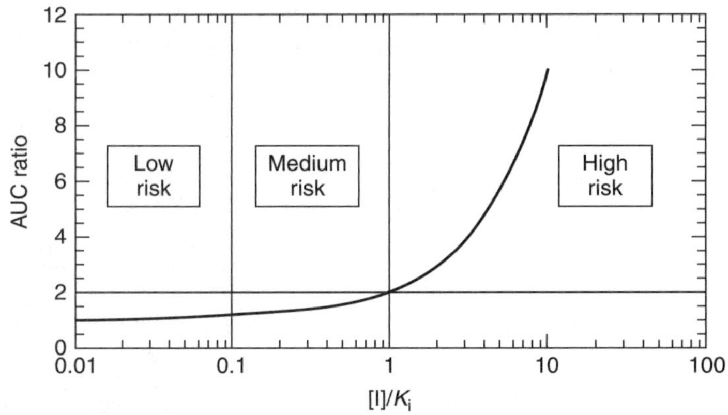

FIGURE 13.1 Relationship between AUC ratio and the ratio of [I]/K_i.

per the FDA current guidance (17), the qualitative risk boundaries are defined as follows: [I]/$K_i \leq 0.1$, remote; [I]/K_i between 0.1–1.0, possible; and [I]/$K_i \geq 1$, likely for risk of clinically significant DDI. The value for [I] was, according to consensus, supposed to be the steady-state peak plasma concentration (C_{max}) of the perpetrator in humans (10). However, in spite of the consensus opinion, some have found that C_{max} (i.e., $[I]_{max}$) does not always give the best prediction of the actual AUC ratio in humans.

Indeed, Houston and colleagues at the University of Manchester in the United Kingdom, who have been instrumental in establishing and refining the prediction of DDIs in humans from *in vitro* K_i data, have shown that the predictive accuracy, which they refer to as *qualitative zoning* (e.g., low, moderate, and high risks), will vary depending on not only which value is selected for [I] but also which value of [I] is used even for a given CYP enzyme. In evaluating nearly 200 DDIs for which the actual human AUC ratio data had been determined in clinical trials and for which they had *in vitro* K_i data, they were able to actually pair experimentally derived AUC ratios with [I]/K_i data (16).

Matching [I]/K_i data with actual AUC ratios provided an assessment for the accuracy of the entire predictive strategy discussed above. Briefly, Houston et al. tabulated the true positive predictions (defined as all the data points for which the AUC ratio was $\geq$2 and the [I]/K_i ratio was $\geq$1) and the true negative predictions (i.e. AUC ratios $\leq$1.1 and [I]/K_i ratios $\leq$0.1) for inhibitory DDIs involving three distinct CYPs, namely, CYP3A4, CYP2D6, and CYP2C9. Their abridged findings are shown in Table 13.2.

The bottom row in Table 13.2 is merely the sum of the percentages of true positives and true negatives, that is, the overall accuracy. Note, for example, for CYP3A4, the highest total percent correct AUC ratios that were predicted from [I]/K_i ratios (83%) occurred when $[I]_{in}$ was used, that is, the computed concentration of perpetrator drug in the hepatic portal vein (Table 13.1). In contrast, the highest percent correct (sum of true positive plus true negative) predictions for

TABLE 13.2 The Percentage of True Positive and True Negative Drug–Drug Interactions (i.e., Accuracy) with CYP 3A4, 2D6, and 2C9 Predicted *In Vitro* Using Different Values for [I]

	CYP3A4				CYP2D6				CYP2C9			
	$[I]_{ave}$	$[I]_{ave,u}$	$[I]_{max}$	$[I]_{in}$	$[I]_{ave}$	$[I]_{ave,u}$	$[I]_{max}$	$[I]_{in}$	$[I]_{ave}$	$[I]_{ave,u}$	$[I]_{max}$	$[I]_{in}$
% Positives predicted	22	7	24	50	20	0	20	29	23	16	23	31
% Negatives predicted	43	43	42	33	65	71	65	25	46	44	35	23
% Correctly predicted	65	50	66	83	85	71	85	54	69	60	58	54

$[I]_{ave}$, total (bound + unbound); $[I]_{ave}$, unbound or $[I]_{ave} \cdot f_u$; $[I]_{max}$, total (bound + unbound); and $[I]_{intraportal}$, total (bound + unbound).
Source: Data abstracted from Reference 16.

CYP2D6 occurred when either $[I]_{max}$ or $[I]_{ave}$ was used (85% for both). Interestingly, for DDIs ascribed to CYP2C9 inhibition, the predictive accuracy of an AUC ratio ≥ 2 or ≤ 1 was highest when $[I]_{ave}$ was used in the $[I]/K_i$ ratio; however, it only reached 69%.

Additional sources of variability can obviously be introduced by the determination of K_i. As noted earlier, the K_i value may depend on the substrate (i.e., victim) selected, the matrix used (microsomes or hepatocytes vs heterologously expressed enzymes), and the kinetics of the enzyme reaction. Rodrigues and Lin (18) point out how the increasingly prevalent phenomenon of atypical enzyme kinetics of the substrate drug can confound the best estimates of kinetic constants. One problem that has plagued the drug metabolism community until recently is the apparent lack of correlation between the clearance rates of multiple probes for the same CYPs. In part, at least, these poor correlations can now be ascribed to the allosteric kinetics of the CYP enzymes. For example, a single substrate for CYP3A4 may be capable of binding at either of two different locations in the broader binding site (19). Two molecules of substrate might bind simultaneously at two different locations within the broader binding site, which could lead to homotropic activation of the enzyme. Two different substrates might bind simultaneously at two different locations within the broader binding site, leading to heterotropic activation of the enzyme. Allosteric kinetics occur among CYPs other than CYP3A, and they give rise to atypical enzyme kinetics that are not described by the simple Michaelis–Menten equation. Multiple binding arrangements for different CYP3A4 substrates may contribute to K_i values for a single inhibitor that vary by more than 10-fold, depending on the substrate used (18). Carefully structured *in vitro* studies with more than one substrate and a range of substrate concentrations have demonstrated partial inhibition, cooperative inhibition, and concentration-dependent inhibition, but no mutual inhibition of CYP3A4 was demonstrated when combinations of nifedipine, midazolam, felodipine, and testosterone were used simultaneously (20). Clearly, the failure to account for atypical kinetics in *in vitro* experiments could give rise to errant values for K_m and K_i. The EUFEPS Conference Report recommended that the issue of enzyme cooperativity for CYP3A4 and CYP2C9 be addressed, at least partially, by the determination of IC_{50} values using at least two low (therapeutic) concentrations of at least two substrates, one of which is known to exhibit homotropic cooperativity, and by full characterization of activation kinetics when defining *in vitro* CL_{int} values.

Despite the vexing and variable issues such as the selection of *in vivo* perpetrator concentrations ([I]), the experimental system for estimating K_i, the issue of nonspecific victim and perpetrator binding, and the sometimes atypical kinetics of CYP enzymes associated with multiple substrate binding sites, $[I]/K_i$ predictions of AUC ratios can be surprisingly good. As discussed previously and shown in Table 13.2 for DDIs involving inhibition of CYP2C9, CYP3A4, and CYP2D6, the general accuracy of predicting the clinical significance of DDIs from $[I]/K_i$ ratios was approximately 70%, 83%, and 85%, respectively, regardless of the source of [I].

If a perpetrator inhibits a CYP enzyme competitively or noncompetitively, then the *in vivo* AUC ratio can often be well predicted from an appropriately selected value for [I] and for an accurately determined K_i, in accordance with Equations 13.2–13.4. However, the model for estimating the effects on AUC must be expanded to account for enzyme inhibition that occurs by irreversible or quasi-irreversible mechanisms.

13.2.2 Mechanism-Based Inhibition

The previous discussion focused on predicting interactions caused by molecules that inhibit catalysis directly in a reversible, concentration-dependent manner. It has long been recognized that compounds may irreversibly or quasi-irreversibly inactivate CYP enzymes during their metabolism through the formation of reactive intermediates that bind covalently to apoprotein or heme or through the ligation of the heme iron. This mechanism-based inhibition is different from the formation of more potent, directly inhibitory metabolites that would also appear to be produced in a timely and metabolism-dependent manner. For mechanism-based inhibition of a CYP, the AUC ratio becomes

$$\text{AUC ratio} = \frac{k_{deg} + \left(\dfrac{[I] \times k_{inact}}{[I] + K_{iapp}} \right)}{k_{deg}} \tag{13.5}$$

where k_{inact} is the apparent inactivation rate constant for the enzyme (21), K_{iapp} is the concentration of inhibitor (perpetrator) that elicits a $k_{inact}/2$, and k_{deg} is the degradation rate constant for the enzyme. However, it may be possible to apply Equation 13.4 even in the case of mechanism-based inhibition so long as the inhibiting drug is preincubated (e.g., for 30 min) before the addition of the substrate (17). Mechanism-based enzyme inhibition may be one of the more important types of DDIs, and examples of drugs that inhibit CYP enzymes in this manner include paroxetine that inhibits CYP2D6 (22), verapamil that inhibits CYP3A4 (23), and protease inhibitors such as ritonavir that inhibit CYP3A4 (24). Other drugs that cause mechanism-based inhibition of CYP enzymes include erythromycin, fluvoxamine, and ethinyl estradiol (25). Some mechanism-based inhibitors cause irreversible inhibition by forming a metabolite intermediate complex (MIC) with the heme of the CYP (26). Primary, secondary, or tertiary amines, or methylenedioxy constituents, in the molecule are prerequisites for compounds that chelate the heme of the CYP (26). More recent studies with metabolites of molecules such as indinavir and nelfinavir that lack these functional groups yet still display MIC formation may indicate other chemical moieties that are also involved (27).

13.2.3 Predicting AUC Ratios When a Substrate is Metabolized by More Than One Enzymatic Pathway

If a drug is metabolized by multiple enzymes, then to what extent must an individual enzyme be responsible for the substrate's metabolism before enzyme inhibition by a perpetrator is likely to be clinically significant? Benchmarks have been set as low as 30%, but a good case can be made for 50%. For a drug that is cleared in part by metabolism via a CYP enzyme, its clearance in the presence of an inhibitor (perpetrator) of that enzyme can be defined as follows:

$$\mathrm{CL}_{\mathrm{int,i}} = \mathrm{CL}_{\mathrm{int}}(1 - \mathrm{fm}_{\mathrm{CYP}}) \tag{13.6}$$

where $\mathrm{CL}_{\mathrm{int}}$ represents all clearance routes. Accordingly, the maximum impact that could occur on the AUC ratio if the metabolic contribution to drug clearance were completely stopped by a perpetrator (inhibitor) would be given by Ito et al. (28)

$$\text{AUC ratio} = \frac{1}{1 - \mathrm{fm}_{\mathrm{CYP}}} \tag{13.7}$$

Thus, only if $\mathrm{fm}_{\mathrm{CYP}}$ were 0.5 would the AUC ratio double by CYP inhibition. However, as described above, the extent of clearance inhibition associated with enzyme inhibition actually depends on the K_{i} value for the perpetrator. If a victim drug is metabolized by a single CYP enzyme and if metabolism accounts for $\geq$50% of the victim drug's total clearance, then the AUC ratio arising from inhibition of that enzyme by a perpetrator is given by the Rowland–Matin equation:

$$\text{AUC ratio} = \frac{1}{\dfrac{\mathrm{fm}_{\mathrm{CYP}}}{1 + \dfrac{[\mathrm{I}]}{K_{\mathrm{i}}}} + (1 - \mathrm{fm}_{\mathrm{CYP}})} \tag{13.8}$$

If a victim drug (i.e., substrate) is metabolized by different enzymes accounting for more than 50% of the total clearance and if a perpetrator inhibits those enzymes in quantitatively different ways, another layer of complexity is added to the estimation of the AUC ratio from $[\mathrm{I}]/K_\mathrm{i}$ data. If two enzymes are inhibited by a particular compound, the fraction of victim drug's metabolism by each of the two pathways is known, and the K_i values for the perpetrator on each of the two pathways is known, then the AUC ratio can be described by

$$\text{AUC ratio} = \frac{1}{\left(\dfrac{\mathrm{fm_{CYP1}}}{1 + \dfrac{[\mathrm{I}]}{K_{\mathrm{i},1}}}\right) + \left(\dfrac{1 - \mathrm{fm_{CYP2}}}{1 + \dfrac{[\mathrm{I}]}{K_{\mathrm{i},2}}}\right)} \tag{13.9}$$

Most often fm is determined experimentally using microsomes or hepatocytes and CYP-selective inhibitors. While heterologously expressed isoforms can be used to indicate qualitatively which enzymes may metabolize a given compound, the quantitative estimation of fm requires the presence of all pathways.

13.2.4 Accounting for Cooperativity in Enzyme Inhibition

As discussed earlier, while most DDI studies of inhibition utilize the enzyme kinetic model of reversible, mutually exclusive binding of victim and perpetrator, atypical enzyme kinetics are becoming increasingly more prominent. Two identical molecules binding at the same time (homotropic) can be either positively or negatively cooperative, while two different molecules binding at the same time (heterotropic) can be positively or negatively cooperative. For an enzyme that can accommodate two molecules at the same time, the AUC ratio elicited by a perpetrator can be described as follows:

$$\text{AUC ratio} = \frac{\left(1 + \dfrac{[\mathrm{I}]}{K_\mathrm{i}}\right)^2}{1 + \dfrac{\gamma[\mathrm{I}]}{\delta K_\mathrm{i}}} \tag{13.10}$$

where γ is an interaction factor that denotes the change in catalytic efficacy in the presence of inhibitor and δ is an interaction factor that denotes the change in binding affinity in the presence of inhibitor (29). When these two interaction factors are equivalent, Equation 13.10 simplifies to Equation 13.4.

13.2.5 Drug Interactions Occurring at the Level of Transporters

Some DDIs presumed to occur at the level of CYP enzymes may actually represent interactions occurring either at transporters or simultaneously at transporters and CYP enzymes. The likely role of transporters in DDIs is becoming increasingly appreciated (30–34).

According to Lin and Yamazaki (35), the role of intestinal P-glycoprotein (P-gp) in the bioavailability of drugs may be much less important than is generally thought because of the high luminal (apical) concentration of drugs and the sink effect of the mucosal circulation. Examples of important interactions do however exist, as illustrated by the observation that the P-gp inhibitor verapamil significantly increased the AUC and the C_{max} of fexofenadine, a poorly metabolized P-gp substrate, in healthy volunteers (36). Moreover, some surfactants have been shown to significantly increase the oral absorption of paclitaxel and docetaxal, which are anticancer agents with limited water solubility, and it has been suggested that this effect is a function of their potent inhibition of intestinal P-gp (37). The apical-to-basal flux of P-gp substrates in enterocytes in the presence of P-gp inhibitors depends not only on the extent of P-gp inhibition but also on the permeability-surface-area products for apical influx, apical efflux, basal influx, and basal efflux (37).

In addition to P-gp (encoded by *ABCB1*), a wide array of transport proteins may play a role in the hepatic processing of xenobiotics. For example, fexofenadine is transported across the basolateral membrane of hepatocytes by OATP1A2 (encoded by *SLCO1A2*) (38). Verapamil, too, is subject to transporter-mediated uptake (39). Thus, the fexofenadine interaction described above might also be explained by inhibition of hepatic influx rather than inhibition of intestinal efflux of fexofenadine. Both grapefruit juice and apple juice have been shown to significantly reduce the AUC of fexofenadine, and this phenomenon is likely due to Organic anion-transporting polypeptide (OATP) inhibition (40).

Other xenobiotic transporters on the basolateral side of human hepatocytes include but are not limited to OATP1B1 (*SLCO1B1*), OATP1B3 (*SLCO1B3*), OATP2B1 (*SLCO2B1*), OAT2 (*SLC22A7*), OAT4 (*SLC22A11*), OCT1 (*SLC22A1*), OCT3 (*SLC22A3*), MRP1 (*ABCC1*), MRP3 (*ABCC3*), and MRP4 (*ABCC4*) (38). P-gp is classified as a primary active transporter driven by ATP hydrolysis, whereas the OATPs are classified as secondary active transporters (37). Human hepatic canalicular transport proteins for which xenobiotics are substrates and that mediate xenobiotic excretion into the bile include P-gp, MRP2 (*ABCC2*), and BCRP (*ABCG2*) (38). In fact, more than two dozen transporters have been implicated in small-molecule tissue fluxes (cellular uptake and efflux), and their substrate preferences, tissue distribution, inhibitors, probe substrates, and *in vitro* characterization methods have been recently described (41). Quantitating transporter function and the effects of inhibitors (or inducers) poses challenges beyond those described above for IVIV prediction of enzyme-related DDIs because in cell-based assays, one must account for vectorial movement of substrates (basolateral to apical or apical to basolateral). Thus, cells must be seeded onto Transwell™ inserts. Isolated hepatocytes appear unsuitable for studying efflux transporters, which are internalized after isolation (42). However, sandwich-cultured hepatocytes can be used to study both efflux as well as uptake since they essentially retain hepatocyte architecture (43).

While microsomes or recombinant enzymes appeared to be the favored systems for making DDI predictions at the EUFEPS conference, Wu and Benet (39) have suggested that intact human cells may be preferred for detecting synergistic or nulling effect associated with the simultaneous inhibition of transporters and drug-metabolizing enzymes. The potential impact of simultaneous inhibition of transporters and drug-metabolizing enzymes, notably CYPs, has been characterized by Benet et al. (39, 44). The combined inhibition of efflux transporters (e.g., P-gp) and CYP activity may lead to different AUC outcomes, depending on which of those tissues is the site for an inhibitory interaction since the topographical relationships between efflux transporters and drug-metabolizing enzymes is different in the enterocyte than in the hepatocyte. Inhibition of P-gp and CYP3A4 activity in enterocytes is predicted to cause a substantial increase in the *in vivo* AUC of substrate, whereas the same inhibition of both processes in hepatocytes may increase the substrate AUC, decrease the AUC, or cause no net change in the AUC (null effect) (39). Thus, if a drug is a substrate for both hepatic P-gp and CYP3A4 and if a drug or new chemical entity (NCE) inhibits both, then the failure of the NCE to elicit significant inhibition of substrate metabolism in, say, a microsomal or recombinant enzyme system will not be proof positive that an *in vivo* DDI will not occur. On the other hand, although some CYP3A inhibitors are also P-gp inhibitors, neither the mechanisms of inhibition nor the K_i values for inhibition need be similar, and it is indeed possible that only the inhibition of CYP3A activity will be responsible for any PK changes, for example, AUC, observed *in vivo* (45).

The quantitative study of P-gp-related inhibitory DDIs is not as mature as that of CYP-related inhibitory DDIs. Nevertheless, a number of IC_{50} and K_i values for P-gp inhibitors derived from *in vitro* studies have been published (37), and kinetic data for transporters as well as other data are available in databases such as the one maintained by Fujitsu [see http://www.fqs.pl/pl/life_science/adme_db/service_ovrview], the human membrane transporter database (HMTD) (46), Transport DB (47), TP-Search (48), and the Transporter Classification Database (TCDB) (49). The roles of other small-molecule transporters in the liver and intestine (41) in DDIs are less well defined. The prediction of *in vivo* clearance of xenobiotics as well as clinically significant DDIs mediated by transporters will improve as the role of each transporter in intestinal epithelial cells and hepatocytes becomes more clearly defined (with regard to individual substrates, both qualitatively and quantitatively, along with the effects of inhibitors and inducers), as the modeling of sequential processing of substrates by transporters and enzymes in the liver and intestines improves, and as *in vitro* systems evolve that retain the full processing capabilities of all transporters and enzymes.

13.3 *IN SILICO* MODELS FOR ADME/TOX PROPERTIES

Computational or *in silico* methods for predicting molecular properties have been widely used alongside the generation of *in vitro* data both to create the model and to validate it (50, 51). These models have been used in the discovery and optimization of novel molecules to clarify ADME/Tox properties and the physicochemical characterization of these molecules. Although we take the calculation of properties such as log P, log D, solubility, and other descriptors for granted, there have been many published models for aspects of ADME/Tox, such as rule-based methods for metabolism and ligand-based methods for blood–brain barrier penetration, bioavailability, and transport (52). These include rule-based models for predicting the likelihood of absorption (53), methods that have more graphical outputs for predicting binding to CYPs (3), and quantitative structure–activity relationship (QSAR) methods for ADME/Tox (54). Such calculations can be performed with very large numbers of molecules or virtual molecules to act as a molecule selection filter.

Computational approaches have been used to model individual enzymes, transporters, ion channels, and receptors important to ADME/Tox (54), while more complex processes such as intrinsic clearance, volume of distribution, and renal clearance (described below) have all been the subject of recent QSAR studies. This section focuses on predictive models for key areas in metabolism and kinetics, CYP inhibition, cardiovascular safety, and biopharmaceutics.

13.3.1 Predictive Models for Complex Pharmacokinetic Properties

Complex PK properties have been modeled computationally (Fig. 13.2), and these include measures of a molecule's clearance, which, together with the volume of distribution, determine the elimination half-life. The models generated for complex processes have thus far been quite small when compared with those generated for physicochemical properties (e.g., clog P or water solubility). As discussed previously, the intrinsic clearance, a measure of the enzyme activity toward a compound, may involve several drug-metabolizing enzymes. An early comparative molecular field analysis (CoMFA) model for this property was developed and utilized for the CYP-mediated metabolism of chlorinated volatile organic compounds (55). More diverse sets of molecules with clearance data derived from human hepatocytes have been used to predict human *in vivo* clearance using multiple linear regression (MLR), principal component analysis (PCA), partial least squares (PLS), or neural networks (NNs) (56). Microsomal and hepatocyte clearance data sets have also been used to generate pharmacophores and to predict each other (57). Similar approaches have been taken with a narrow series of compounds such as 11 corticosteroids with stepwise MLR (58), 23 cephalosporins with PLS (59), and 44 antimicrobials with *k*-nearest-neighbor method (60). More recently, the largest model to date used 503 molecules (398 training and 105 testing) with multiple, different descriptors and the support vector machine (SVM); *k*-nearest neighbor; and general regression NN methods. The majority of these models had mean fold error values of less than 2 for approximately 70% of the molecules (61).

A second complex PK property is the volume of distribution (V_d), which is a function of the extent of drug partitioning into tissue versus plasma, and there

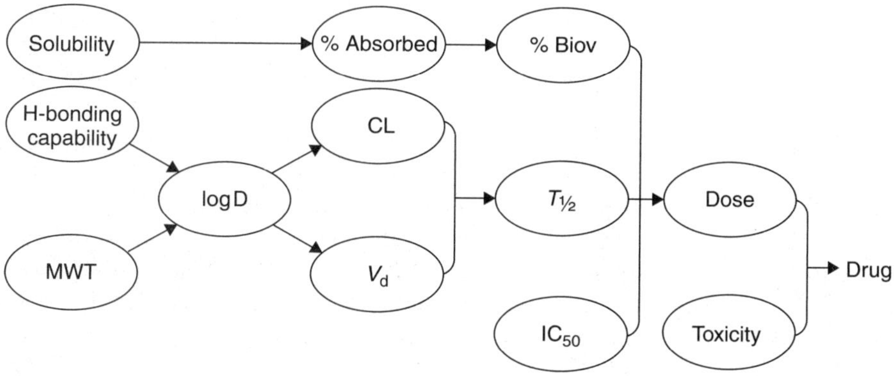

FIGURE 13.2 Relationships between molecule properties. CL, clearance; V_d, volume of distribution; MWT, molecular weight; $t_{1/2}$, half-life; % Biov, bioavailability. *Source:* Modified from Reference 62.

have been several attempts at modeling this property (63, 64). This property, along with the plasma half-life, determines the appropriate dose of a drug. One study has used 11 corticosteroids with stepwise MLR (58), while another modeled the volume of distribution for 23 cephalosporins with PLS (59), and a third used 44 antimicrobials with the *k*-nearest-neighbor method (60). Another extensive study used 253 diverse drugs from the literature with 8 molecular descriptors and the Sammon and Kohonen mapping methods. These models appeared to classify correctly 80% of the compounds (65). Recently, a set of 384 drugs with literature values for the volume of distribution at steady-state data was used with a mixture discriminant analysis–random forest method and 31 molecular descriptors (66). The model was tested with 23 molecules and resulted in a geometric mean fold error of 1.78-fold, which was comparable to more complex predictions for V_d from animal *in vitro* and other methods (66). Additional models for V_d utilized 129 molecules and stepwise regression, producing mean fold error results ranging from 1.79–2.17 (67). Another approach compared Bayesian NNs, classification trees, and PLS using data for humans (199 molecules) and rats (2086 molecules) V_{ss} (the volume of distribution at steady state) which were partitioned into training and test sets; these models were used to predict test sets for the other species. A consensus of the three approaches performed the best in terms of model statistics (68).

A third PK property, the plasma half-life ($t_{1/2}$), has also been modeled using data for 458 drugs from the literature and 4 molecular descriptors with Sammon and Kohonen maps. Like the previously described volume of distribution models, these models could classify correctly 80% of the compounds (65).

A fourth complex PK property is renal clearance, the elimination of unchanged drug by the kidney. In one published model, 130 molecules were used with 62 VolSurf or 37 Molconn-Z descriptors, and when the models were tested with 20 molecules and one model using SIMCA and Molconn-Z descriptors, they correctly classified 85% of the compounds between the two boundary conditions (i.e., 0–20% and 20–100%) (69).

Although many of these studies represent individual models, very few have actually generated multiple PK properties for the same data set, the exceptions being the study of 11 corticosteroids, which modeled clearance, volume of distribution, fraction unbound in plasma, and percent oral absorption (58), and the study with 23 cephalosporins, which modeled clearance and volume of distribution (59). Although developing predictive models for each PK parameter is important, developing integrated models that can accurately predict the interplay of clearance, distribution, and excretion as they determine the plasma-concentration-time profile for new molecules remains a key objective.

13.3.2 Predicting the Formation of Metabolite Inhibitor Complexes (MICs) with CYP3A4

As described in Section 13.2.1, mechanism-based enzyme inhibition may be one of the more important types of DDIs and MIC formation may be of concern. Often, the potential for MIC formation is suggested by molecular structure and the chemical subsitutents; however, the presence of an amine function, for example, does not always imply that a molecule will form an MIC (70). For example, 3-hydroxy tamoxifen and 4-hydroxy tamoxifen do not form MIC, while tamoxifen and *n*-desmethyltamoxifen form an MIC *in vitro* (71). Some macrolide antibiotics such as clarithromycin form MICs (72) while miocamycin does not (73), which suggests that steric hindrance may be important for correct orientation in the CYP3A4 ($+b_5$) binding site. It would appear other molecular properties besides primary, secondary, and tertiary amines or methylenedioxyphenyl features are important. Pharmacophore, recursive partitioning, and logistic regression computational methods have all been used to aid in discriminating between CYP3A4 ($+b_5$) MIC-forming compounds and those that do not form an MIC (74). The qualitative approach of common feature pharmacophore alignments for selected compounds suggests that CYP3A4 ($+b_5$) MIC formation requires at least four hydrophobic interactions and a hydrogen bond acceptor interaction. Conversely, non-MIC-forming compounds appear to possess fewer hydrophobic pharmacophore features with hydrogen bond donor rather than acceptor interactions. Recursive partitioning approaches also highlighted hydrophobic and hydrogen bond acceptor interactions (74) while a logistic regression model used a hydrogen bond acceptor descriptor and possessed the highest prediction accuracy. Differentiating between MIC-forming and non-MIC-forming compounds is complex, likely requiring numerous hydrophobic and hydrogen bond acceptor features for MIC formation. An *in vitro* model has been proposed that can predict the *in vivo* clinically observed inhibition of CYP3A4 (72). On the basis of the results with the computational MIC models, it is likely these could compliment *in vitro* testing, representing a preliminary approach to predicting CYP3A4 ($+b_5$) MIC formation, an important mechanism for DDIs.

13.3.3 Predicting Interactions with the Cardiac Ion Channel hERG

QT interval prolongation and severe ventricular arrhythmias have been associated with different classes of drugs including some antipsychotics (75, 76), and blockade of one or more cardiac potassium channels is the most likely explanation. A recent study characterized the effects of multiple antipsychotics against human cardiac ion channels (I_{to}, I_{Na}, I_{sus}, I_{K1} and human ether-a-gogo-related gene (hERG)) (77). The I_{to}, I_{sus}, I_{K1}, and I_{Na} recorded from human cardiac myocytes were relatively insensitive to the agents tested. All the drugs tested were found to block hERG current in a concentration-dependent manner, with affinities in the nanomolar concentration range (77). Using data from clinical studies (78, 79), plasma concentrations of haloperidol, olanzapine, risperidone, thioridazine, and ziprasidone and changes in the corrected QT (QTc) interval were evaluated for a potential correlation (Table 13.3). A reduction in hERG current amplitude, varying from approximately 14.0% to approximately 43.6%, was found to correlate with steady-state C_{max} plasma concentrations, and the extent of these QTc interval increases varied from relatively modest to more pronounced (78, 79). A significant positive association was observed between the estimated hERG blockade at the steady-state C_{max} plasma concentrations and the increase in QTc (78, 79).

Many computational studies have therefore generated QSAR or pharmacophore model for hERG (54, 86) to aid in understanding those structural features within molecules that are important for hERG binding. In our study of antipsychotics described above (77), a pharmacophore was generated and tested with olanzapine and two metabolites, correctly rank ordering them. This would suggest that it could be possible to use an *in silico* model before the *in vitro* model to predict potential *in vivo* effect. Although blockade of repolarizing

TABLE 13.3 Correlation Between the Increase in QTc Interval and hERG Blockade

Compound	hERG IC_{50} (μM)[a]	Dose (mg/day)[b]	Mean C_{max} (ng/ml)[b]	Mean Free Plasma C_{max} (nM)[c]	Mean ↑ QTc (ms)[d]	% hERG Blockade
Haloperidol	0.0268	15	16.3	3.5	7.1	15.1
Olanzapine	0.231	20	55.1	12.3	1.7	14.0
Risperidone	0.148	16	58.7	14.3	3.6	17.0
Thioridazine	0.033	300	765	20.6	30.1	43.6
Ziprasidone	0.125	160	171	4.1	15.9	22.7

[a]Data abstracted from Reference 80.
[b]Data abstracted from References 77 and 78.
[c]Calculated as mean C_{max} multiplied by the free (unbound) fraction (haloperidol: 8% (81), olanzapine: 7% (82), risperidone: 10% (82), thioridazine: 1% (83–85), and ziprasidone: 1% (82)) and divided by the molecular weight multiplied by 10^{-3}.
[d]Mean changes in QTc were calculated by the baseline correction method (77, 78).
Source: Modified from Reference 76.

currents, such as hERG, can lead to delayed repolarization and a prolongation of the QT interval, the clinical consequences of these effects can be off set if a drug that blocks hERG also blocks depolarizing currents mediated by the sodium channel I_{Na} and/or calcium channel I_{Ca}. As an example of this phenomenon, verapamil (87, 88) and others (89) are hERG blockers but also block I_{Ca} at pharmacologically relevant concentrations yet do not frequently produce torsade de pointes, a severe and often fatal ventricular arrhythmia. In the future, if we are to develop noncardiac drugs that avoid hERG interaction, we must understand the structural requirements for binding to this channel and other ion channels if reliable predictions for *in vivo* consequences are to be made.

13.3.4 Computational Approaches to the Biopharmaceutics Drug Disposition Classification System (BDDCS)

Another key goal in pharmaceutical research is to correlate *in vitro* drug release and dosage form performance to *in vivo* drug disposition. In addition to the goals of the pharmaceutical industry, reliably linking dosage form performance *in vitro* with *in vivo* behavior offers regulatory authorities an opportunity to evaluate and approve formulation changes without the need for additional clinical studies. The Biopharmaceutics Classification System (BCS) has been a helpful guide and had considerable impact on drug regulatory process and practice based on the classification of molecules into four categories on the basis of their aqueous solubility and gastrointestinal permeability characteristics (90). This classification system has been expanded by Wu and Benet (39) to incorporate routes of elimination (metabolism vs excretion unchanged) and the effects of influx and efflux transporters. Their categorization, with the acronym BDDCS (Biopharmaceutics Drug Disposition Classification System), emphasizes that the clinical impact of efflux transporters in modulating oral absorption and drug PK is most applicable to class 2 molecules (i.e., those with low solubility and high permeability) and possibly class 4 compounds (i.e., those with low solubility and low permeability). Many of the key parameters (e.g., permeability) can be determined *in vitro* or estimated using computer programs (91). A recent study used *in vitro* dissolution data as input in the GastroPlus™ software to successfully predict *in vivo* PK data (92) We have recently investigated computational methods to produce predictive models for BDDCS classes using molecular descriptors alone for a data set consisting of 165 training and 56 test set molecules. Using machine learning methods such as SVM along with readily interpretable descriptors, as well as complex descriptor sets from commercial vendors such as Molconn-Z (93) and VolSurf (94), we have been able to assign molecules into the four classes. Combinations of 2D and 3D descriptors are important for drugs of classes 2–4, whereas 2D descriptors alone are relevant for class 1 predictions. An earlier BCS model (95, 96) used a smaller number of compounds with partitioned total surface area (PTSA, a conformation-dependent property requiring a 3D structure for calculations) and solubility and permeability parameters, which were calculated independently. Computational approaches for rapidly assigning drugs to BDDCS classifications would provide valuable additional insight into bioavailability aspects of a drug, which could be applied in drug discovery to identify promising molecules and to highlight those that may have future developability issues.

13.4 SUMMARY

The age of the *in silico* ADME/Tox field compared to other computational methods is quite young, but already many databases and modeling approaches have been used (97) and we are seeing simple rules of thumb (98) developed analogous to the Rule of 5 put forth by Lipinski and colleagues (53) to aid medicinal chemists view common molecular properties as they relate to desirable drug performance characteristics. However, as with any rule of thumb, there will always be some exceptions. The advantages of *in silico* ADME/Tox methods include the capacity to focus more chemical and biological resources to the promising chemicals emerging from virtual screening, a greater speed in the discovery process, and decreased animal and reagent usage, while the limitations include an inability to account for molecular and protein flexibility and the relative lack of predictive models for animal PK properties. There has been some discussion of local versus global models and how the predictions for test molecules should also provide a measure of similarity compared with the training set to understand the chemical space covered for optimal predictions (i.e., the applicable prediction space) (99–101). Similarity measures in the form of Euclidean or Tanimoto distances calculated between pairs of molecules or PCA analysis are applicable for this purpose (80, 102).

In the examples above, we have indicated how *in silico* methods could be of use either to predict likely *in vivo* ADME/Tox properties or, at the very least, to provide a filter for *in vitro* testing before *in vivo* testing. We have already suggested a synergy between *in silico* ADME/Tox approaches and PKPD simulations that might facilitate IVIVC (52). While we might consider that in future we will see a closer integration of *in silico*, *in vitro*, and *in vivo* methods, computational ADME/Tox will undoubtedly have its biggest impact if it can reliably predict *in vivo* disposition. In the near future, it should be possible to consider *in silico in vivo* correlations (ISIVCs) for some of the available models for drug disposition.

REFERENCES

1. Ekins S, Swaan PW. Computational models for enzymes, transporters, channels and receptors relevant to absorption, distribution, metabolism, excretion and toxicology. Rev Comput Chem 2004;20:333–415.
2. Pritchard JF, Jurima-Romet M, Reimer ML, Mortimer E, Rolfe B, Cayen MN. Making better drugs: decision gates in non-clinical drug development. Nat Rev Drug Discov 2003;2:542–553.

3. Ekins S, de Groot M, Jones JP. Pharmacophore and three dimensional quantitative structure activity relationship methods for modeling cytochrome P450 active sites. Drug Metab Dispos 2001;29:936–944.
4. Ekins S, Berbaum J, Harrison RK. Generation and validation of rapid computational filters for CYP2D6 and CYP3A4. Drug Metab Dispos 2003;31:1077–1080.
5. Jorgensen WL. The many roles of computation in drug discovery. Science 2004;303:1813–1818.
6. Cheng A, Diller DJ, Dixon SL, Egan WJ, Lauri G, Merz KMJ Jr. Computation of the physico-chemical properties and data mining of large molecular collections. J Comput Chem 2002;23:172–183.
7. Tucker GT, Houston JB, Huang SM. EUFEPS conference report. Optimising drug development: strategies to assess drug metabolism/transporter interaction potential—towards a consensus. European Federation of Pharmaceutical Sciences. Eur J Pharm Sci 2001;13:417–428.
8. Tucker GT, Houston JB, Huang SM. Optimizing drug development: strategies to assess drug metabolism/transporter interaction potential—toward a consensus. Pharm Res 2001;18:1071–1080.
9. Tucker GT, Houston JB, Huang SM. Optimizing drug development: strategies to assess drug metabolism/transporter interaction potential-toward a consensus. Clinical Pharmacol Therap 2001;70:103–114.
10. Bjornsson TD, Callaghan JT, Einolf HJ, Fischer V, Gan L, Grimm S, et al. The conduct of in vitro and in vivo drug-drug interaction studies: a pharmaceutical research and manufacturers of America (PhRMA) perspective. Drug Metab Dispos 2003;31:815–832.
11. Bjornsson TD, Callaghan JT, Einolf HJ, Fischer V, Gan L, Grimm S, et al. The conduct of in vitro and in vivo drug-drug interaction studies: a PhRMA perspective. J Clin Pharmacol 2003;43:443–469.
12. Galetin A, Brown C, Hallifax D, Ito K, Houston JB. Utility of recombinant enzyme kinetics in prediction of human clearance: impact of variability, CYP3A5, and CYP2C19 on CYP3A4 probe substrates. Drug Metab Dispos 2004;32:1411–1420.
13. Rodrigues AD. Integrated cytochrome P450 reaction phenotyping: attempting to bridge the gap between cDNA-expressed cytochromes P450 and native human liver microsomes. Biochem Pharmacol 1999;57:465–480.
14. Proctor NJ, Tucker GT, Rostami-Hodjegan A. Predicting drug clearance from recombinantly expressed CYPs: intersystem extrapolation factors. Xenobiotica 2004;34:151–178.
15. Obach RS. Nonspecific binding to microsomes: impact on scale-up of in vitro intrinsic clearance to hepatic clearance as assessed through examination of warfarin, imipramine, and propranolol. Drug Metab Dispos 1997;25:1359–1369.
16. Ito K, Brown HS, Houston JB. Database analyses for the prediction of in vivo drug-drug interactions from in vitro data. Br J Clin Pharmacol 2004;57:473–486.
17. FDA USA. Guidance for Industry:Drug interaction studies—study design, data analysis, and implications for dosing and labeling. 2006;1–52.
18. Rodrigues AD, Lin JH. Screening of drug candidates for their drug–drug interaction potential. Curr Opin Chem Biol 2001;5:396–401.
19. Ekins S, Stresser DM, Williams JA. In vitro and pharmacophore insights into CYP3A enzymes. Trends Pharmacol Sci 2003;24:191–196.

20. Galetin A, Clarke SE, Houston JB. Multisite kinetic analysis of interactions between prototypical CYP3A4 subgroup substrates: midazolam, testosterone, and nifedipine. Drug Metab Dispos 2003;31:1108–1116.
21. Ghanbari F, Rowland-Yeo K, Bloomer JC, Clarke SE, Lennard MS, Tucker GT, et al. A critical evaluation of the experimental design of studies of mechanism based enzyme inhibition, with implications for in vitro-in vivo extrapolation. Curr Drug Metab 2006;7:315–334.
22. Bertelsen KM, Venkatakrishnan K, von Moltke LL, Obach RS, Greenblatt DJ. Apparent mechanism-based inhibition of human CYP2D6 in vitro by paroxetine: comparison with fluoxetine and quinidine. Drug Metab Dispos 2003;31:289–293.
23. Wang YH, Jones DR, Hall SD. Prediction of cytochrome P450 3A inhibition by verapamil enantiomers and their metabolites. Drug Metab Dispos 2004;32:259–266.
24. von Moltke LL, Durol AL, Duan SX, Greenblatt DJ. Potent mechanism-based inhibition of human CYP3A in vitro by amprenavir and ritonavir: comparison with ketoconazole. Eur J Clin Pharmacol 2000;56:259–261.
25. Zhou S, Chan E, Lim LY, Boelsterli UA, Li SC, Wang J, et al. Therapeutic drugs that behave as mechanism-based inhibitors of cytochrome P450 3A4. Curr Drug Metab 2004;5:415–442.
26. Franklin MR. Inhibition of mixed-function oxidations by substrates forming reduced cytochrome P-450 metabolic-intermediate complexes. Pharmacol Ther 1977;2:227–245.
27. Ernest CS 2nd, Hall SD, Jones DR. Mechanism-based inactivation of CYP3A by HIV protease inhibitors. J Pharmacol Exp Ther 2005;312:583–591.
28. Ito K, Hallifax D, Obach RS, Houston JB. Impact of parallel pathways of drug elimination and multiple cytochrome P450 involvement on drug-drug interactions: CYP2D6 paradigm. Drug Metab Dispos 2005;33:837–844.
29. Galetin A, Ito K, Hallifax D, Houston JB. CYP3A4 substrate selection and substitution in the prediction of potential drug-drug interactions. J Pharmacol Exp Ther 2005;314:180–190.
30. Yu DK. The contribution of P-glycoprotein to pharmacokinetic drug-drug interactions. J Clin Pharmacol 1999;39:1203–1211.
31. Suzuki H, Sugiyama Y. Role of metabolic enzymes and efflux transporters in the absorption of drugs from the small intestine. Eur J Pharm Sci 2000;12:3–12.
32. Chiou WL, Chung SM, Wu TC. Potential role of P-glycoprotein in affecting hepatic metabolism of drugs. Pharm Res 2000;17:903–905.
33. Kim RB. Drugs as P-glycoprotein substrates, inhibitors, and inducers. Drug Metab Rev 2002;34:47–54.
34. Endres CJ, Hsiao P, Chung FS, Unadkat JD. The role of transporters in drug interactions. Eur J Pharm Sci 2006;27:501–517.
35. Lin JH, Yamazaki M. Clinical relevance of P-glycoprotein in drug therapy. Drug Metab Rev 2003;35:417–454.
36. Yasui-Furukori N, Uno T, Sugawara K, Tateishi T. Different effects of three transporting inhibitors, verapamil, cimetidine, and probenecid, on fexofenadine pharmacokinetics. Clin Pharmacol Ther 2005;77:17–23.
37. Mizuno N, Niwa T, Yotsumoto Y, Sugiyama Y. Impact of drug transporter studies on drug discovery and development. Pharmacol Rev 2003;55:425–461.

38. Chandra P, Brouwer KL. The complexities of hepatic drug transport: current knowledge and emerging concepts. Pharm Res 2004;21:719–735.

39. Wu CY, Benet LZ. Predicting drug disposition via application of BCS: transport/absorption/elimination interplay and development of a biopharmaceutics drug disposition classification system. Pharm Res 2005;22:11–23.

40. Dresser GK, Bailey DG, Leake BF, Schwarz UI, Dawson PA, Freeman DJ, et al. Fruit juices inhibit organic anion transporting polypeptide-mediated drug uptake to decrease the oral availability of fexofenadine. Clin Pharmacol Ther 2002;71:11–20.

41. Xia CQ, Milton MN, Gan LS. Evaluation of drug-transporter interactions using in vitro and in vivo models. Curr Drug Metab 2007;8:341–363.

42. Hoffmaster KA, Turncliff RZ, LeCluyse EL, Kim RB, Meier PJ, Brouwer KL. P-glycoprotein expression, localization, and function in sandwich-cultured primary rat and human hepatocytes: relevance to the hepatobiliary disposition of a model opioid peptide. Pharm Res 2004;21:1294–1302.

43. Liu X, LeCluyse EL, Brouwer KR, Lightfoot RM, Lee JI, Brouwer KL. Use of Ca2+ modulation to evaluate biliary excretion in sandwich-cultured rat hepatocytes. J Pharmacol Exp Ther 1999;289:1592–1599.

44. Benet LZ, Cummins CL, Wu CY. Transporter-enzyme interactions: implications for predicting drug-drug interactions from in vitro data. Curr Drug Metab 2003;4:393–398.

45. Yasuda K, Lan L-B, Sanglard D, Furoya K, Schuetz JD, Schuetz EG. Interaction of cytochrome P450 3A inhibitors with P-glycoprotein. J Pharm Exp Ther 2002;303:323–332.

46. Yan Q, Sadee W. Human membrane transporter database: a web-accessible relational database for drug transport studies and pharmacogenomics. AAPS PharmSci 2000;2:E20.

47. Schapira M, Raaka BM, Samuels HH, Abagyan R. Rational discovery of novel nuclear hormone receptor antagonists. Proc Natl Acad Sci USA 2000;97: 1008–1013.

48. Ozawa N, Shimizu T, Morita R, Yokono Y, Ochiai T, Munesada K, et al. Transporter database, TP-Search: a web-accessible comprehensive database for research in pharmacokinetics of drugs. Pharm Res 2004;21:2133–2134.

49. Saier MH Jr, Tran CV, Barabote RD. TCDB: the transporter classification database for membrane transport protein analyses and information. Nucleic Acids Res 2006;34:D181–186.

50. Ekins S, Mestres J, Testa B. In silico pharmacology for drug discovery: applications to targets and beyond. Br J Pharmacol 2007;152:21–37.

51. Ekins S, Mestres J, Testa B. In silico pharmacology for drug discovery: methods for virtual ligand screening and profiling. Br J Pharmacol 2007;152:9–20.

52. Ekins S, Ecker GF, Chiba P, Swaan PW. Future directions for drug transporter modelling. Xenobiotica 2007;37:1152–1170.

53. Lipinski CA, Lombardo F, Dominy BW, Feeney PJ. Experimental and computational approaches to estimate solubility and permeability in drug discovery and development settings. Adv Drug Del Rev 1997;23:3–25.

54. Ekins S, Swaan PW. Computational models for enzymes, transporters, channels and receptors relevant to ADME/TOX. Rev Comput Chem 2004;20:333–415.

55. Waller CL, Evans MV, McKinney JD. Modeling the cytochrome P450-mediated metabolism of chlorinated volatile organic compounds. Drug Metab Dispos 1996;24:203–210.
56. Schneider G, Coassolo P, Lave T. Combining in vitro and in vivo pharmacokinetic data for prediction of hepatic drug clearance in humans by artificial neural networks and multivariate statistical techniques. J Med Chem 1999;42:5072–5076.
57. Ekins S, Obach RS. Three dimensional-quantitative structure activity relationship computational approaches of prediction of human in vitro intrinsic clearance. J Pharmacol Exp Ther 2000;295:463–473.
58. Mager DE, Jusko WJ. Quantitative structure-pharmacokinetic/pharmacodynamic relationships of corticosteroids in man. J Pharm Sci 2002;91:2441–2451.
59. Karalis V, Tsantili-Kakoulidou A, Macheras P. Quantitative structure-pharmacokinetic relationships for disposition parameters of cephalosporins. Eur J Pharm Sci 2003;20:115–123.
60. Ng C, Xiao Y, Putnam W, Lum B, Tropsha A. Quantitative structure-pharmacokinetic parameters relationships (QSPKR) analysis of antimicrobial agents in humans using simulated annealing k-nearest-neighbor and partial least-square analysis methods. J Pharm Sci 2004;93:2535–2544.
61. Yap CW, Li ZR, Chen YZ. Quantitative structure-pharmacokinetic relationships for drug clearance by using statistical learning methods. J Mol Graph Model 2006;24:383–395.
62. Nyberg G, Axelsson R, Martensson E. Binding of thioridazine and thioridazine metabolites to serum proteins in psychiatric patients. Eur J Clin Pharmacol 1978;14:341–350.
63. Lombardo F, Obach RS, Shalaeva MY, Gao F. Prediction of volume of distribution values in humans for neutral and basic drugs using physicochemical measurements and plasma protein binding. J Med Chem 2002;45:2867–2876.
64. Lombardo F, Obach RS, Shalaeva MY, Gao F. Prediction of human volume of distribution values for neutral and basic drugs. 2. Extended data set and leave-class-out statistics. J Med Chem 2004;47:1242–1250.
65. Balakin KV, Ivanenkov YA, Savchuk NP, Ivaschenko AA, Ekins S. Comprehensive computational assessment of ADME properties using mapping techniques. Curr Drug Discov Tech 2005;2:99–113.
66. Lombardo F, Obach RS, Dicapua FM, Bakken GA, Lu J, Potter DM, et al. A hybrid mixture discriminant analysis-random forest computational model for the prediction of volume of distribution of drugs in human. J Med Chem 2006;49:2262–2267.
67. Ghafourian T, Barzegar-Jalali M, Dastmalchi S, Khavari-Khorasani T, Hakimiha N, Nokhodchi A. QSPR models for the prediction of apparent volume of distribution. Int J Pharm 2006;319:82–97.
68. Gleeson MP, Waters NJ, Paine SW, Davis AM. In silico human and rat Vss quantitative structure-activity relationship models. J Med Chem 2006;49:1953–1963.
69. Doddareddy MR, Cho YS, Koh HY, Kim DH, Pae AN. In silico renal clearance model using classical Volsurf approach. J Chem Inf Model 2006;46:1312–1320.
70. Kajita J, Keiko I, Eiichi F, Kuwabara T, Kobayashi H. Effects of olopatadine, a new antiallergic agent, on human liver microsomal cytochrome P450. Drug Metab Dispos 2002;30:1504–1511.

71. Zhao X-J, Jones DR, Wang Y-H, Grimm SW, Hall SD. Reversible and irreversible inhibition of CYP3A enzymes by tamoxifen and metabolites. Xenobiotica 2002;32:863–878.
72. Mayhew BS, Jones DR, Hall SD. An in vitro model for predicting in vivo inhibition of cytochrome P450 3A4 by metabolic intermediate complex formation. Drug Metab Dispos 2000;28:1031–1037.
73. Kasahara M, Suzuki H, Komiya I. Studies on the cytochrome P450 (CYP)-mediated metabolic properties of miocamycin:evaluation of the possibility of a metabolic intermediate complex formation with CYP, and identification of the human CYP isoforms. Drug Metab Dispos 2000;28:409–417.
74. Jones DR, Ekins S, Li L, Hall SD. Computational approaches that predict metabolic intermediate complex formation with CYP3A4 (+b5). Drug Metab Dispos 2007;35:1466–1475.
75. Welch R, Chue P. Antipsychotic agents and QT changes. J Psychiatry Neurosci 2000;25:154–160.
76. Mehta D, Mehta S, Petit J, Shriner W. Cardiac arrhythmia and haloperidol. Am J Psychiatry 1979;136:1468,1469.
77. Crumb WJ Jr, Ekins S, Sarazan D, Wikel JH, Wrighton SA, Carlson C, et al. Effects of antipsychotic drugs on Ito, INa, Isus, IK1, and hERG: QT prolongation, structure activity relationship, and network analysis. Pharm Res 2006;23:1133–1143.
78. Harrigan EP, Miceli JJ, Anziano R, Watsky E, Reeves KR, Cutler NR, et al. A randomized evaluation of the effects of six antipsychotic agents on QTc, in the absence and presence of metabolic inhibition. J Clin Psychopharmacol 2004;24:62–69.
79. Pfizer briefing document for Zeldox Capsules (Ziprasidone HCl). 2000 Jul 19.
80. Ekins S, Balakin KV, Savchuk N, Ivanenkov Y. Insights for human Ether-a-Go-Go-Related Gene Potassium Channel inhibition using recursive partitioning, Kohonen and Sammon mapping Techniques. J Med Chem 2006;49:5059–5071.
81. Ekins S, Crumb WJ, Sarazan RD, Wikel JH, Wrighton SA. Three-dimensional quantitative structure-activity relationship for inhibition of human ether-a-go-go-related gene potassium channel. J Pharmacol Exp Ther 2002;301:427–434.
82. Hardmand JG, Limbird LE, editors. The Pharmacological Basis of Therapeutics. New York: McGraw-Hill; 2001.
83. Physicians' Desk Reference, Medical economics, thompson healthcare. Montvale (NJ) 2004.
84. Belpaire FM, Vanderheeren FA, Bogaert MG. Binding of thioridazine and some of its metabolites to human serum protein and human albumin. Arzneimittelforschung 1975;25:1969–1971.
85. Dinovo EC, Pollak H, Gottschalk LA. Partitioning of thioridazine and mesoridazine in human blood fractions. Methods Find Exp Clin Pharmacol 1984;6:143–146.
86. Aronov AM, Balakin KV, Kiselyov A, Varma-O'Brien S, Ekins S. Applications of QSAR methods to ion channels. In: Ekins S, editor. Computational toxicology: risk assessment for pharmaceutical and environmental chemicals. Hoboken (NJ): John Wiley and Sons; 2007. pp. 353–389.
87. Crumb W, Cavero I. QT interval prolongation by non-cardiovascular drugs: issues and solutions for novel drug development. Pharm Sci Technol Today 1999;2:270,280.

88. Zhang S, Zhou Z, Gong Q, Makielski JC, January CT. Mechanism of block and identification of the verapamil binding domain to HERG potassium channels. Circ Res 1999;84:989–998.
89. Kang J, Chen XL, Wang H, Ji J, Reynolds W, Lim S, et al. Cardiac ion channel effects of tolterodine. J Pharmacol Exp Ther 2004;308:935–940.
90. Amidon GL, Lennernas H, Shah VP, Crison JR. A theoretical basis for a biopharmaceutic drug classification: the correlation of in vitro drug product dissolution and in vivo bioavailability. Pharm Res 1995;12:413–420.
91. Chang C, Bahadduri PM, Polli JE, Swaan PW, Ekins S. Rapid identification of P-glycoprotein substrates and inhibitors. Drug Metab Dispos 2006;34:1976–1984.
92. Wei H, Lobenberg R. Biorelevant dissolution media as a predictive tool for glyburide a class II drug. Eur J Pharm Sci 2006;29:45–52.
93. Kier LB, Hall LH. Molecular molecular connectivity in structure-activity analysis. Letchworth: Wiley, Research Studies Press; 1986.
94. Cruciani G, Crivori P, Carrupt P-A, Testa B. Molecular fields in quantitative structure-permeation relationships: the VolSurf approach. J Mol Struct 2000;503:17–30.
95. Bergstrom CA. In silico predictions of drug solubility and permeability: two rate-limiting barriers to oral drug absorption. Basic Clin Pharmacol Toxicol 2005;96:156–161.
96. Bergstrom CA, Strafford M, Lazorova L, Avdeef A, Luthman K, Artursson P. Absorption classification of oral drugs based on molecular surface properties. J Med Chem 2003;46:558–570.
97. Jónsdóttir SO, Jørgensen FS, Brunak S. Prediction methods and databases within chemoinformatics: emphasis on drugs and drug candidates. Bioinformatics 2005;21:2145–2160.
98. Gleeson MP. Generation of a set of simple, interpretable ADMET rules of thumb. J Med Chem 2008;51:817–834.
99. Dimitrov S, Dimitrova G, Pavlov T, Dimitrova N, Patlewicz G, Niemela J, et al. A stepwise approach for defining the applicability domain of SAR and QSAR models. J Chem Inf Model 2005;45:839–849.
100. Tetko IV, Bruneau P, Mewes HW, Rohrer DC, Poda GI. Can we estimate the accuracy of ADME-Tox predictions? Drug Discov Today 2006;11:700–707.
101. Sheridan RP, Feuston BP, Maiorov VN, Kearsley SK. Similarity to molecules in the training set is a good discriminator for prediction accuracy in QSAR. J Chem Inf Comput Sci 2004;44:1912–1928.
102. Khandelwal A, Bahadduri P, Chang C, Polli JE, Swaan P, Ekins S. Computational models to assign biopharmaceutics drug disposition classification from molecular structure. Pharm Res 2007;24:2249–2262.

14

IN VITRO–IN VIVO CORRELATIONS IN DRUG DISCOVERY AND DEVELOPMENT: CONCEPTS AND APPLICATIONS IN TOXICOLOGY

REX DENTON, KIMBERLY BRANNEN, AND BRUCE D. CAR

14.1 INTRODUCTION

Pharmaceutical industry productivity, as measured by the annual federal registration of new chemical entities (NCEs), has progressively declined since the early 1990s. Reasons cited for this decline include reduction in scientific innovation, heightened safety standards, and progressively more challenging, less well-characterized drug targets (1). Although a highly complex combination of these reasons and others exist for low productivity, considering the plethora of technologic advancements and 100-fold increase in drug targets (480 pregenome draft, 5000–10,000 postgenome draft), waning scientific innovation is likely a trivial component in declining productivity.

Historically, hepatotoxicity and nephrotoxicity dominated as the most significant causes of attrition because of safety. The withdrawal of drugs such as terfenadine and astemizole have more recently increased the focus on the potential for cardiac liabilities, with commensurately higher regulatory agency standards relative to the benefit–risk ratio for cardiovascular liabilities. Strategies to eliminate compounds carrying serious liabilities before their advancement from discovery to development depend on robustly validated and cost-efficient assays capable of utilizing low quantities of drug. This chapter outlines several

Predictive Approaches in Drug Discovery and Development: Biomarkers and In Vitro/In Vivo Correlations, First Edition. Edited by J. Andrew Williams, Jeffrey R. Koup, Richard Lalonde, and David D. Christ.

approaches for the early detection of potential liabilities and therefore enhancing informed decision making relative to compound progression from the discovery environment to drug Development. Specifically, approaches for the evaluation of unwanted ancillary pharmacology, mitochondrial toxicity, hepatotoxicity, cardiotoxicity, and teratogenicity are presented, and general considerations in the interpretation of predictive toxicity assay results are provided throughout.

14.2 PREDICTION OF MECHANISMS OF TOXICITY AND THE DEVELOPMENT OF COUNTERSCREENS

In vitro toxicity assessments and counterscreening activities test the relationship between safety liabilities and well-understood physicochemical characteristics of new molecules. These assays are typically inexpensive and outline known or potential risks of new molecules to cause undesired or adverse changes within living systems. Reliable *in vitro* target-organ assessment and screening panels identify and attempt to accurately predict toxicity, and are often used to outline safety risks and roughly define safety margins of a chemical series well before candidate drugs enter development. Appropriate *in vitro* screening of off-target counterscreening strategies reduces risk and cost, and streamlines chemical optimization of novel druglike molecules. These relatively inexpensive, preliminary experiments are typically based on predefined criteria, and are completed in a few days to weeks.

14.2.1 High Throughput, *In Vitro* Screening Panels

High throughput, direct ligand-binding methods can detect potential adverse reactions, assign activity to general categories (i.e., "bin"), or both, based on the interaction of a molecule of interest with a panel of enzymes and receptors typically associated with known effects on biological systems. *In vitro* binding screens typically orient the researcher to potential nonspecific interactions and potential *in vivo* effects, potential toxicities, or atypical effects that impact optimization and further development of the molecule or chemical series. Many binding screens are predictive of known risks and are widely evaluated using high throughput assessment and preliminary optimization of most novel chemical entities. The identification of off-target binding interactions are used to either eliminate structural liabilities from a molecule or chemical series, or if the interactions are persistent or unavoidable, to suggest an appropriate counterscreening strategy.

A typical high throughput binding panel includes receptors, enzymes, and ion channels that evaluate the selectivity and potency of compounds for specific assay targets with the goal of predicting the potential for adverse findings *in vivo*. The relevance of data derived from screening with ligand-binding assays is fundamentally dependent on high affinity, reversible (or lack thereof), and saturable binding of the test compound to the binding panel targets and the consideration of concentrations that may be produced *in vivo*. Typically, a higher throughput, initial rough-cut analysis is performed using a single concentration (typically 1 or 10 μM) of the test molecule combined with a known concentration of a tracer ligand possessing a high affinity for the target. Following incubation under appropriate conditions specific for each assay, the magnitude of the inhibition of the high affinity probe binding by the test compound, relative to control (a sample lacking the test compound), is expressed as a percent of the control binding. Positive "hits" occur when a threshold value (typically 50%) inhibition of control binding is exceeded. Because the result from a single concentration cannot be used to determine a relative affinity or characterize allosteric binding interactions for the unknown molecule, "hits" on the panel are typically followed up with a more rigorous binding assay performed at a variety of concentrations. The resulting binding isotherm is then used to calculate the concentration that effectively inhibits 50% of the probe binding (IC_{50}), and determine the overall level of affinity the test molecule relative to known ligands for the evaluated target. Even more experiments would be necessary to determine the nature of the inhibition (e.g., competitive or other) and the true inhibition constant, K_i.

14.2.2 Selection of the Panel

Examples of targets and generally accepted liabilities are shown in Table 14.1. These binding targets are typically receptors, enzymes, and ion channels associated with acute effects, known toxicities, or frequently occurring off-target (ancillary pharmacology) effects.

14.2.3 Key Factors Affecting the Extrapolation of *In Vitro* Results to *In Vivo* Effects

14.2.3.1 Protein Binding. Several factors should be considered when interpreting data from *in vitro* binding screens. For any assessment performed *in vitro*, including binding panels, test compounds are evaluated using defined buffers and media that do not reflect conditions *in vivo*, especially those relating concentration to effect. Many drugs bind reversibly to plasma proteins *in vivo*, and according to the free drug hypothesis, it is only the free (unbound) drug that can bind to extracellular receptors and channels or pass through cell membranes to interact with intracellular enzymes and other targets (2). Since *in vitro* binding screens often use little if any additional protein (other than the target preparation), the IC_{50} values typically determined represent the free or unbound concentration. Therefore, binding panel results commonly overstate the potential for interaction, as the concentrations of unbound drug *in vivo* may be much lower than those tested in the screening assays. Determining the protein binding and describing the potency of a novel drug to interact with the intended target in terms of unbound concentrations better anticipates the interaction of the drug with the panel target under *in vivo* conditions. The science and impact of specific and nonspecific interactions of drugs with plasma proteins and the potential effects on pharmacologic and pharmacokinetic behavior has been extensively reviewed (3–5).

TABLE 14.1 Typical Screening Targets and Associated Biological Activity

Target	Organ System and Potential Activity and Liability
Cholinesterase, acetyl	CNS effects, acute toxicity
Monoamine oxidase MAO-A/B	Drug–drug interactions with SSRI's, CNS toxicity
Phosphodiesterase PDE3	Hemodynamic effects, oocyte development
Phosphodiesterase PDE4	Emesis, systemic vasculitis
Others, phosphodiesterases	Profiling of other PDEs indicated if potent hits against PDE3/4
Adenosine α	Hemodynamic, common off-target of kinase inhibitors
Adrenergic $\alpha 2$	Pulmonary bronchodilation, vascular smooth muscle contraction, platelet aggregation, presynaptic autoreceptor norepinephrine release
Adrenergic $\alpha 1$	Lipolysis, tachycardia, vascular smooth muscle contraction
Adrenergic β	Cardiovascular effects, tachycardia, hemodynamic effects
Androgen (testosterone)	Endocrine, androgeny, hirsuitism
Calcium channel L-type	Electrocardiographic parameters, hemodynamic effects
hERG	Cardiovascular, QT prolongation, torsades de pointes
Dopamine D_1	CNS. Stereotypy, ataxia, dependence, psychosis, abuse potential parathyroid hormone release, memory impairment
Dopamine D_2	CNS. Stereotypy, ataxia, dependence, psychosis, abuse potential, parathyroid hormone release
Dopamine D_3	CNS. Stereotypy, ataxia, dependence, abuse potential
Estrogen	Endocrine, androgenesis, reproductive effects
GABA	CNS, sedation, seizure, reproductive effects
Glucocorticoid	Immunological and inflammatory effects, apoptosis, adrenal effects, immunosuppression
Glutamate, NMDA	CNS toxicity
Histamine H1	Various. Sedation, bronchoconstriction, gastric secretion, gut contractility
Muscarinic, nonselective	CNS effects
Nicotinic acetylcholine	CNS effects
Opiate μ (OP3, MOP)	Various. Sedation, proprioception, abuse potential, gastrointestinal stasis
Opiate, nonselective	Sedation, proprioception, abuse potential
Progesterone PR-B	Endocrine, reproductive effects
Serotonin 5-HT	CNS effects. Profiling against several individual 5-HT receptors is suggested if binding to one or more receptors is identified in the initial screen. Agonism versus antagonism should be determined.

TABLE 14.1 ***(Continued)***

Target	Organ System and Potential Activity and Liability
Serotonin 5-HT 2B	Cardiovascular. Valvulopathies (agonist), cardiomyopathy (antagonist—rodents). Antagonism generally not relevant to human risk
Cardiac sodium channel	Cardiovascular. Conduction and electrocardiographic parameters (QRS, PR interval)
Somatostatin sst2	Endocrine. Regulation of secretion (pituitary gland, the pancreas, and the gastrointestinal tract)
Transporter, choline	Development and reproduction. Neural tube defects, terata, subsequent behavioral abnormalities
Transporter, dopamine (DAT)	CNS effects
Transporter, norepinephrine (NET)	CNS effects
Transporter, serotonin 5-HT (SERT)	CNS effects

Abbreviations: CNS, central nervous system; 5-HT, 5-hydroxytryptamine; MAO, monoamine oxidase; SSRI, selective serotonin re-uptake inhibitor.

14.2.3.2 Indices of Exposure* In Vivo*: Maximum Plasma Concentrations (C_{max}) and Area under the Plasma Concentration versus Time Curve (AUC). Toxicologists and other scientists frequently default to area under the drug concentration versus time curve (AUC) determined in toxicokinetic studies with animals as the key associative or causative parameter in toxicity. In general, this is appropriate when applied to target-organ toxicities such as hepatocellular toxicity; however, when considering ancillary pharmacology targets, the inhibition of which is a concentration–threshold event, the peak drug concentration or C_{max} is more relevant.

Significant inhibition of certain targets *in vivo*, especially the cardiac ion channels such as hERG, would be expected when the maximum concentration is high, and exceeds the value determined *in vitro*. If these threshold concentrations (especially of free or unbound drug as discussed previously) are exceeded, even for brief periods, the risk for biological stimulation or inhibition may be significant. Determining C_{max} experimentally is greatly dependent on study design, especially when absorption, distribution, and clearance are very rapid, and early data points may miss or underestimate actual C_{max}, rendering accurate safety projections for target effects such as hERG inhibition occasionally difficult.

14.2.3.3 Compound Solubility. "Negative" results, or data that suggest a lack of interaction between the test molecule and screening targets, must also be scrutinized carefully. Molecules that are highly insoluble in the aqueous solutions used for *in vitro* screening assessments often return data that suggest a low risk since little drug will be in solution and capable of interacting with target proteins. If multiple concentrations demonstrate a low plateau of saturable binding, with or without a lack of linearity, the data may suggest that the test molecule is insoluble in the test system buffer. As a result, when performing any *in vitro* assessment, all negative results should be evaluated for solubility in aqueous solutions and reaction buffers. Traditionally, "equilibrium" solubility was the most suitable analytical method; however, newer, multiwell compatible laser nephelometry methods have been used to overcome throughput issues with the traditional methods and utilize the DMSO solutions that are typically prepared for the liquid handling of stock solutions.

While *in vitro* ligand binding data are important, they are often inadequate by themselves to determine potential for toxicity, and further characterization of "hits" from the screening panel is typically performed to place the binding results into context. As discussed previously, this usually requires an in-depth understanding of the pharmacology of the interaction, the agonist and antagonist properties of the interaction, and drug disposition characteristics such as protein binding. Additionally, specialized *in vitro* or *in vivo* experiments are generally performed to place the screening data into appropriate perspective. These experiments and systems are too numerous to list, but can include intact cell (culture, coculture, or suspension) or *ex vivo* organ systems, or *in vivo* studies to directly determine potential toxicity and verify the relevance of the *in vitro* screening data in an expeditious manner.

14.3 COUNTERSCREENING STRATEGIES AND EXAMPLES

In general, high chemical selectivity for the desired pharmacologic target is associated with improved tolerability and safety relative to less selective molecules. During early drug discovery, a chemical series will often demonstrate a high affinity for unwanted, off-target interactions. Minimizing or eliminating these interactions and improving selectivity are important goals, especially when the "hit" in the screen is to a closely related target.

A chemical series may demonstrate poor selectivity or show unwanted adverse effects related to unforeseen interaction with closely related targets or transduction pathways that arise because of (i) evolutionary duplication and subsequent divergence, with unanticipated expression in other tissues or (ii) undesirable activity and subsequent toxicity in additional biological pathways. Consequently, many biologically active molecules share high affinity for, and are capable of binding to, other closely related molecular targets. For example, in the antidiabetic area, it was shown recently that the highly selective DPP4 inhibitor (over DPP2, DPP8, and DPP9) sitagliptin (Januvia®), carried fewer liabilities in nonclinical toxicology studies than vildagliptin, which may in part have contributed to the more tortuous registrational path for vildagliptin (6).

14.3.1 Strategic Considerations for Counterscreening Against Conserved Protein Targets

Conservation, evolutionary duplication, and divergence within the genome are reflected in shared activities for different receptors, channels, and enzymes and shared affinities for substrates and drugs. Thus, promiscuous drug binding and related pharmacology with additional, closely related target receptors, channels, or enzymes may predict toxic effects. Families of proteins are often included during *in vitro* assessments when target sequence conservation is high. In such instances, direct counterscreening against the conserved protein target(s) provides information that enables development of strategies to retain the desired pharmacology while eliminating potential toxicity. In such cases, target pharmacology and toxicity are typically governed by the kinetic characteristics of the interaction of the pharmacophore for the pharmaceutical target, and toxic effects are anticipated by understanding target pharmacology and tissue expression patterns. Determining the subjective *in vitro* boundary or parameter that distinguishes a desired pharmacological effect from a potentially toxic liability can be attempted when comparing data to suitable *in vivo* models. Often, the degree of receptor expression and occupancy in target tissues is an informative measure of the potential to cause a desired change at a target organ while avoiding undesirable effects in other tissues. Pharmacokinetic descriptions are widely used to determine efficacy boundaries and avoid toxicity, and special emphasis is placed on understanding the expression patterns of proteins in various target tissues.

14.3.2 PDE4 is an Example of a Conserved Sequence Target Counterscreen

Phosphodiesterase (PDE) 4 is an example where a pharmacological strategy was used to mitigate adverse central nervous system (CNS) effects *in vivo* (7). PDE enzymes degrade and terminate the cyclic nucleotide-mediated signal transduction produced by the second messengers cAMP and cGMP. Nine PDE4 families encoded by four distinct genes differ only by preference for specific substrates, sensitivity to endogenous activators and inhibitors, and tissue distribution of the gene products (8). The conservation of PDE family genes suggests an evolutionary duplication and subsequent divergence that resulted in specific physiological responses in immune and inflammatory cells, bronchi, and the CNS.

As a result of sequence conservation and multiple conformations of unique expression products, there are 11 different PDE families, representing over 50 different cAMP and cGMP PDE variants. PDE4 activity alone is provided by a large family (>20 isoforms), which is encoded by four genes (PDE4A, 4B, 4C, and 4D) (9, 10). These gene products can interact with each other and also serve to influence the function of the PDE4 catalytic unit where they orchestrate the functional consequence of phosphorylation of the PDE4 catalytic unit by ERK, and of phosphorylation of the N-terminus by PKA (11). As these differentially targeted and regulated isoforms are cell specific, it is likely that PDE4 isoforms have specific functional roles in different tissues. Correspondingly, inhibitors of PDE4, such as rolipram (Fig. 14.1), show a variety of physiological effects.

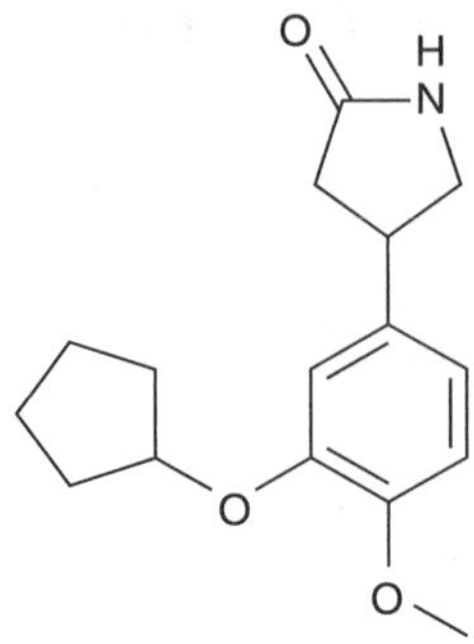

FIGURE 14.1 Structure of rolipram.

Rolipram-induced elevation of intracellular cAMP causes increased synthesis and release of norepinephrine in peripheral tissues, enhanced central noradrenergic transmission, and suppressed expression of proinflammatory cytokines and other mediators of inflammation. The pharmacological effects include attenuation of depression and inflammation, and both effects are potentially relevant clinically. While the anti-inflammatory, immunomodulatory, and smooth muscle relaxant activities of PDE4 inhibitors support the therapeutic value of PDE4 inhibition for treatment of asthma and other inflammatory diseases (demonstrated with a variety of *in vitro* and *in vivo* systems, e.g., isolated human bronchi), first-generation PDE4 inhibitors such as rolipram demonstrated dose-limiting CNS side effects (centrally mediated nausea and vomiting, increased gastric acid secretion, and potential psychotropic activity) thereby limiting potential clinical effectiveness (12–15).

In the case of centrally mediated emesis and nausea, the PDE4-binding characteristics suggested that improvement in the selectivity of PDE4 inhibition of human monocyte-derived PDE4 catalytic activity could be gained, and interactions with the higher affinity (nanomolar) [^{3}H]rolipram-binding sites that predominate in the CNS could be minimized. Although the analysis of low and high affinity activity is complicated by conserved protein sequences, discrete kinetic analysis and radioligand binding studies on full-length and truncated forms of recombinant human PDE4 suggested that acute toxicity may be differentiated from therapeutically useful interactions by ligand binding to distinct high affinity forms of the enzyme in the CNS (PDE or Sr) versus binding to a lower affinity form (LPDE or Sc) on the peripheral proinflammatory and bronchial smooth muscle cells. The stereoisomers of rolipram bind to the catalytic sites of the two conformers with different affinities. For example, *R*-rolipram binds to Sr with a K_d of 1–5 nM but has approximately 100-fold weaker binding to Sc. Furthermore, the R-(−) enantiomer of rolipram demonstrates an approximately 10- to 20-fold selectivity over the S-(+) enantiomer in binding to HPDE, whereas only a two- to fourfold selectivity is seen at LPDE. These two conformational binding states are exhibited by all four PDE4 enzymes (16, 17). The rank order of potency of a variety of PDE4 inhibitors for this binding site differs from their rank order of potency for inhibition of cAMP hydrolysis (16).

Because the desired and undesired activity could be assigned to differential isoform binding, counterscreening efforts focused on maximizing the therapeutic potential of inhibitors of the low K_m cAMP-specific PDE (PDE4), while attempting to minimize affinity for the high K_m cAMP-specific site, which predominates in the CNS. Inhibitors such as *R*-rolipram and RS25344, which bind with high affinity, are significantly less active against the truncated enzyme compared to the full-length enzyme, whereas second generation inhibitors show equal potency against both forms of the enzyme. Such "second generation PDE inhibitors" show improved potency to the lower affinity form, while not affecting high affinity binding site, which is predominant in the CNS. This counterscreening strategy demonstrates the importance of understanding the basic pharmacological properties of a target to optimize desired interactions while minimizing adverse effects.

14.4 MITOCHONDRIA AS A TARGET FOR TOXICITY

Mitochondria are complex, critical organelles, present in all cells, and are increasingly associated with drug-mediated toxicities. In addition to generating adenosine triphosphate (ATP), mitochondria carry maternally transmitted genetic material distinct from nuclear heterochromatin, and are able to replicate and express their primitive genome independently of cell division. Other functions of mitochondria are related to the utilization of this ATP energy in different cell types.

Maintenance of mitochondrial function and the production of ATP, the critical source of cellular energy, depend intimately on the architecture and components of the two membranes defining this organelle. Mitochondrial membranes consist of a unique lipid bilayer. The outer membrane is generally more permeable to ions and solutes than the inner layer. The inner membrane encloses a water-containing compartment (called the *matrix*), containing the maternal mitochondrial DNA and most of the enzymes involved in oxidative phosphorylation. This inner membrane is less permeable and contains special membrane "pore" proteins that are involved with transport of energy containing molecules as well as drugs into the core matrix. This pore complex is often a target for mitochondrial toxicants, which in general terms, typically increase permeability, inhibit normal transport, or both. The inner membrane's lipid composition is distinguished from typical eukaryotic cellular membrane lipid bilayers by the presence of comparatively little cholesterol, and large amounts of cardiolipin. Cardiolipin is a phospholipid that serves as a "scaffold" for anchoring the specific electron chain transport proteins. By virtue of the unique charge characteristics of cardiolipin, some drugs (e.g., anthraquinones such as adriamycin and doxorubicin) bind preferentially, resulting in potentially toxic concentrations of the molecules within in the inner mitochondrial membrane (18). Thus, function of mitochondria is physiologically regulated on a cell-specific basis and is as diverse as the production of heat to steroid synthesis (19).

Ultimately, drug-induced mitochondrial impairment and toxicity can result from the formation or cellular liberation of excessive Ca^{2+}, free radical formation, disruption of the mitochondrial membrane potential, inhibition of mitochondrial replication, or disruption of the electron transport process. When a cell is exposed to elevated Ca^{2+} concentrations (arising from either cellular deterioration or toxicity), to preserve cellular energy, the negative membrane potential, established as a result of electron transport, deteriorates rapidly, resulting in the release of apoptotic factors. Other examples of how mitochondrial function can be impaired acutely include the impairment of the exchange of requisite substrates via membrane transporters, inhibition of metabolic pathways that fuel respiration, and direct effects of drugs on cardiolipin. Such acute effects accelerate electron transport component oxidation yielding oxygen-centered and nitrogen-centered free radicals. In contrast, drugs that impair DNA replication or protein synthesis will diminish mitochondrial and bioenergetic capacity over longer time periods. In addition, many drugs can precipitate irreversible mitochondrial failure by induction of the "permeability transition pore," ultimately leading to release of proapoptotic factors that include cytochrome *c*. Drugs can also change the equilibrium between proapoptotic and antiapoptotic proteins, such as Bak/Bax and Bcl-2, among many others, causing mitochondrial toxicity.

Mitochondrial toxicity, produced by the inhibition of mammalian DNA polymerase-γ, was found to be the basis of numerous toxicities following the introduction of nucleoside reverse transcriptase inhibitors (NRTIs) for the treatment of HIV. NRTI's by design, prevent mitochondrial replication by the inhibition of mtDNA. Thus, any molecule demonstrating significant antipolymerase activity can reasonably be expected to cause toxicity related to impaired mitochondrial function at some concentration, as mitochondria replicate depend on cellular requirement for energy production (and other critical mitochondrial functions). As a result, this inhibition by NRTI's and other molecules ultimately results in fewer mitochondria. While NRTI toxicity may be manifested in a variety of toxic sequelae, it is especially apparent in tissues with high oxidative demands, such as liver and muscle (20, 21). Additionally, sequelae related to the impaired transport of lipids (i.e., lipodystrophy, lipoatrophy, lipidoses) are often evident in certain tissues. Other drugs may produce mitochondrial-mediated ototoxicity and nephrotoxicity, including aminoglycoside antibiotics that impair mitochondrial protein synthesis (22). In addition to these longer-term effects, other drugs, such as some thiazolidinones (23) and fibrates (24), directly inhibit the electron transport chain components in mitochondria. Compounds such as dinitrophenol and nonsteroidal anti-inflammatory drugs (NSAIDs) such as diclofenac, aspirin, nimesulide, meloxicam, piroxicam, and indomethacin (25) uncouple electron transport from ATP synthesis and break down the membrane gradient (26). Disruption of electron transport dissipates the membrane potential ($\Delta\Psi$), and ATP production is diminished, or abolished, depending on the severity of the disruption.

Mitochondrial toxicity can also be produced by drugs that impose oxidative stress via redox cycling or glutathione depletion, reactive oxygen species

(ROS) generation via CYP2E1 metabolism, or reactive metabolite formation. Free radicals can directly inactivate several portions of the transport complexes. Some thiazolidinediones and statins induce mitochondrial permeability transition, leading to irreversible collapse of the transmembrane potential and release of proapoptotic factors. Such impairment can be prevented by Bcl-xL overexpression (27). In addition, inhibition of metabolic pathways that fuel oxidative phosphorylation, such as fatty acid oxidation and the Krebs cycle, will impair ATP production as will inhibition of any of the many mitochondrial membrane transporters that exchange metabolites across the impermeable inner membrane.

14.4.1 Prediction of Drug-Induced Mitochondrial Toxicity

Drug-induced mitochondrial toxicity is actively evaluated as part of the general pharmaceutical screening process. The evaluation of mitochondrial toxicity using conventional cell culturing methods in immortalized cell lines is, however, generally not well suited for the assessment of mitochondrial toxicity. This is because of the high glucose concentration in the medium, and the cells therefore deriving the majority of their ATP through glycolytic conversion, rather than lipid oxidation and oxidative phosphorylation. As a result, certain facets of mitochondrial function are poorly reflected in typical cell culture media. There have been efforts to make this *in vitro* system more reflective of the conditions *in vivo*. Marroquin et al. (28) replaced glucose in the growth media with the less easily metabolized galactose, resulting in a net negative equivalent of ATP production within the cell. These cells must then use mitochondrial oxidative phosphorylation induced by additional substrates to derive additional ATP necessary to maintain cellular energy requirements. As a result of enhanced mitochondrial activity, cells grown in galactose are susceptible to mitochondrial toxicants.

Traditionally, mitochondrial status is monitored via biological oxygen uptake or consumption, measured using low throughput polarography instrumentation with either intact cells or, more commonly, isolated mitochondria. Data from this assay indicate that many drugs with various adverse events are acute mitochondrial toxicants, and potency tracks well with severity and frequency of such events. Newer methods generally consist of dedicated, arrayed, phosphorescent oxygen-selective probes. These probes are compatible with a microtiter plate assay format, and can be integrated with standard fluorescence plate reader instrumentation, thereby overcoming the throughput limitations of the traditional poloragraphic measurements. Measurement of this oxygen consumption enables the rapid and specific detection of mitochondrial dysfunction in both isolated mitochondria and whole cells, thereby providing a simple yet highly sensitive means of assessing the impact of a compound on mitochondrial function.

Other techniques have been developed and utilized for mitochondrial toxicity screening. Measurements of mitochondrial membrane potential ($\Delta\Psi$) are easily adapted to multiwell, plate-based formats. These assays, however, do not distinguish the specific mode of mitochondrial toxicity, (i.e., uncoupling, inhibition, or mitochondrial permeability transition pore induction). Mitochondrial permeability transition is a hallmark of cytotoxic molecules and other anticancer molecules, and may contribute to toxicity in organs with high metabolic activity levels. Detecting electron transport inhibition by drug molecules, such as fibrates and biguanides, and respiratory chain activity is now possible using immunocapture technology in higher throughput capabilities, or with traditional radiolabeled deoxynucleotide triphosphates (dNTPs). Mitochondrial impairment may also be evaluated using isolated organ models. For example, clofibrate decreases oxygen consumption by 40% within 15 min and selectively kills periportal hepatocytes in an *ex vivo* liver model (24).

14.5 ASSAYS FOR THE PREDICTION OF HEPATOTOXICITY

Hepatotoxicity still ranks highly amongst potential causes of postmarketing drug withdrawal, and is a leading cause of drug failure in nonclinical and clinical development before registration. Prediction of hepatoxicity in the pharmaceutical discovery environment has been reviewed extensively (29).

Cellular targets within the liver (in order of diminishing importance) include hepatocytes, the biliary tree, and Kupffer cells. The prediction of *hepatocellular* toxicity, presumed largely to result from the biotransformation of parent drug to metabolites that are directly or indirectly cytotoxic, is a robust science, with fewer compounds now being advanced with significant unpredicted metabolic liabilities. (Predictive biotransformation approaches are discussed elsewhere in this text.) This section illustrates one well-validated *in vitro* assay for the prediction of hepatocellular toxicity *in vivo*. It is important to recognize that currently neither biotransformation assays nor assays for hepatocellular toxicity are consistently useful for the prediction of biliary toxicity (such as biliary hyperplasia or events leading to biliary cirrhosis) or for toxicity initiated in Kupffer cells.

Historically, cell lines derived from hepatocytes such as HepG2 cells have been used in screens and assigned predictive value for hepatotoxicity. When primary human hepatocytes or liver transcriptomic profiles are compared to hepatoma-derived cells such as HepG2, there are striking differences in RNA expression (William Foster, personal communication, 2008). Specifically, it has been known for some time that the genes associated with oxidative metabolism of xenobiotics by cytochrome *P*450 (CYP) enzymes, and transporters both at apical and basolateral hepatocyte surfaces, are markedly reduced in cell lines and nonoptimized or longer-term cultures of primary hepatocytes. As biotransformation events generally underpin the pathogenesis of most xenobiotic-induced hepatocellular toxicity, the absence of CYPs and transporters render these cell lines of less utility in predicting potential hepatotoxicity *in vivo*.

To address these deficiencies, a completely automated, transformed hepatocyte-based predictive tool for assessment of hepatotoxic potential was developed. To restore expression, key metabolic enzymes such as CYP 3A4, 1A1, 1A2, 2B, 2E1, 2C9, and 2C19, were stably transfected (SV40) into human hepatocytes (THLE-5 cells ATCC registration CRL-11113). Five of these cell lines are used in routine screening assessments (29). The hepatotoxic predictivity of the parent cell line and transfected sub-lines was validated with a set of over 700 xenobiotics, delivering approximately 80% sensitivity for prediction of hepatoxicants, and an overall fidelity of approximately 60%. Toxicities that were not predicted well were included biliary or other mechanisms not expected to. Predictive assays for biliary toxicity will likely require the full expression of the three-dimensional differentiation of the biliary tree (30). The development of these predictive models is currently hampered by technical complexity and restrictive patents, thereby constraining a broader evaluation of potential biliary toxicants *in vitro*.

14.6 PREDICTION OF POTENTIAL CARDIAC ION CHANNEL, CARDIOMYOPATHY, AND HEMODYNAMIC EFFECTS

In a retrospective analysis of the causes of drug candidate attrition, cardiovascular safety issues ranked as the most frequent cause, responsible for approximately 30% of early nonclinical and clinical drug termination (1). Highly predictive assays of the potential for drugs to interact with cardiac myocytes, cardiac conduction system, and vasculature have been developed (31) in response to this high attrition rate and the close scrutiny of cardiovascular safety by regulatory agencies.

Cardiac myocytes and the Purkinje fiber conduction system of the heart express several key ion channels, the most importance of which for drug development are the K^+ or hERG channel (I_{Kr}), Na^+ channel (SCN5A), and L-type Ca^{++} channel (I_{Ca}). Drugs inhibiting the function of hERG delay ventricular repolarization, prolong QT interval and predispose subjects to a rare and often fatal arrhythmia known as *torsades de pointes*. Inhibition of the Na^+ channel results in prolongation of PR and QRS intervals, delayed conduction and the potential for primary heart conduction block. Inhibition of the L-type Ca^{++} channel results in decreased myocardial contractility and other secondary hemodynamic changes.

Drug interactions with each of these ion channels may be screened for using ligand-binding assays with isolated proteins or electrophysiological (EP) assays in which transmembrane effects on ion currents are determined with stably transfected cell lines or primary explants of cardiac myofiber or Purkinje fibers. Ligand-binding assays usually comprise the first tier of screening, with EP assays typically representing more robust, but lower throughput tests. The correlation between the outcome of these assays and effects observed when electrocardiogram (ECG) or blood pressure measurements are performed in animals is perhaps the highest of all *in vitro* screens and can be used in the rank ordering for compounds and elimination of compounds with severe safety risks.

As for many other parameters, inhibition of ion channels, endothelial receptors, or enzymes correlates best with the unbound fraction of drug (32), a topic discussed previously in section 14.2.3.1. *In vitro* ligand binding and EP assays are typically conducted in the absence of added plasma proteins. The attenuating effects of plasma binding are most pronounced, for example, in the case of hERG inhibitors, for drugs with high plasma protein binding (>95%) and hence little unbound or free drug *in vivo*. As IC_{10} effects of hERG inhibition may still produce measurable QT prolongation, one must not assume that the potential for QT prolongation at free drug concentrations in plasma at or less than the IC_{50} values typically reported represents an *in vitro–in vivo* assay discordance. The correlation of QT prolongation to free fraction for hERG inhibitors is less clear for drugs with substantial unbound fractions.

In addition to effects on the heart, both cardiac and peripheral vascular effects may result in altered systolic and/or diastolic pressures. These result from unwanted drug interactions with one or several vascular smooth muscle and endothelial receptors or enzymes. The most common of these interactions are readily detected in ancillary pharmacology screens.

Before advancing a compound from discovery to clinical development, it is recommended that at least one *in vivo* evaluation be undertaken to ensure cardiovascular liabilities such as QT prolongation or increased or decreased blood pressure are well understood and clearly characterized. Anesthetized rabbits and conscious guinea pigs are typically used to assess the potential for QT prolongation. Rats with surgically implanted sensors that can be monitored telemetrically are highly effective at detecting potential effects on blood pressure, and avoid the complicating effects of anesthesia. Conscious telemetered macaque and dog studies are generally performed in late discovery or early development. Reconciling findings in these assays with the multiple cardiovascular screens undertaken in discovery is important for both human risk assessment and drug backup strategies.

14.7 PREDICTIVE TERATOLOGY

A significant proportion of safety-related compound attrition in the pharmaceutical industry results from teratology findings in regulatory reproductive toxicology studies (33). With the relatively late timing of segment II studies in the drug development process, the cost of teratogenicity-based attrition is especially high. For at least the last 30 years, identifying and implementing alternatives to the standard mammalian *in vivo* teratology models have been sought for their utility in screening compounds earlier in drug discovery and development for teratogenicity. These fall into the basic categories of cell culture, organ culture, or whole organism models. The assays discussed herein are by no means an exhaustive list, but many of the most popular and best tested methods are briefly reviewed.

14.7.1 Cell Culture Models of Teratology

Cultures of primary cells isolated from embryos are well established for teratology screening and mechanistic studies. Probably the best studied and validated example is the micromass (MM) assay. Flint and Orton (34) developed the limb bud MM assay in which compounds are tested for their ability to inhibit the differentiation of cells isolated from rat embryonic forelimb buds into chondrocytes *in vitro*. This assay has been thoroughly validated (34, 35). Recently the

European Centre for the Validation of Alternative Methods (ECVAM) Scientific Advisory Committee recommended that the limb bud MM assay be used in regulatory safety testing in the European Union (36). A similar assay with cells isolated from embryonic midbrain explants has also been described (34).

Compared to primary cell culture, permanent cell lines have the advantage of eliminating euthanasia of adult and embryonic animals, thereby reducing animal usage. The mouse embryonic stem cell test (EST), or similar methods using pluripotent embryonic stem (ES) cell lines derived from blastocysts, have surfaced as highly promising assays in recent years. The EST evaluates a compound's effect on differentiation and viability of ES cells and benchmarks those against the toxicity of the compound in 3T3 cells, which are used as a surrogate for maternal or general toxicity. ECVAM performed an extensive validation of the EST for categorizing compounds into three classes according to their teratogenic potency (non-teratogen, weak teratogen, strong teratogen) (37). The concordance of the assay results with documented *in vivo* mammalian teratogenicity data was 78% in this study, and the EST is recommended as an assay for regulatory safety testing in the EU (36). In addition, results from initial attempts to incorporate molecular endpoints into an ES assay suggest that gene expression analysis may make the method even more powerful (38–42). Another potentially valuable application of the EST in risk assessment would be to adapt the assay to use with human ES cells (43).

14.7.2 Organ Explant Culture Models of Teratology

As described below, the development of a number of embryonic tissues cannot be studied *in vitro* with whole embryos. In some cases, these tissues can be explanted from embryos and grown in organ cultures. Examples include limb buds (44, 45), palatal shelves (46, 47), mandibles (48–50), and hearts (51, 52). Embryo–fetal organ cultures are probably of limited utility in general screening for teratogenicity, but they can be quite useful in studying mechanisms of development and teratogenesis.

14.7.3 Whole Organism Models of Teratology

A number of assays exist that use whole developing organisms for predictive teratology, with some involving the removal of mammalian embryos from the uterus of the mother and culturing them *in vitro*. Others models take advantage of the ease of use and observation of rapidly developing non-mammalian species. Whole embryo approaches have the distinct advantage of including all embryonic tissues and aspects of development, and some make it possible to study all periods of development. However, clearly none of these models includes an inherent ability to model maternally mediated effects, whether protective or toxic.

The best accepted and validated of the whole embryo models is the postimplantation rodent whole embryo culture (WEC) technique. In the 1970s, Dennis New and his colleagues refined methods first described in the 1930s and developed the method currently used for growing postimplantation mammalian embryos in circulating or rolling cultures (53, 54). This assay produces embryonic development in rats (53) or mice (55) very similar to that which occurs in utero. The method is most successful when used with early organogenesis-stage embryos and when the culture period is limited to 48 h, typically gestation days 9–11 in rats and 8–10 in mice (56). While the time and stage limitations prevent investigation of a number of important developmental processes, the culture period represents one of the windows of greatest sensitivity to teratogens (57). WEC has been used for the last three decades in predictive teratology screening, and the method validated for regulatory use in Europe by ECVAM, with a level of concordance to *in vivo* data similar to that found for the EST (58). Attempts have been made to adapt WEC techniques to other mammalian species. For most, success has been quite limited, but significant progress has been made recently in the rabbit embryo model (59, 60).

Amphibian embryos have been used for many years in developmental biology research. Developing frogs are also a valuable model that allow for rapid, inexpensive teratogenicity screening in the whole organism throughout organogenesis. The Frog Embryo Teratogenesis Assay—*Xenopus* (FETAX) was originally developed by Dumont and colleagues (61, 62) in the 1980s and extensively validated for teratogenicity screening in the 1990s (63–66). Results of the FETAX are largely predictive of mammalian teratogenicity, and the assay is used frequently for hazard assessment in environmental toxicology studies. Most work with the FETAX was done with *Xenopus laevis*. More recently, *Xenopus tropicalis* was shown to be an effective species in this assay with a number of practical advantages over *X. laevis* (67). The assay involves culturing frog embryos from the blastula stage for 96 or 48 h, respectively, depending on the species used, and evaluating morphological, growth, and viability endpoints following waterborne exposure to test compounds or environmental samples.

Much like the advantages of the FETAX assay, a zebrafish embryo assay provides testing throughput similar to cell culture methods, using an *in vivo* whole organism model with intact pharmacokinetics and pharmacodynamics capabilities (68). Zebrafish are emerging as a promising new model for teratology screening, with interest expressed within academia, industry, and government (69–73). No thorough validation of the method as a teratogenicity screen or comprehensive analysis of the most appropriate experimental design or conditions has, however, been reported to date. Unresolved considerations include determining at which developmental stage(s) embryo–larvae should be exposed or evaluated, which zebrafish strain to use, how to analyze and interpret the data, and what endpoints to include. One published validation study described the use of a method referred to as the *Dar*T assay, which had an 88% concordance with outcomes in mammals, and compounds that were correctly categorized included some that required metabolic activation for mammalian teratogenicity (74).

Invertebrates, such as hydra, are also potentially useful species for teratogenicity screening. The hydra developmental toxicity assay was developed by

Marshall Johnson (75–78), and it compares the effects of a test compound on adult polyp hydra and artificial "embryos" created by dissociating and aggregating adult cells. The concentrations of test agent that cause adult (A) and developmental (D) toxicity are compared in an A/D ratio; high A/D ratios indicate selective developmental hazards. The assay is very inexpensive, has been tested in a number of laboratories, and the results reviewed by Collins (79); however, it is useful mainly for hazard identification, as the endpoints included in the assay do not provide information useful for exploring teratogenic mechanisms or characterizing target tissues. In addition, there is generally significant skepticism regarding the relevance of an invertebrate assay such as this in predicting mammalian teratogenicity.

REFERENCES

1. Car BD. Enabling technologies in reducing drug attrition due to safety failures. Am Drug Discov 2006;1:53–56.
2. Trainor GL. The importance of plasma protein binding in drug discovery. Expert Opin Drug Discov 2007;2:51–64.
3. Copeland RA. Determination of serum protein binding affinity of inhibitors from analysis of concentration-response plots in biochemical activity assays. J Pharm Sci 2000;89:1000–1007.
4. Copeland RA. Mechanistic considerations in high-throughput screening. Anal Biochem 2003;320:1–12.
5. Copeland RA. Evaluation of enzyme inhibitors in drug discovery: A guide for medicinal chemists and pharmacologists. Methods Biochem Anal 2005;46:1–265.
6. Lankas GR, Leiting B, Roy RS, Eiermann GJ, Beconi MG, Biftu T, Chan CC, Edmondson S, Feeney WP, He HB, Ippolito DE, Kim D, Lyons KA, Ok HO, Patel RA, Petrov AN, Pryor KA, Qian XX, Reigle L, Woods A, Wu JK, Zaller D, Zhang XP, Zhu L, Weber AE, Thornberry NA. Dipeptidyl peptidase IV inhibition for the treatment of type 2 diabetes: potential importance of selectivity over dipeptidyl peptidases 8 and 9. Diabetes 2005;54:2988–2994.
7. Torphy TJ. Phosphodiesterase isozymes: molecular targets for novel antiasthma agents. Am J Respir Crit Care Med 1998;157:351–370.
8. Beavo JA, Conti M, Heaslip RJ. Multiple cyclic nucleotide phosphodiesterases. Mol Pharmacol 1994;46:399–405.
9. Bolger G, Michaeli T, Martins T, St. John T, Steiner B, Rodgers L, Riggs M, Wigler M, Ferguson K. A family of human phosphodiesterases homologous to the dunce binding learning and memory gene products of Drosophila melanogaster are potential targets for antidepressant drugs. Mol Cell Biol 1991;13:6558–6571.
10. Richter W, Jin SL, Conti M. Splice variants of the cyclic nucleotide phosphodiesterase PDE4D are differentially expressed and regulated in rat tissue. Biochem J 2005;388:803–811.
11. Essayan DM. Cyclic nucleotide phosphodiesterases. J Allergy Clin Immunol 2001;108:671–680.

12. Heaslip RJ, Evans DY. Emetic, central nervous system and pulmonary activities of rolipram in the dog. Eur J Pharmacol 1995;286:281–290.
13. Bertolino A, Crippa D, Di Dio S, Fichte K, Musmeci G, Porro V, Rapisarda V, Sastre-y-Hernández M, Schratzer M. Rolipram versus imipramine in inpatients with major, "minor" or atypical depressive disorder: a double-blind doubledummy study aimed at testing a novel therapeutic approach. Int Clin Psychopharmacol 1988;3:245–253.
14. Hebenstreit GF, Fellerer K, Fichte K, Fischer G, Geyer N, Meya M, Sastre-y-Hernández W, Schöny M, Schratzer W, Soukop W, Trampitsch E, Varosanec S, Zawada E, Zöchling R. Rolipam in major depressive disorder: results of a double-blind comparative study with imipramine. Pharmacopsychiatry 1989;22:156–160.
15. O'Connolly M, Dierdorf D, Greb WH, Mayer M-ER, D Wolf. Efficacy of denbufylline in patients with multi-infarct dementia. Drug Dev Res 1988;14:195–198.
16. Souness JE, Rao S. Proposal for pharmacologically distinct conformers of PDE4 cyclic AMP phosphodiesterases. Cell Signal 1997;9:227–236.
17. Manning CD, Burman M, Christensen SB, Cieslinski LB, Essayan DM, Grous M, Torphy TJ, Barnette MS. Suppression of human inflammatory cell functionby subtype selective PDE4 inhibitors correlates with inhibition of PDE4A and PDE4B. Br J Clin Pharmacol 1999;128:1393–1398.
18. Wallace KB, Starkov AA. Mitochondrial targets of drug toxicity. Annu Rev Pharmacol Toxicol 2000;40:353–388.
19. Skulachev VP. *Membrane Bioenergetics*. New York: Springer-Verlag; 1988.
20. Lai K, Gang DL, Zawacki JK, Cooley TP. Fulminant hepatic failure associated with 2′,3′-dideoxyinosine (ddI). Ann Intern Med 1991;115:283–284.
21. Arnaudo E, Dalakas M, Shanske S, Moraes CT, DiMauro S, Schon EA. Depletion of muscle mitochondrial DNA in AIDS patients with zidovudine- induced myopathy. Lancet 1991;337:508–510.
22. Fischel-Ghodsian N. Genetic factors in aminoglycoside toxicity. Pharmacogenomics 2005;6:27–36.
23. Masubuchi Y, Kano S, Horie T. Mitochondrial permeability transition as a potential determinant of hepatotoxicity of antidiabetic thiazolidinediones. Toxicology 2006;222:233–239.
24. Brunmair B, Lest A, Staniek K, Gras F, Scharf M, Roden M, Nohl H, Waldhäusl W, Fürnsinn C. Fenofibrate impairs rat mitochondrial function by inhibition of respiratory control complex I. J Pharmacol Exp Ther 2004;311:109–114.
25. Nulton-Persson AC, Szweda LI, Sadek HA. Inhibition of cardiac mitochondrial respiration by salicylic acid and acetylsalicylate. J Cardiovasc Pharmacol 2004;44:591–595.
26. Moreno-Sánchez R, Bravo C, Vásquez C, Ayala G, Silveira L, Martínez-Lavín M. Inhibition and uncoupling of oxidative phosphorylation by nonsteroidal anti-inflammatory drugs: study in mitochondria, submitochondrial particles, cells, and whole heart. Biochem Pharmacol 1999;57:743–752.
27. Blanco-Colio LM, Justo P, Daehn I, Lorz C, Ortiz A, Egido J. Bcl-xL overexpression protects from apoptosis induced by HMG-CoA reductase inhibitors in murine tubular cells. Kidney Int 2003;64:181–191.

28. Marroquin LD, Hynes J, Dykens JA, Jamieson JD, Will Y. Circumventing the Crabtree effect: replacing media glucose with galactose increases susceptibility of HepG2 cells to mitochondrial toxicants. Toxicol Sci 2007;97:539–547.

29. Dambach DM, Andrews BA, Moulin F. New technologies and screening strategies for hepatotoxicity: use of in vitro models. Toxicol Pathol 2005;33:17–26.

30. Marion TL, Leslie EM, Brouwer KL. Use of sandwich-cultured hepatocytes to evaluate impaired bile acid transport as a mechanism of drug-induced hepatotoxicity. Mol Pharm 2007;4:911–918.

31. Hanson LA, Bass AS, Gintant G, Mittelstadt S, Rampe D, Thomas K. ILSI-HESI cardiovascular safety subcommittee initiative: evaluation of three non-clinical models of QT prolongation. J Pharmacol Toxicol Methods 2006;54:116–129.

32. Webster R, Lesihman D, Walker R. Towards a drug concentration effect relationship for QT prolongation and torsades de pointes. Curr Opin Drug Discov Devel 2002;5:116–126.

33. Guengerich FP, MacDonald JS. Applying mechanisms of chemical toxicity to predict drug safety. Chem Res Toxicol 2007;20:344–369.

34. Flint OP, Orton TC. An *in vitro* assay for teratogens with cultures of rat embryo midbrain and limb bud cells. Toxicol Appl Pharmacol 1984;76:383–395.

35. Spielmann HE, Genschow E, Brown NA, Piersma AH, Verhoef M, Spanjersberg MQ, Huuskonen H, Paillard F, Seiler A. Validation of the rat limb bud micromass test in the international ECVAM validation study on three *in vitro* embryotoxicity tests. Altern Lab Anim 2004;32:245–274.

36. Spielmann H. Predicting the risk of developmental toxicity from *in vitro* assays. Toxicol Appl Pharmacol 2005;207:375–380.

37. Genschow E, Spielmann H, Scholz G, Pohl I, Seiler A, Clemann N, Bremer S, Becker K. Validation of the embryonic stem cell test in the international ECVAM validation study on three in vitro embryotoxicity tests. Altern Lab Anim 2004;32:209–244.

38. zur Nieden NI, Ruf LJ, Kempka G, Hildebrand H, Ahr HJ. Molecular markers in embryonic stem cells. Toxicol In Vitro 2001;15:455–461.

39. Buesen R, Visan A, Genschow E, Slawik B, Spielmann H, Seiler A. Trends in improving the embryonic stem cell test (EST): an overview. ALTEX 2004;21:15–22.

40. Seiler A, Visan A, Buesen R, Genschow E, Spielmann H. Improvement of an *in vitro* stem cell assay for developmental toxicity: the use of molecular endpoints in the embryonic stem cell test. Reprod Toxicol 2004;18:231–240.

41. zur Nieden NI, Kempka G, Ahr HJ. Molecular multiple endpoint embryonic stem cell test–a possible approach to test for the teratogenic potential of compounds. Toxicol Appl Pharmacol 2004;194:257–269.

42. Seiler A, Buesen R, Hayess K, Schlechter K, Visan A, Genschow E, Slawik B, Spielmann H. Current status of the embryonic stem cell test: the use of recent advances in the field of stem cell technology and gene expression analysis. ALTEX 2006;23:393–399.

43. Adler S, Pellizzer C, Hareng L, Hartung T, Bremer S. First steps in establishing a developmental toxicity test method based on human embryonic stem cells. Toxicol In Vitro 2008;22:200–211.

44. Lessmollmann U, Hinz N, Neubert D. *In vitro* system for toxicological studies on the development of mammalian limb buds in a chemically defined medium. Arch Toxicol 1976;36:169–176.
45. Friedman L. Teratological research using in vitro systems. II. Rodent limb bud culture system. Environ Health Perspect 1987;72:211–219.
46. Whitby KE. Teratological research using in vitro systems. III. Embryonic organs in culture. Environ Health Perspect 1987;72:221–223.
47. Abbott BD, Buckalew AR. Embryonic palatal responses to teratogens in serum-free organ culture. Teratology 1992;45:369–382.
48. Slavkin HC, Bringas P Jr, Sasano Y, Mayo M. Early embryonic mouse mandibular morphogenesis and cytodifferentiation in serumless, chemically defined medium: a model for studies of autocrine and/or paracrine regulatory factors. J Craniofac Genet Dev Biol 1989;9(2):185–205.
49. Slavkin HC, Sasano Y, Kikunga S, Bessem C, Bringas P Jr, Mayo M, Luo W, Mak G, Rall L, Snead ML. Cartilage, bone and tooth induction during early embryonic mouse mandibular morphogenesis using serumless, chemically-defined medium. Connect Tissue Res 1990;24:41–51.
50. Chai YP, Bringas P Jr, Shuler C, Devaney E, Grosschedl R, Slavkin HC. A mouse mandibular culture model permits the study of neural crest cell migration and tooth development. Int J Dev Biol 1998;42:87–94.
51. Kimler BF, Rethorst RD, Cox GG. Investigation of the interaction of cardiotoxic anticancer agents using the fetal mouse heart organ culture system. Invest New Drugs 1986;4:127–133.
52. Porter GA Jr, Rivkees SA. Ontogeny of humoral heart rate regulation in the embryonic mouse. Am J Physiol Regul Integr Comp Physiol 2001;281: R401–R407.
53. New DA, Coppola PT, Cockcroft DL. Comparison of growth in vitro and in vivo of post-implantation rat embryos. J Embryol Exp Morphol 1976;36:133–144.
54. New DA. Whole-embryo culture and the study of mammalian embryos during organogenesis. Biol Rev Camb Philos Soc 1978;53:81–122.
55. Sadler TW. Culture of early somite mouse embryos during organogenesis. J Embryol Exp Morphol 1979;49:17–25.
56. New DA. The culture of postimplantation embryos. Hum Reprod 1991;6:58–63.
57. Rogers JM, Kavlock RJ. Developmental toxicology. In Klaassen CD, editor. *Casarett and Doul's Toxicology: The Basic Science of Poisons*. New York: McGraw-Hill; 2008.
58. Piersma AH, Genschow E, Verhoef A, Spanjersberg MQ, Brown NA, Brady M, Burns A, Clemann N, Seiler A, Spielmann H. Validation of the postimplantation rat whole-embryo culture test in the international ECVAM validation study on three *in vitro* embryotoxicity tests. Altern Lab Anim 2004;32:275–307.
59. Pitt JA, Carney EW. Development of a morphologically-based scoring system for postimplantation New Zealand White rabbit embryos. Teratology 1999;59:88–101.
60. Carney EW, Tornesi B, Keller C, Findlay HA, Nowland WS, Marshall VA, Ozolins TR. Refinement of a morphological scoring system for postimplantation rabbit conceptuses. Birth Defects Res B Dev Reprod Toxicol 2007;80:213–222.
61. Dumont JN, Schultz TW, Newman SM. A Frog Embryo Teratogenesis Assay: Xenopus (FETAX)—A model for teratogen screening. Teratology 1982;25: 37A–38A.
62. Dumont JN, Schultz TW, Eppler RG. The response of the FETAX model to mammalian teratogens. Teratology 1983;27: 39A–40A.

63. Bantle JA, Burton DT, Dawson DA, Dumont JN, Finch RA, Fort DJ, Linder G, Rayburn JR, Buchwalter D, Maurice MA, Turley SD. Initial interlaboratory validation study of FETAX: phase I testing. J Appl Toxicol 1994;14:213–223.

64. Bantle JA, Finch RA, Burton DT, Fort DJ, Dawson DA, Linder G, Rayburn JR, Hull M, Kumsher-King M, Gaudet-Hull AM, Turley SD. FETAX interlaboratory validation study: phase III—Part 1 testing. J Appl Toxicol 1996;16:517–528.

65. Fort DJ, Stover EL, Bantle JA, Rayburn JR, Hull MA, Finch RA, Burton DT, Turley SD, Dawson DA, Linder G, Buchwalter D, Dumont JN, Kumsher-King M, Gaudet-Hull AM. Phase III interlaboratory study of FETAX, Part 2: interlaboratory validation of an exogenous metabolic activation system for frog embryo teratogenesis assay—Xenopus (FETAX). Drug Chem Toxicol 1998;21:1–14.

66. Bantle JA, Finch RA, Fort DJ, Stover EL, Hull M, Kumsher-King M, Gaudet-Hull AM. Phase III interlaboratory study of FETAX. Part 3. FETAX validation using 12 compounds with and without an exogenous metabolic activation system. J Appl Toxicol 1999;19:447–472.

67. Fort DJ, Rogers RL, Thomas JH, Buzzard BO, Noll AM, Spaulding CD. Comparative sensitivity of Xenopus tropicalis and Xenopus laevis as test species for the FETAX model. J Appl Toxicol 2004;24:443–457.

68. Zon LI, Peterson RT. *In vivo* drug discovery in the zebrafish. Nat Rev Drug Discov 2005;4:35–44.

69. Parng C, Seng WL, Semino C, McGrath P. Zebrafish: a preclinical model for drug screening. Assay Drug Dev Technol 2002;1 (1Pt 1):41–48.

70. Parng C. In vivo zebrafish assays for toxicity testing. Curr Opin Drug Discov Devel 2005;8:100–106.

71. Hill AJ, Teraoka H, Heideman W, Peterson RE. Zebrafish as a model vertebrate for investigating chemical toxicity. Toxicol Sci 2005;86:6–19.

72. Jacobs A. Use of nontraditional animals for evaluation of pharmaceutical products. Expert Opin Drug Metab Toxicol 2006;2:345–349.

73. Rubinstein AL. Zebrafish assays for drug toxicity screening. Expert Opin Drug Metab Toxicol 2006;2:231–240.

74. Nagel R. DarT: The embryo test with the Zebrafish Danio rerio—a general model in ecotoxicology and toxicology. ALTEX 2002;19 (Suppl 1):38–48.

75. Johnson EM. A subvertebrate system for rapid determination of potential teratogenic hazards. J Environ Pathol Toxicol 1980;4:153–156.

76. Johnson EM, Gorman RM, Gable BE, George ME. The Hydra attenuata system for detection of teratogenic hazards. Teratog Carcinog Mutagen 1982;2:263–276.

77. Johnson EM, Gabel BE. An artificial 'embryo' for detection of abnormal developmental biology. Fundam Appl Toxicol 1983;3:243–249.

78. Johnson EM, Dansky LA. A hydra assay as a pre-screen for teratogenic potential. Food Chem Toxicol 1987;25:637–638.

79. Collins TF. Teratological research using in vitro systems. V. Nonmammalian model systems. Environ Health Perspect 1987;72:237–249.

15

ASSESSING THE POTENTIAL FOR INDUCTION OF CYTOCHROME *P450* ENZYMES AND PREDICTING THE *IN VIVO* RESPONSE

JIUNN H. LIN

15.1 INTRODUCTION

One of the intriguing aspects of cytochrome *P*450 (CYP) enzymes is that some of the CYPs are inducible. CYP induction-mediated drug–drug interactions are one of the major concerns in clinical practice and for the pharmaceutical industry. The major issue associated with CYP induction is reduction in therapeutic efficacy of coadministered medications. Because CYP induction is a metabolic liability in drug therapy, it is highly undesirable to develop new drug candidates that are potent CYP inducers. For this reason, many drug companies today include the assessment of CYP induction at the stage of drug discovery as part of the selection processes of new drug candidates for further clinical development.

CYPs, the most important drug-metabolizing enzyme system, play a crucial role in the oxidative metabolism of endogenous and exogenous substances. In humans, more than 50 CYP isoforms have been identified. Collectively, the CYP1A, CYP2C, CYP2D, and CYP3A subfamilies are responsible for the oxidative metabolism of most clinically useful drugs.

Unlike CYP inhibition, which produces an almost immediate response, CYP induction is a slow biochemical process involving several cellular regulatory steps. Induction takes time to reach a higher steady-state enzyme level as a result

Predictive Approaches in Drug Discovery and Development: Biomarkers and In Vitro/In Vivo Correlations, First Edition. Edited by J. Andrew Williams, Jeffrey R. Koup, Richard Lalonde, and David D. Christ.

of a new balance between the rate of biosynthesis and degradation. After discontinuing treatment with a CYP inducer, it also takes time to return to the enzyme's basal level. The time-dependent process of CYP induction is best exemplified by the interaction study of verapamil and rifampicin (1). Because of the nature of time-dependent process, CYP induction may complicate drug-dosing regimens in chronic drug therapy. The addition of any potent inducer to or withdrawal of a potent inducer from an existing drug-dosing regimen may cause pronounced changes in drug concentrations, leading to failure of drug therapy or adverse effects.

Several *in vitro* models have been established to assess the potential of CYP induction, including liver slices, immortalized cell lines, and primary or cryopreserved hepatocytes (2–5). Among these models, primary cultures of human hepatocytes are probably the most predictive model for evaluating CYP induction.

The purpose of this chapter is to review the molecular mechanisms of CYP induction and the pharmacokinetic consequences. In addition, factors that affect the magnitude of CYP induction and extrapolation of *in vitro* CYP induction data to *in vivo* situations are also discussed. Finally, assessment of the potential of CYP3A4 induction at the drug discovery and development stage is discussed.

15.2 MOLECULAR MECHANISMS OF CYP INDUCTION

Aryl hydrocarbon receptor (AhR), constitutive androstane receptor (CAR), and pregnane X receptor (PXR) are the major transcription factors that mediate CYP induction. Both CAR and PXR belong to the gene family (family NR1) and share a common heterodimerization partner, the retinoid X receptor (XRX) (6, 7). On the other hand, the *AhR* genes belong to Per-Arnt-Sim (PAS) family of transcription factors and require AhR nuclear translocator (Arnt) as its heterodimerization partner (8).

15.2.1 Aryl Hydrocarbon Receptor (AhR)

The human CYP1A subfamily consists of two members: CYP1A1 and CYP1A2. While CYP1A2 is one of the major CYPs in human liver, accounting for approximately 10% of the total amount of hepatic CYPs, CYP1A1 is mainly expressed in human lung, placenta, and lymphocytes in much less relative abundance (8). Initial evidence of AhR involvement in the induction of CYP1A1 and CYP1A2 enzymes was provided by the identification of a hepatic cytosolic protein that exhibited stereospecific and high affinity binding to 2,3,7,8-tetrachlorodibenzo-*p*-dioxin (TCDD), an environmental toxin (9). Studies in three mouse hepatoma cell lines, wild-type, AhR-defective, and Arnt-defective cell lines, led to the identification of Arnt, a second regulatory protein in CYP1A induction (10). The AhR-defective variant cells contained an altered AhR, which had low affinity for TCDD and responded poorly to TCDD. While TCDD binding to the AhR in Arnt-defective cells was almost identical to that in wild-type cells, the Arnt-defective cells failed to respond to TCDD. Together, these data demonstrated that both AhR and Arnt are required for CYP1A induction.

The AhR/Arnt heterodimer (complex) activates *CYP1A1* gene by interacting with responsive elements of the gene. Transfection experiments revealed that the DNA upstream of the *CYP1A1* gene contains two types of transcriptional elements. The more distal element has the functional properties of AhR-dependent transcriptional enhancer, spanning about 400 base pairs (bp) in about 1000 bp upstream of the CYP1A1 transcription start site, while the more proximal element functions as a transcriptional promoter, spanning about 150 bp immediately near the transcriptional start site (11, 12). The enhancer contains at least three binding sites for the AhR/Arnt complex, while in contrast the promoter has no binding sites for the Ahr/Arnt complex. Together with the identification of AhR and Arnt, the discovery of transcriptional elements in the *CYP1A* gene provides the basis for our understanding of how the cells recognize inducers of different chemical structure and how the induction signal is conveyed to the transcriptional machinery.

15.2.2 Constitutive Androstane Receptor (CAR)

CAR was first isolated in 1994 by Baes et al. (13) through screening of a cNDA library. The functional role of CAR in CYP2B induction has been clearly demonstrated with transgenic mice. Studies utilizing *CAR*-null mice have provided direct evidence that CAR is an essential mediator of the transcriptional response to the classical CYP-inducer phenobarbital (PB). While treatment with PB significantly induced the *Cyp2b10* gene in $CAR^{(+/+)}$ mice, PB had no inductive effect in $CAR^{(-/-)}$ mice (14). In contrast to the classical nuclear receptors that are activated by their cognate ligands, the CAR forms a heterodimer with XRX that binds to retinoid acid response elements (RAREs) and transactivates the target genes of RAREs in a constitutive manner in the absence of ligands. Because CAR can transactivate the target genes in a constitutive manner without ligands, CAR is referred to as *constitutive androstane receptor*. The unique nature of CAR has led to an intensive search of endogenous ligands of the receptor. While searching for potential ligands of CAR, androstane metabolites, such as androstanol and andostenol, were found to effectively inhibit the constitutive activity of CAR (15). Forman et al. (16) showed that the intrinsic activity of CAR was completely inhibited by endogenous steroid androstanol and androstenol, with an IC_{50} value of approximately 400 nM. These androstane ligands are good examples of naturally occurring inverse agonists that reverse transcriptional activation by nuclear receptors.

The CAR/XRX heterodimer transcriptionally activates the *CYP2B* genes by interacting with the PB-responsive DNA elements [PB-responsive enhancer module (PBREM)]. A major breakthrough in CYP2B induction was made by Trottier et al. (17), who identified a 163-bp PBREM in the rat *CYP2B2* 5′-flanking region situated at −2318 to −2155 bp upstream of the transcription start site. The identification of PBREM made a very significant contribution to our understanding of the role of CAR in the induction of *CYP2B* genes by PB and a number of PB-like inducers.

CAR is predominately expressed in the liver and to a lesser extent in the intestine (13). Similar to AhR, CAR is located in the cytoplasma of hepatocytes of uninduced mice (18). Western blot as well as immunohistochemical analyses confirmed that CAR is completely absent in the liver nuclei of untreated mice. After the incubation of PB with primary hepatocytes, CAR is accumulated rapidly in the nuclei, suggesting that CAR nuclear translocation is the first activation step in response to PB (18). However, even with sensitive *in vitro* binding assays, PB has not been shown to bind directly to either human CAR or mouse CAR. Therefore, ligand binding does not appear to be critical for CAR nuclear translocation, and the question of how PB translocates CAR into the nucleus is still under extensive investigation. Although the mechanism involved in CAR nuclear translocation is still not fully understood, it seems to involve a specific and sensitive dephosphorylation event. In mouse primary hepatocyte cultures, the phosphatase inhibitor okadaic acid was found to effectively inhibit PB-mediated CAR translocation and *Cyp2b10* induction, providing evidence that dephosphorylation of mouse CAR is required for the nuclear translocation process (19). In addition, CAR nuclear translocation appears to be dependent on a leucine-rich region near the C-terminus of the receptor (20).

15.2.3 Pregnane X Receptor (PXR)

Although CYP3A enzymes have long been known to be inducible, the molecular mechanism remained unknown until the identification of PXR. Database searching and subsequent cDNA cloning resulted in the discovery of a new mouse orphan nuclear receptor designated PXR by Kliewer et al. in 1998 (6). Direct evidence that PXR mediates the induction of *CYP3A* genes was provided by transgenic mice. Although pregnenolone-16α-carbonite (PCN) and dexamethasone (DEX) significantly induced *Cyp3a11* gene in wild-type $PXR^{(+/+)}$ mice, mice with the $PXR^{(-/-)}$ failed to induce *Cyp3a11* gene expression when challenged with PCN or DEX (21). Soon after the discovery of mouse PXR, human SXR (hPXR or steroid and xenobiotic receptor) was identified by Bertilsson et al. (22). SXR is expressed predominately in human liver and to a lesser extent in small intestine. SXR is a promiscuous nuclear receptor, which can be activated by numerous structurally diverse xenobiotics and drugs, and is referred as the *master* regulator of CYP enzymes. The number of compounds identified as hPXR ligands and CYP3A inducers continue to grow; however, there appears no obvious quantitative structure–activity relationship among these ligands.

Activation of PXR appears to be simpler compared with activation of CAR. Negishi and coworkers (23) have shown that PXR is located in the cytoplasma of untreated mouse liver cells and is concentrated in the nucleus after treatment with PCN, suggesting translocation of PXR into the nucleus. Using hsp90 antibody, these investigators further demonstrated that PXR formed a complex with endogenous cytoplasmic CAR retention protein (CCRP) and hsp90 in HepG2 cells. On the basis of these results, the investigators concluded that nuclear translocation of PXR occurs and the PXR-CCRP-Hsp90 complex maintains the receptor in the cytosol.

Like many other nuclear receptors, PXR contains two functional domains: a ligand-binding domain (LBD) and a highly conserved DNA-binding domain (DBD). The highly conserved DBD, characterized by two C4-type zinc fingers, links the PXR to the specific regions of the target genes (24). Crystal structure analyses suggest that the LBD of the human PXR is highly hydrophobic and flexible. The unique structure of the ligand pocket not only allows PXR to bind a diverse set of substrates of different molecular size but also permits a single molecule to dock in multiple orientations (25), thus explaining why PXR can be activated by various structurally diverse ligands. The LBD serves as the binding site for ligands and also contains transcriptional activation domains such as the activation function-2 (AF-2) helix. The binding of a ligand to the LBD results in a conformational change in AF-2 helix, leading to the recruitment of coactivators and cointegrators, and in turn transactivation of the target genes (24).

PXR interacts with its cognate response elements in the 5′-flanking region of the target genes as a heterodimer with RXR. Three DNA-binding sites have been identified for human *CYP3A4* gene. One site is located at approximately −150 bp region of the proximal promoter and another two sites are found between −7733 and −7672 bp of the distal enhancer region of CYP3A4 gene, termed as *xenobiotic-responsive-enhancer module* (XREM) (7). The XREM is composed of two hexameric half-sites (AG(G/T)TCA) organized as a direct repeat with a three-nucleotide spacing (DR-3), and an everted repeat separated by 6 bp (ER-6), while the promoter element contains a ER-6 motif (26). Mutation studies suggest that all three DNA-binding sites are required for PXR activation; a mutation of either one of these three sites alone decreased the response activity only by approximately 30%, while mutation of all three sites almost completely abolished the response activity (7).

15.3 SPECIES DIFFERENCES IN CYP INDUCTION

To date, our understanding of the molecular mechanisms regarding the role of nuclear receptors in the induction of *CYP* genes is based mainly on *in vitro* models and animal studies. Although it is believed that CYP induction is regulated in humans in the same fashion as in animals, inductive response to inducers is markedly different, both quantitatively and qualitatively, among animal species. For example, omeprazole, a gastric acid-suppressing drug, is a robust CYP1A enzyme inducer in humans but has little inductive effect in mice or rabbits (27, 28). Similarly, significant species differences in CYP3A induction were also observed. Rat CYP3A enzymes are readily induced by PCN, whereas neither rabbit nor human CYP3A enzymes are induced by PCN. In contrast, rifampicin is a potent inducer for CYP3A enzymes in rabbits and humans, while it has little inductive effect on CYP3A enzymes in rats (29, 30).

Marked species differences in CYP induction even occur between rodents. Degawa et al. (31) reported that significant species differences in CYP1A1/2 induction were observed among mice, rats, hamsters, and guinea pigs when treated with 2-amino-1-methyl-6-phenyl-imidazole[4,5-*b*]pyridine (PhIP). Treatment with PhIP caused significant increases in the levels of CYP1A1/2 in the liver of rats, while CYP1A1/2 was not induced by PhIP in mice, hamsters, and guinea pigs. Species-dependent induction between mice and rats was also reported for CYP2B and CYP3A after treatment with tamoxifen (32). Treatment with tamoxifen resulted in significant increases in both CYP2B and CYP3A proteins as well as their enzyme activities in rats, while no induction was observed in mice. In contrast, treatment with 4-vinyl-1-cyclohexene caused significant induction of CYP2A and CYP2B in mice but not in rats (33).

The molecular basis for the observed species differences in CYP induction has been intensively studied. A transfection study was conducted by Barwick et al. (34) to determine whether the observed species difference in CYP3A induction was due to species differences in the sequences of the promoters in the *CYP3A* genes. The CYP3A promoters from rat and rabbit *CYP3A* genes were transfected separately into primary hepatocytes from rats and rabbits. The rat *CYP3A1* promoter was transcriptionally activated by PCN in rat hepatocytes but not in rabbit hepatocytes. Conversely, the rabbit *CYP3A6* promoter was transcriptionally activated by rifampicin in rabbit hepatocytes but not in rat hepatocytes. These results suggest that the observed species differences in CYP3A induction by xenobiotics is not due to the structural differences in the gene promoter.

Subsequently, PXR was identified as the cellular factor responsible for species differences in CYP3A induction. Jones et al. (35) compared the CYP3A induction produced by a variety of different compounds, such as DEX, RU486, clotrimazole, PB, and rifampicin, in rat and rabbit hepatocytes with their ability to activate the rat and rabbit PXR. There were marked differences in the induction of *CYP3A* gene expression between rat and rabbit hepatocytes. For example, rifampicin markedly increased CYP3A6 mRNA in rabbit hepatocytes but had no effect on CYP3A1 mRNA in rat hepatocytes. In contrast, PCN significantly increased CYP3A1 mRNA in rat hepatocytes but not in rabbit hepatocytes. Consistent with the observations in hepatocytes, a good inducer in hepatocytes of a given species was also a good activator of PXR for the same species. For instance, PCN is a good activator for rat PXR but not for rabbit PXR, while rifampicin is an efficacious activator for rabbit PXR but not for rat PXR (35). These results suggest that structural difference in PXR is the molecular basis for the observed species differences in CYP3A induction by xenobiotics.

The notion that structural difference in PXR is responsible for the observed species differences in CYP3A induction is further supported by the study with humanized PXR mice (21). Although *Cyp3a11* gene was markedly induced in mice by PCN and DEX, targeted disruption of the mouse *PXR* gene resulted in loss of PCN- and DEX-mediated *Cyp3a11* gene induction. The disruption of *PXR* alleles was confirmed by the absence of PXR expression in the liver and small intestine, two principal PXR-expressing tissues. The transgenic mPXR-null mice ($PXR^{(-/-)}$) were then humanized by introduction of human PXR. Interestingly, the humanized mice responded to the human CYP3A4-specific inducer, rifampicin, but not PCN. Therefore, humanization of PXR is sufficient to convert the induction characteristics from mice to humans. Similarly, the conversion of species-specific inducibility of CYP3A was observed *in vitro*, when human PXR was transfected into rat hepatocytes (21). In control rat hepatocytes, CYP3A23 was strongly induced by PCN, while rifampicin and clotrimazole had little or no inductive effect on CY3A23. When human PXR was cotransfected into rat hepatocytes, however, there was a significant induction of CYP3A23 by rifampicin and clotrimazole. These results provide direct evidence that species differences in CYP3A induction are due mainly to the structural differences in PXR between species.

Structurally, PXR consists of a DBD and a LBD in the same fashion across animal species. Structural comparison of PXR from different species revealed that there is greater than 95% sequence homology in the DBD regions but only 76–83% homology for LBD regions (24). Because the DBD is highly conserved among the PXR across species, the species differences in CYP3A induction is more likely due to the structural differences in the LBD. Indeed, mutation studies indicated that alteration of four amino acids in the LBD of mouse PXR to human PXR (Arg 203 Leu, Pro 205 Ser, Gln 404 His and Gln 407 Arg) was sufficient to switch the mouse PXR that responds to PCN to a human-like PXR, which is activated effectively by SR-12813 but not by PCN (25).

Similar to the PXR, comparison of CAR amino acid sequence revealed that the human and rat CAR share only about 70% amino acid homology in their LBD and demonstrates marked species differences in responses to xenobiotics (36). For example, clotrimazole is an efficacious deactivator of human CAR but has little or no effect on mouse CAR. Conversely, TCPOBOP (1,4-bis[z-(3,5-dichloropyridyloxyl)] benzene) is a potent deactivator for mouse CAR but lacks any activity on human CAR (24). Similarly, there is a significant species difference in CYP2B induction between rats and mice. TCPOBOP, the most potent ligand for mouse CAR, has little effect on rat CAR (24). Like PXR, it is reasonable to assume that the structural difference in the LBD of nuclear receptors is the major cellular factor responsible for species differences in CYP2B induction (36). Collectively, it becomes evident that it is difficult to make the prediction of CYP induction in humans using animal models, because of these marked species differences.

15.4 EFFECTS OF CYP INDUCTION ON PHARMACOKINETICS

Direct assessment of CYP induction *in vivo* by measuring enzyme amount and activity in the liver is difficult in clinical settings because of ethical considerations and practical limitations. A simple but indirect way of assessing the effect of CYP induction in clinical settings is the comparison of the plasma AUC of a drug before and after coadministration of an inducer. However, as discussed below, the magnitude of changes in the plasma AUC after inducer treatment is highly dependent on the route of administration and the pharmacokinetic characteristics of the affected drug. Therefore, the changes in plasma AUC of a drug after treatment with an inducer may not directly reflect the changes in enzyme activity quantitatively, that is, a 10-fold decrease in the plasma AUC by an inducer does not necessarily reflect a 10-fold increase in enzyme activity.

Moreover, the changes in the AUC of a victim drug after inducer treatment may reflect the sum of induction of multiple CYP enzymes in various tissues. Because CYP enzymes are not only expressed in the liver but also present in other tissues responsible for drug metabolism, including intestine, lung, and kidney, it is expected that CYP induction also occurs in various tissues beside the liver. For example, after treatment of mice with TCDD (10 μg/kg), a 20- to 40-fold CYP1A1 induction was observed in liver, lung, and skin in mice (37). The enzyme level of CYP1A1 in the lung was about 4% of that in the liver before the TCDD treatment, while the pulmonary CYP1A1 enzyme level was increased to about 10–15% of that in the liver after the TCDD treatment. Therefore, interpretation of the effect of CYP induction *in vivo* based on the pharmacokinetic analysis of AUC data is not as straightforward as generally believed.

15.4.1 Pharmacokinetics in the Induced State Depend on the Route of Drug Administration

Depending on the pharmacokinetic characteristics of drugs, the route of drug administration may have a significant impact on the AUC changes caused by CYP induction (38). For high clearance drugs metabolized by the liver, a marked decrease in AUC during CYP induction is expected following oral administration, whereas CYP induction has little effect on the AUC after intravenous (IV) administration. Kinetically, the AUC of a drug after IV administration (AUC_{iv}) is determined mainly by the systemic clearance. For drugs whose high hepatic clearance is due to metabolism (e.g., CYP-mediated oxidation), an increase in metabolic (intrinsic) clearance caused by CYP induction will have little effect on the AUC_{iv} because the systemic clearance of high hepatic clearance drugs is determined by hepatic blood flow and is insensitive to the changes in enzyme activity. In contrast, the AUC after oral administration (AUC_{po}) for these drugs is determined by both systemic clearance and bioavailability. For drugs that are well absorbed and extensively metabolized in the liver, oral bioavailability will be limited by presystemic (first-pass) metabolism in the liver. Although their systemic clearance is not sensitive to the changes in enzyme activity, their bioavailability is very sensitive to the changes in enzyme activity, because of changes in first-pass metabolism.

The pentobarbital–alprenolol interaction is a good example in support of the notion that the magnitude of the plasma AUC changes for a drug after CYP induction is highly dependent on the drug's pharmacokinetic characteristics and the route of drug administration. The metabolism of alprenolol, a high hepatic clearance drug with a clearance of 1200 ml/min (a value approaching to the hepatic blood flow of 1500 ml/min in humans), is known to be significantly induced by pentobarbital in humans. Alprenolol was administered orally or intravenously to five healthy subjects on two different occasions before and after 10–14 daily doses of 100-mg pentobarbital (39). The AUC_{po} of alprenolol after a 200-mg oral dose decreased dramatically from an average of 706 to 154 ng h/ml after barbiturate treatment. In contrast, the AUC_{iv} of alprenolol following a 5-mg IV dose decreased slightly from an average of 67 to 58 ng h/ml following pentobarbital treatment. Thus, there was about a fivefold decrease in the AUC_{po} of alprenolol after oral dosing and only a 15% decrease in the AUC_{iv} of alprenolol after IV administration. Interestingly, there was no significant change in the elimination half-life (~2 h) of alprenolol before and after pentobarbital treatment, independent of the route of administration, supporting the argument that CYP induction has little effect on the systemic clearance of high hepatic clearance drugs. The marked decrease in the AUC_{po} of alprenolol is due mainly to a decrease in bioavailability resulting from an increase in first-pass metabolism.

Similarly, treatment with rifampicin had little effect on the kinetics of nifedipine (a high hepatic clearance drug) after IV administration, whereas rifampicin greatly reduced the AUC_{po} of nifedipine following oral administration (40). After 7 days of rifampicin treatment (600 mg/day), the AUC_{po} of nifedipine decreased from an average of 280 to 18 ng h/ml, when a 20-mg oral dose of nifedipine was given. In contrast, the AUC_{iv} of nifedipine only slightly reduced from an average of 38 to 27 ng h/ml, when the drug was dosed intravenously at 20-µg/kg body weight. There was about a 16-fold decrease in the AUC_{po} of nifedipine after oral administration, while only 30% decrease in the AUC_{iv} after IV dosing. As with alprenolol, rifampicin treatment had little effect on the elimination half-life (~2 h) of nifedipine after either oral or IV administration.

This route-dependent effect has also been reported for another high hepatic clearance drug, verapamil. Verapamil is administered as a racemate and the fate of the individual enantiomers was determined. Treatment of rifampicin had little effect on the AUC_{iv} of (*S*)- and (*R*)-verapamil after IV administration of verapamil, whereas it caused a dramatic decrease in the AUC_{po} of (*S*)- and (*R*)-verapamil after oral administration of verapamil by 30- to 50-fold (1). After rifampicin treatment for 16 days (600 mg/day), the AUC_{po} of (*S*)-verapamil decreased from an average of 151 to 5 ng h/ml after oral administration (120 mg), while the AUC_{iv} of (*S*)-verapamil decreased from 77 to 61 ng h/ml following IV dosing (10 mg). Overall, these results from the above examples clearly demonstrated that for high hepatic clearance drugs, CYP induction has little effect on the AUC_{iv} after IV administration, while it causes a marked decrease in the AUC_{po} after oral dosing.

Unlike high hepatic clearance drugs, enzyme induction produces significant effects on both the systemic clearance and AUC for low clearance drugs metabolized by the liver, independent of the route of administration. The systemic clearance of low hepatic clearance drugs is limited by enzyme activity and sensitive to changes in enzyme activity. Because low hepatic clearance drugs are generally not subject to significant first-pass metabolism and have relatively high bioavailability, both the AUC_{iv} and AUC_{po} are determined mainly by the systemic clearance. Therefore, an increase in the intrinsic metabolic clearance caused by CYP induction will have significant effect on both the AUC_{iv} and AUC_{po} of low hepatic clearance drugs, and it would be expected that the magnitude of the changes in the AUC of the low hepatic clearance drugs will be quantitatively similar between oral and IV administration. In addition, CYP induction will also have a significant effect on the elimination half-life of low hepatic clearance drugs, regardless of the route of administration.

The methadone–rifampicin interaction is a good example illustrating the point that both the AUC_{iv} and AUC_{po} of the low hepatic clearance drugs are sensitive to the changes in enzyme activity caused by CYP induction. Clinically, the effect of rifampicin (600 mg/day for 5 days) on the pharmacokinetics of methadone has been studied with 12 healthy volunteers (41). Methadone, a low hepatic clearance drug with a clearance of 140 ml/min, is used to treat opiate addiction by preventing opiate withdrawal syndrome. After an IV dose (4.5 mg), the AUC_{iv} and half-life of methadone decreased from an average of 816 ng h/ml and 38 h before rifampicin treatment to 259 ng h/ml and 18 h after rifampicin treatment, respectively. Similarly, the AUC_{po} and half-life of methadone decreased from an average of 1128 ng h/ml and 33 h before rifampicin treatment to 262 ng h/ml and 25 h, respectively, after a 10-mg oral dose of methadone. The fold decrease in the AUC after oral dose was similar to that following IV administration.

The elimination of alprazolam, a low hepatic clearance drug, was investigated in two groups of healthy volunteers with and without rifampicin treatment at 450-mg daily dose for 4 days (42). The AUC_{po} of alprazolam was about eightfold lower in the rifampicin treatment group (28.4 ng h/ml) compared to the control group without rifampicin (224 ng h/ml) after a 1-mg oral dose of alprazolam. Similar to the AUC, the half-life was also decreased dramatically from an average of 14 h in the control group to 2.6 h in the rifampicin-treated group. In another clinical study, the disposition of diazepam, a low hepatic clearance drug, was investigated in 21 healthy subjects before and after 7 days of administration of 600-mg rifampicin. In this study, diazepam was given orally at a dose of 10 mg. After treatment with rifampicin, there was a significant decrease in the AUC and half-life of diazepam (43). The AUC of diazepam decreased from an average of 4430 ng h/ml before rifampicin treatment to 1040 ng h/ml after treatment. In addition, the elimination half-life of diazepam decreased from an average of 52 h before rifampicin treatment to 15.8 h after the treatment. Overall, these examples clearly demonstrate that CYP induction has a similar effect on the AUC_{iv} and the AUC_{po} for low hepatic clearance drugs. In addition, CYP induction also has a significant effect on the half-life of low hepatic clearance drugs. Unlike high hepatic clearance drugs, the changes in plasma AUC of low hepatic clearance drugs after treatment with an inducer may more accurately reflect the changes in enzyme activity.

Another intriguing aspect of the effect of CYP induction on the pharmacokinetics of drugs is that the magnitude of changes in the AUC_{po} after oral dosing tends to be greater for high hepatic clearance drugs compared to that for low hepatic

TABLE 15.1 Effect of Rifampicin (RIF) on the Oral AUC of Drugs that are Metabolized Predominately by CYP3A4

Drug	Type of Clearance (CL)[a]	Rifampicin (mg/day) Days Treated	AUC (ng h/ml) Before RIF	AUC (ng h/ml) After RIF	Fold Induction[b]	References
Cyclosporine	Low	600 mg 11 days	8986	2399	3.7	48
Tacrolimus	Low	600 mg 18 days	351	112	3.1	49
Methadone	Low	600 mg 5 days	1128	262	4.3	41
Alprazoam	Low	450 mg 4 days	224	28	8.0	42
Diazepam	Low	600 mg 7 days	4430	1040	4.2	43
Zolpidem	Low	600 mg 5 days	1202	336	3.6	44
Zopiclone	Low	600 mg 5 days	473	86	5.5	45
Quinidine	Moderate	600 mg 7 days	8000	910	8.8	50
Midazolam	Moderate	600 mg 5 days	612	25	24.0	46
Triazolam	Moderate	600 mg 5 days	14.8	0.74	20.0	47
Nifedipine	High	600 mg 7 days	280	18	15.5	40
Indinavir	High	600 mg 8 days	18.8[c]	1.2[c]	16.0	51
S-Verapamil	High	600 mg 12 days	152	5	30.0	1
R-Verapamil	High	600 mg 12 days	724	14	52.0	1

[a]Type of clearance: low clearance less than 200 ml/min; moderate clearance greater than 500 ml/min; high clearance greater than 1000 ml/min.
[b]Fold induction is defined as the ratio of AUC before:after rifampicin.
[c]In μg h/ml.

clearance drugs. The clearance of methadone, alprazolam, diazepam, zolpidem, and zolpiclone (<200 ml/min) is substantially smaller than hepatic blood flow (1500 ml/min), while the clearance of nifedipine and verapamil (>1000 ml/min) approaches the hepatic blood flow, and the former group are defined as low hepatic clearance drugs, while those in the later group are considered as high hepatic clearance drugs (1, 40, 44, 45). As shown in Table 15.1, the magnitude of decrease in the AUC_{po} is smaller for the low hepatic clearance drugs (4- to 8-fold) than that for high hepatic clearance drugs (15- to 50-fold). Similar to the high clearance drugs, the moderate hepatic clearance drugs (~500 ml/min) such as midazolam and triazolam are also subject to a significant decrease in the AUC_{po} after oral dosing during CYP induction (46, 47).

CYP2C8 and CYP2C9 can also be induced by rifampicin (52, 53); however, the effect of rifampicin on drugs that are predominately metabolized by CYP2C8 or CYP2C9 appears to be less significant compared to that on drugs metabolized by CYP3A4 (Table 15.2). For example, rifampicin treatment only caused two- to fourfold changes in both the AUC and half-life of warfarin, a low clearance drug, which is metabolized predominately by CYP2C9 (54). After treatment with rifampicin at 300 mg twice daily for 4 days, the AUCs and half-life of (*R*)-warfarin decreased from an average of 159 μg h/ml and 43 h to 47 μg h/ml and 17 h, respectively, while the AUCs and half-life of (*S*)-warfarin decreased from an average of 232 μg h/ml and 26 h to 60 μg h/ml and 13 h, respectively, in

TABLE 15.2 Effect of Rifampicin (RIF) on the Oral AUC of Drugs that are Metabolized Predominately by CYP2C8 or CYP2C9 in Humans

Drug	Type of Clearance (CL)[a]	Rifampicin (mg/day) Days Treated	AUC (ng h/ml) Before RIF	AUC (ng h/ml) After RIF	Fold Induction[b]	References
Rosiglitazone	Low	600 mg 6 days	2676	988	2.7	56
Glimepiride	Low	600 mg 5 days	287	190	1.5	58
Gliclazide	Low	600 mg 6 days	44[c]	15[c]	2.9	59
Glyburide	Low	600 mg 5 days	324	198	1.6	60
Glipizide	Low	600 mg 5 days	801	621	1.3	61
S-Warfarin	Low	600 mg 4 days	220[c]	59[c]	3.7	54
R-Warfarin	Low	600 mg 4 days	159[c]	48[c]	3.3	54

[a]Type of clearance: low clearance less than 200 ml/min.
[b]Fold induction is defined as the ratio of oral AUC before:after rifampicin treatment.
[c]In μg h/ml.

healthy volunteers receiving an oral dose of racemic warfarin (0.75 mg/kg body weight). Rosiglitazone, an insulin-sensitizing agent for the treatment of patients with type 2 diabetes, is eliminated predominately by CYP2C8 (55). After dosing with rifampicin 600 mg once daily for 6 days in healthy volunteers, the AUC_{po} of rosiglitazone after an oral dose of 8 mg decreased from an average of 2676 to 988 ng h/ml (56). The AUC_{po} of a 1-mg oral dose of glimepiride, a new sulfonylurea antidiabetic agent eliminated predominately by CYP2C9-mediated metabolism (57), was decreased by only a 1.5-fold after dosing with 600 mg of rifampicin once daily for 5 days (58), emphasizing the modest effect that induction of CYP2C9 may have on the pharmacokinetics of drugs metabolized by this enzyme. Similarly, rifampicin produced a modest effect on the AUC_{po} of gliclazide, glyburide, and glipizide, which are eliminated predominately via metabolism by CYP2C9 (59, 60). The changes in the AUC_{po} after the rifampicin treatment ranged from 1.3- to 2.9-fold for gliclazide, glyburide, and glipizide (Table 15.2).

All the CYP2C8/CYP2C9 drugs listed in Table 15.2 can be defined as low hepatic clearance drug, that is, clearance values of less than 200 ml/min in humans. The relatively small changes in the AUC_{po} of these CYP2C8/CYP2C9 drugs after rifampicin treatment are consistent with the notion that the magnitude of changes in the AUC_{po} after CYP induction tends to be lower for low hepatic clearance drugs compared to that for high hepatic clearance drugs. However, even for the class of low hepatic clearance drugs, the effect of rifampicin appears to be less significant for CYP2C8/CYP2C9 drugs compared to CYP3A4 drugs. As shown in Tables 15.1 and 15.2, the changes in AUC_{po} were generally smaller for CYP2C8/CYP2C9 drugs (1.3- to 3.5-fold) than that for CYP3A4 drugs (3.5- to 8-fold). The differences in the magnitude of changes in the AUC_{po} between CYP3A4 drugs and CYP2C8/CYP2C9 drugs may be due to differential induction between different *CYP* genes by rifampicin, suggesting that the *CYP2C8* and *CYP2C9* genes are less sensitive to rifampicin as compared with *CYP3A4* gene.

The argument that *CYP2C8* and *CYP2C9* genes are less sensitive to rifampicin is supported by data from *in vitro* studies. In a study with human hepatocytes, CYP3A4 mRNA increased about 25-fold, while the CYP2C8 and CYP2C9 mRNA only increased 3- to 4-fold after incubation for 24 h with rifampicin (61). In another *in vitro* study with cultured human hepatocytes, enzyme activity and protein levels of CYP2C8 and CYP2C9 were induced by 3- to 5-fold after rifampicin treatment, while the protein and activity of CYP3A4 was induced by 10-fold after rifampicin treatment (62). Similar observations of differential induction of CYP genes by rifampicin have also been reported by other investigators (54, 63). Although the PXR response elements of *CYP2C8* and *CYP2C9* genes have not been identified, the differential induction between *CYP2C* and *CYP3A* genes by rifampicin may possibly reflect the differences in the affinity of PXR/RXR complex to the responsive elements between *CYP2C8/CYP2C9* genes and *CYP3A4* genes (64–66).

15.5 THE EFFECTS OF TIME AND DOSE ON CYP INDUCTION

Enzyme induction is a slow regulatory process, involving biosynthesis of mRNA and protein, and therefore, CYP induction is expected to be a time- and concentration (dose)-dependent process. The time- and concentration-dependent induction of CYP1A1 and CYP1A2 by TCDD has been demonstrated in studies using human splenic lymphocytes and the colon carcinoma cell line LS180 (67, 68). In the LS180 cell cultures, maximal induction of CYP1A1 and CYP1A2 mRNA occurred between 6 and 24 h after the treatment with TCDD, while maximum CYP1A1 and CYP1A2 protein levels were reached at 48 h and between 48 and 72 h, respectively. These results clearly suggest that there is a lag time between the biosynthesis of mRNA and protein synthesis of new enzyme.

Consistent with *in vitro* observations, time- and dose-dependent TCDD-mediated CYP1A1/2 induction has also been demonstrated in rats (69). The protein expression of CYP1A1 and CYP1A2 was induced in a dose-dependent manner after a single oral dose of TCDD to rats over a dose range of 0.01–30 μg/kg. The ED_{50} for TCDD-induced CYP1A1 and CYP1A2 protein expression was estimated to be 0.22 and 0.4 μg/kg, respectively. Although a significant increase in the expression of TCDD-induced CYP1A1 and CYP1A2 protein was observed at 24 h, both CYP1A1 and CYP1A2 protein reached their maximal levels 7 days after a single oral dose of TCDD (10 μg/kg). After reaching the peak, protein levels declined slowly via protein degradation, and even 35 days after TCDD administration, they remained slightly elevated. The persistence of the expression of TCDD-induced CYP1A1 and CYP1A2 protein may be related to the slow hepatic elimination of TCDD since hepatic concentrations of TCDD reached a maximum about 8 h after TCDD administration and were slowly eliminated with a half-life of approximately 10 days. A physiologically based pharmacokinetic model incorporating tissue retention of TCDD and its binding affinity to AhR has been successfully developed to describe the persistence of CYP1A induction in rats and mice for TCDD and its analogs (70–72).

Dose-dependent induction of CYP1A1/2 by omeprazole has been observed in humans (28). In a clinical study with six healthy volunteers, endoscopic tissue specimens were analyzed for mRNA, and enzyme activity was measured by de-ethylation of ethoxyresorufin before and after treatment with 20 mg/day of omeprazole for 1 week. Large interindividual variations of CYP1A induction were observed, ranging from no change in one subject to a sixfold increase in another subject. The individual who did not respond initially after dosing with 20 mg/day had a marked increase in both mRNA and enzyme activity after receiving 60 mg of omeprazole daily for 1 week, suggesting that enzyme induction is dose dependent. Similarly, dose-dependent induction of CYP1A2 by omeprazole has been demonstrated between 40 and 120 mg doses, as measured by the ^{13}C-[*N*-3-methyl]-caffeine breath test (73). In another clinical study, CYP1A2 was induced in poor metabolizers (PMs) of CYP2C19 but not in extensive metabolizers (EMs) after 7 days of treatment with omeprazole at 40 mg/day (74). Because omeprazole is eliminated predominately by CYP2C19-mediated metabolism, the plasma AUC of omeprazole in PMs of CYP2C19 was approximately fivefold higher than that in EMs after an oral dose of 40 mg. Therefore, the differential effect of omeprazole on CYP1A induction between PMs and EMs of CYP2C19 can be explained by concentration-dependent induction.

Cigarette smoking is known to induce CYP1A2 enzyme and the smoking-induced CYP1A2 returns to the basal level via protein degradation after cessation of smoking. The degradation half-life of the smoking-induced CYP1A2 has been studied in heavy cigarette smokers after cessation of smoking (75). This study was conducted with 8 men and 4 women, who smoked 20 cigarettes or more per day. Subjects were phenotyped for CYP1A2 activity at 6, 4, and 1 days before smoking cessation and at 0, 1, 2, 3, 6, 8, 10, and 13 days thereafter by measuring caffeine clearance. The CYP1A2 activity decreased as a function of time and a maximal decrease was observed at 6 or 8 days after cessation. The degradation half-life of CYP1A2 activity was estimated to be about 36 h in these subjects.

The time- and dose-dependent induction of CYP3A4 by rifampicin has also been demonstrated in primary cultures of human hepatocytes (76, 77). In human hepatocytes treated with rifampicin, the induction of CYP3A4 activity (measured by testosterone 6β-hydroxylation) was concentration dependent, and the EC_{50} for rifampicin was estimated to be 0.3–0.5 μM (78). However, there are no clear-cut clinical examples of dose-dependent CYP3A4 induction reported. Clinically, the effect of rifampicin on the pharmacokinetics of diazepam (a low hepatic clearance drug) was not significantly different between oral rifampicin doses of 600 and 1200 mg (43). The AUC of diazepam decreased from an average of 4430 to 1040 ng h/ml after the 600-mg dose of rifampicin and from an average of 4170 to 1130 ng h/ml after 1200 mg of rifampicin. Similarly, the effect of 600-, 900-, or 1200-mg daily doses of rifampicin on the disposition of propranolol (a high hepatic clearance drug) was also not significantly different (79). Together, these results suggest that the inducing effect of rifampicin at the dosage of 600 mg is near maximal.

Time-dependent CYP3A4 induction by rifampicin has, however, been demonstrated in humans. Clinically, the time course of CYP3A4 induction was evaluated by measuring the trough concentrations of verapamil before, during, and after a 12-day treatment with 600 mg of rifampicin once daily (1). The maximal effect of rifampicin on CYP3A4 activity was observed at 8 days after starting the rifampicin treatment and returned to the baseline activity about 2 weeks after discontinuing rifampicin treatment. The half-life for this increase in CYP3A4 activity was estimated to be about 0.9 day for (*R*)-verapamil and 1.0 day for (*S*)-verapamil, while the half-life for the decline in CYP3A4 activity was 1.5 days for (*R*)-verapamil and 2.1 days for (*S*)-verapamil.

Consistent with the half-life of CYP3A4 induction of approximately 1 day, significant CYP3A4 induction was observed as early as 8 h after a single rifampicin dose (80). Nifedipine (10 mg) was given orally 8 h after a single dose of rifampicin (1200 mg) in healthy volunteers. The AUC_{po} of nifedipine decreased from an average of 573 ng h/ml before rifampicin treatment to 205 ng h/ml at 8 h after the single rifampicin dose. Although the change in the AUC_{po} of nifedipine after 8 h (2.8-fold) is much less than that produced by 7 days of treatment (15.5-fold as shown in Table 15.1), an 8-h pretreatment with rifampicin is sufficient to bring about significant increase in the activity of CYP3A4.

15.6 INDUCTION OF INTESTINAL AND HEPATIC CYP

Although the small intestine is normally regarded as an absorptive organ in the disposition of orally administrated drugs, it contains various drug-metabolizing enzymes, including CYP enzymes, and therefore is metabolically active. Because of its anatomical position, the small intestine contributes to the overall first-pass metabolism of many drugs and some studies have even suggested that the role of intestinal metabolism is quantitatively more important than that of hepatic metabolism in the overall first-pass effect (40, 81, 82). Much of the evidence for these claims has been derived indirectly from comparisons of the AUCs following IV and oral administration, before and after treatment of rifampicin. In a clinical study, Wu et al. (81) claimed that the intestinal first-pass metabolism for cyclosporine (50%) is approximately twice the hepatic first-pass metabolism (24%). Similarly, Gorski et al. (82) claimed that the intestinal first-pass metabolism of midazolam is quantitatively more important than hepatic first-pass metabolism in a study of healthy volunteers, where intestinal and hepatic first-pass metabolism was estimated to be 53% and 29%, respectively. Moreover, in another clinical study, Holtbecker et al. (40) claimed that intestinal CYP3A4 is more sensitive to induction by rifampicin than hepatic CYP3A4. These investigators estimated that the intestinal first-pass metabolism of nifedipine, based on AUC data, increased from 21.8% to 75.8% (a 3.5-fold increase), while the hepatic first-pass metabolism increased from 47.4% to 67.4% (a 40% increase) after treatment with 600 mg/day of rifampicin for 7 days. From these results, these investigators concluded that the contribution of intestinal metabolism to overall first-pass metabolism is quantitatively more important than hepatic metabolism. In addition, they claimed that the intestinal metabolism is induced preferentially by orally administered inducers compared with hepatic metabolism.

This notion that the small intestine plays a quantitatively more important role in first-pass metabolism has been critically questioned (83). Previous conclusions about the quantitative importance of intestinal first-pass metabolism may have been exaggerated as a result of invalid assumptions and problems inherent in the pharmacokinetic analysis of the AUC data. The argument that the role of small intestine in first-pass metabolism is not quantitatively as important as liver is further supported by the fact that the protein levels (per milligram of microsomal protein) of CYP enzymes, such as CYP2C8, CYP2C9, and CYP3A4, in hepatocytes are 5- to 10-fold greater than that in the enterocytes of small intestinal mucosal epithelium (84–86). This difference is even greater when one considers the total mass of microsomal protein in the intestinal mucosa and liver. Similar to the observations in humans, the protein levels of CYP enzymes are much lower in the small intestine than that in the liver in rats (87, 88). Consistent with the low CYP protein levels of intestinal CYP3A4, Kleinbloesem et al. (89) have demonstrated clinically that the intestinal first-pass metabolism of nifedipine in patients with a portacaval shunt (where portal blood bypasses the liver) was absent, because the bioavailability of nifedipine was almost complete (100%) in these patients. In another clinical study, the intestinal first-pass metabolism of verapamil in patients with a portacaval shunt was also reported to be absent (90). Collectively, these results clearly suggest that the role of small intestine in first-pass metabolism is not as important as some scientists have claimed.

The other widespread misconception that intestinal CYP enzymes, compared with hepatic CYP enzymes, are preferentially induced by compounds dosed orally arises in part from the general belief that there is more direct availability of the orally dosed inducers to the intestine as compared to the liver and in part from the observations that the decreases in the oral AUC_{po} of drugs are generally greater than decreases in the IV AUC_{iv} after induction. However, as discussed previously, the observation of a greater change in the oral AUC_{po} versus the IV AUC_{iv} after induction does not necessarily reflect a greater degree of enzyme induction in the small intestine. Recall that CYP induction has more profound effect on the AUC_{po} than the AUC_{iv}, particularly for high hepatic clearance drugs (1, 40, 81, 82) because the AUC_{iv} is driven primarily by drug delivery, that is, hepatic blood flow, and not enzyme activity. Moreover, although Holtbecker et al. (40) concluded that intestinal CYP3A4 is more sensitive to rifampicin than hepatic CYP3A4, the calculation of the intestinal and hepatic first-pass metabolism was carried out with invalid assumptions regarding AUC.

In fact, the literature data suggest that the degree of CYP3A4 induction in the intestine is generally lower than the degree of hepatic CYP3A4 induction. In a clinical study of 14 patients in which liver biopsies were collected before and after rifampicin treatment of 600 mg/day for 4 days, the protein level of hepatic

CYP3A4 was increased by about 6- to 10-fold (91). On the other hand, dosing six healthy volunteers with 600 mg of rifampicin for 10 days only resulted in a two- to threefold induction of CYP3A4 protein in the small intestine (92). Similarly, intestinal CYP3A4 mRNA, collected via endoscopic biopsies, was increased by approximately threefold in five healthy volunteers treated with 600 mg/day of rifampicin for 7 days (93). Similar to humans, a greater degree of CYP3A induction was observed for the liver compared with the small intestine after the treatment of rats with DEX, another prototypical CYP3A inducer (87). These results support the argument that the magnitude of induction of intestinal CYP enzymes is not greater than the induction of hepatic CYP enzymes.

15.7 *IN VITRO* AND *IN VIVO* ASSESSMENT OF CYP INDUCTION

From an industrial perspective, CYP induction is a highly undesirable drug property, not only because of metabolic liability with respect to the potential for decreased therapeutic efficacy but also because of the concern of marketing competition (94). Therefore, it is highly desirable to develop new drug candidates that are not potent CYP inducers, to avoid the potential of CYP induction-mediated drug interactions. For this reason, many drug companies today routinely include early evaluation of CYP induction at the drug discovery stage as part of the selection processes of drug candidates (95, 96).

While many CYP enzymes are known to be inducible, CYP3A4 induction is probably the most important cause of the documented induction-based drug–drug interactions (97, 98). This is expected because CYP3A4 accounts for roughly 40% of the total CYP in human liver and catalyzes the metabolism of more than 60% of clinically used drugs. Therefore, most drug companies focus mainly and routinely on the assessment of CYP3A4 induction, and assessing the induction potential for other CYP enzymes is evaluated only when there is a need. Although animal models may provide some useful information on the factors that affect CYP induction, significant species differences in the inductive response preclude the use of *in vivo* animal models for the assessment of human CYP3A4 induction for new drug candidates. Therefore, the use of *in vitro* systems is the only means by which the potential for human CYP3A4 induction can be routinely assessed in drug discovery.

15.7.1 *In Vitro* Methods for the Assessment of CYP3A4 Induction

Several *in vitro* models have been established to assess the potential of CYP3A4 induction for new drug candidates, including liver slices, immortalized cell lines, and fresh and frozen primary hepatocytes. Each method has its advantage and disadvantage, and the reader is referred to other review articles for further information on this topic (3–5). Among these models, primary cultures of human hepatocytes have been used extensively by academic and industrial laboratories for evaluating CYP3A4 induction. It is generally accepted that the primary hepatocyte culture is the most predictive *in vitro* model for assessing CYP induction and therefore primary cultures of human hepatocytes have become the "gold standard" for *in vitro* testing of CYP3A4 induction. Moreover, this system is widely accepted by drug companies for assessing the potential CYP3A4 induction of new drug candidates (95, 96).

In addition to assays for enzymatic activity, protein, or mRNA levels, several *in vitro* reporter gene assays have been developed for assessing the induction of the *CYP3A4* gene (99, 100). In one study, 14 commercially available compounds were evaluated and compared for their ability to induce CYP3A4 in human hepatocytes and to activate human PXR in a reporter gene assay (99). Sandwiched primary cultures of human hepatocytes from six donors were used and changes in CYP3A4 were assessed by microsomal testosterone 6β-hydroxylase activity, hepatic CYP3A4 mRNA, and protein levels (analyzed using branched DNA technology/Northern blotting and Western blotting, respectively), in addition to the PXR response. A reasonably good and statistically significant correlation between human PXR activation and CYP3A4 induction in human hepatocytes was observed for these 14 compounds. In another study, 17 compounds, including known CYP3A4 inducers, were evaluated for their PXR activation, using a reporter gene assay cotransfected with PXR and the glucocorticoid receptor (GR) (100). The rank order of the potency of the known CYP3A4 inducers was generally in agreement with the results from the PXR/GR cotransfected reporter gene assay. Collectively, these results suggest that the PXR reporter gene assay can be used as a reliable *in vitro* model for CYP3A4 induction. Because of its robustness and reliability, the PXR reporter assay is particularly useful at the drug discovery stage for high throughput screening.

15.7.2 CYP3A4 Induction and Data Presentation

While enzyme induction is defined as an increase in mRNA and enzyme protein levels, the associated change in the enzyme activity is more clinically relevant than the changes in mRNA or protein level. For example, ritonavir induced CYP3A4 mRNA and protein levels in human hepatocytes in a concentration-dependent manner, but there was a decrease in CYP3A4 enzyme activity because of its strong CYP3A4 inhibition activity (101). In the clinical setting, ritonavir inhibits the CYP3A4-mediated metabolism of drugs and the induction of CYP3A4 activity is offset by its strong inhibitory effect. Therefore, it is desirable to have all three endpoint measurements of a compound's ability to induce CYP3A4, mRNA and protein levels and enzyme activity, at the same time in order to accurately interpret the data. Kinetically, the intrinsic parameters EC_{50} (effective concentration for 50% maximal induction) and E_{max} (maximal CYP induction) of a drug are probably more useful for the interpretation of *in vitro* induction data. The EC_{50} for an inducer allows a direct comparison to be drawn between the plasma concentration (C) of the agent and the degree of induction following drug administration, while the E_{max} predicts the maximal extent of CYP induction by the agent in patients. The relationship between these parameters and the overall induction effect (E) is shown in the following equation:

$$E = \frac{E_{max} \cdot C}{EC_{50} + C} \tag{15.1}$$

The EC_{50} and E_{max} values have been used to compare the capacity of troglitazone and rifampicin to induce CYP3A4 activity in primary human hepatocyte cultures and to predict the induction potential of troglitazone in humans (78). While the E_{max} for troglitazone induction (measured as the rate of testosterone 6β-hydroxylation) was comparable to that of rifampicin, troglitazone's EC_{50}, 5–10 μM, was 10- to 30-fold higher than that of rifampicin, 0.3–0.5 μM. Thus, the intrinsic induction activity (E_{max}/EC_{50}) of troglitazone is only about 10% of the rifampicin value. In addition, the derived EC_{50} of troglitazone for CYP3A4 induction is much higher than the steady-state plasma concentration of troglitazone in clinical studies, 2 μM. Taken together, these results suggest that troglitazone is a weak CYP3A4 inducer at the clinical dose range. In another study with human hepatocytes, the EC_{50} and E_{max} for CYP3A4 induction by SR12813 were estimated to be 0.6–1.0 μM and 5 nmol/min/mg protein, respectively, and the corresponding values for rifampicin were 0.3 μM and 10 nmol/min/mg protein (5, 35). Because the E_{max} of SR12813 is about 50% of the value of rifampicin, significant induction by the compound would be predicted to occur *in vivo* only when plasma concentrations of SR12813 are much greater than 1 μM. From these two examples, it is clear that the potential of a drug to induce CYP3A4 *in vivo* depends not only on its intrinsic activity (E_{max}/EC_{50}), but also on the relationship between its plasma concentration and EC_{50} value.

Determining the EC_{50} and E_{max} may not be very practical for routine, automated screening during the drug discovery or early development stage because of the need for large numbers of human hepatocytes and the requirement of including more concentration points to estimate the enzyme kinetic parameters. Additionally, it may be difficult to obtain an E_{max} value due to the cell toxicity of the test compound at high concentrations. Alternatively, the induction potential, expressed as a fold-change in relation to the basal level, or induction potency, expressed as a percentage of the rifampicin value at 10 μM, can be used to assess CYP3A4 induction in hepatocytes at the drug discovery stage (2).

The primary advantage of the use of the fold-change method is that this approach provides useful information on interindividual variability of CYP3A4 induction in hepatocyte preparations from different donors. However, the use of the fold-change method without proper reference may sometimes lead to inappropriate conclusions (2). There are two reasons for this. First, the basal levels of CYP3A4 in some individuals are so low that they are difficult to accurately quantify, leading to an inaccurate estimation of the fold-change. For example, induction of CYP3A4 enzyme activity by a drug candidate, calculated as a fold-change, varied from three- to eightfold in human hepatocytes from five different human donors. However, when the results were expressed as a percentage of the induction normalized by a prototypical inducer (rifampicin), the range was from 16% to 30% (2). Thus, based on the fold-change of enzyme activity, one may conclude that the drug candidate is a potent CYP3A4 inducer, although it would only be a moderate inducer based on the percentage induction relative to rifampicin. Second, the basal CYP3A4 level and associated enzyme activity can be highly variable between hepatocyte preparations obtained from different donors. It is important to point out that the fold-change in CYP3A4 elicited by a given inducing agent is highly dependent on the basal level of CYP3A4 in hepatocyte preparations; the lower the basal level the higher the fold induction (3, 102). Because relatively small numbers are typically studied (normally $n = 3-4$ individual donors), it is quite possible that hepatocytes may obtained from liver donors that all have low basal CYP3A4 level (or high basal CYP3A4 level).

In conclusion, it would be highly desirable to assess the potential of CYP3A4 induction by measuring the intrinsic parameters EC_{50} and E_{max} with respect to CYP3A4 mRNA, protein levels, and enzyme activity. If the measurement of the intrinsic parameters EC_{50} and E_{max} is difficult due to practical limitations, determination of induction potency as the percentage of the maximal induction by rifampicin, rather than fold-change method, would be preferable.

15.7.3 *In Vitro/In Vivo* Extrapolation of CYP Induction

The ultimate goal of *in vitro* induction studies is to predict induction-based drug interactions in the clinical settings; however, the prediction of *in vivo* CYP induction based on *in vitro* data is quite complicated. As discussed above, CYP induction by drugs is not only a concentration-dependent but also a time-dependent process. In addition, the magnitude of CYP induction depends on the net balance of enzyme biosynthesis and degradation in the time course during the chronic dosing of inducing agent. It would be very difficult to predict the extent of CYP induction as function of time in the short term; however, after a sufficient period of time during chronic dosing, CYP induction should reach a steady state and the process becomes time invariant. According to Equation 14.1, in theory, at steady state, the magnitude (E) of *in vivo* induction can be predicted for a drug candidate using the *in vitro* EC_{50} and E_{max} values and the steady-state concentration of a drug (C).

It is believed that only the free (unbound) fraction of drug, that is, drug that is not bound to protein, would cross cell membranes and activate nuclear receptors (PXR or CAR) and induce the biosynthesis of CYP3A4 in hepatocytes. Therefore, the drug concentration (C) used for predicting CYP3A4 induction should be the unbound (free) drug concentration at the active site of the PXR (or CAR) receptor. Because direct measurement of unbound drug concentration at the active site is almost impossible, the unbound drug concentration in plasma is used in lieu of unbound drug concentration at the active site, based on the assumption in the free drug hypothesis that equilibrium of unbound drug occurs readily between plasma and hepatocytes. Although it is generally accepted that plasma protein binding plays an important role in determining the degree of CYP inhibition (103–105), there is no information on the effect of plasma protein binding on CYP3A4 induction. In fact, the effect of plasma protein binding on CYP3A4 induction is still a debatable issue. Some scientists believe that the role of plasma protein binding in CYP3A4 induction may not be as important as in CYP3A4 inhibition because CYP3A4 induction is a time-dependent process and reflects a cumulative effect, while CYP3A4 inhibition is an immediate effect. Other scientists, however, believe that the role of plasma protein binding in CYP induction is as important as CYP inhibition because there is always a clear concentration-dependent relationship between the magnitude of CYP induction and drug concentration in the *in vitro* hepatocyte system after a sufficient period of incubation time.

Without a full understanding of the role of plasma protein binding in CYP induction, it is difficult to predict the magnitude of CYP induction *in vivo*. Therefore, more efforts should be made to evaluate the effect of plasma and tissue protein binding on the *in vivo* CYP3A4 induction. Another critical issue that remains largely unresolved in predicting *in vivo* CYP3A4 induction is the large variability associated with CYP3A4 induction. Considerable interindividual variability in CYP induction has been observed both *in vitro* and *in vivo* (38, 106), and until all the sources of this biological variability are identified and fully understood, quantitative predictions of *in vivo* CYP induction will remain elusive.

Despite these limitations, attempts have been made to predict *in vivo* induction by comparing *in vitro* and *in vivo* E_{max}. As shown in Equation 15.1, the *in vitro* E_{max} of an inducer can be obtained when the drug concentration of the inducer is much greater than its EC_{50} value, and the *in vivo* E_{max} of the inducer can also be obtained when a high dose of the inducer is given. Although CYP induction is a time-dependent process, in theory, the E_{max} becomes a time-invariant parameter when a steady-state condition is reached (i.e., after sufficient period of time during chronic dosing). There are several good examples of the use of the E_{max} approach to predict CYP induction. For example, the induction of CYP3A enzymes (measured by 6β-testosterone hydroxylation) in rat hepatocytes by DEX was concentration dependent, and the EC_{50} was estimated to be 1.3 μM (107). The mean *in vitro* E_{max} was estimated to be sixfold over its basal activity after 72-h incubation in rat hepatocytes at a high concentration of DEX (15 μM). On the other hand, the mean *in vivo* E_{max} was determined to be about fivefold in rats following administration of DEX at 50 mg/kg/day for 4 days. The *in vivo* E_{max} (expressed as fold-change) was in good agreement with the *in vitro* E_{max} (107). Similarly, a good correlation between the maximal *in vitro* and *in vivo* CYP3A induction for 13 compounds has been reported by Silva et al. (2). In this study, rat hepatocytes were incubated with the test compounds at concentration of 50 μM for 4 days (96 h), while the compounds were given orally to rats at 400 mg/kg/day for 4 days. These results suggest that the *in vivo* E_{max} can be predicted reasonably well using *in vitro* E_{max} data.

The *in vitro/in vivo* E_{max} approach has also been applied for rifampicin in humans. The *in vitro* E_{max} of rifampicin at 10 μM was estimated to be in the range of 10- to 20-fold of the basal CYP3A4 protein level in human hepatocytes (65, 108, 109). The *in vitro* E_{max} (fold-change) of CYP3A4 by rifampicin is in good agreement with the *in vivo* E_{max} reported by Ged et al. (91), where there was an 18-fold increase in CYP3A4 protein levels in liver biopsies after dosing with 600 mg/day of rifampicin for 4 days. It should be noted that the peak plasma concentrations of rifampicin are in the range of 5–15 μM in patients following an oral dose of 600 mg and that the plasma protein binding of rifampicin is about 70–80% in human plasma (110). Thus, maximal CYP3A4 induction by rifampicin would be expected to occur *in vivo* after 600-mg oral doses.

15.8 CONCLUSIONS

Although our understanding of CYP induction has advanced significantly over the past 15 years through the discovery of key nuclear receptors, much still remains to be learned about the molecular mechanisms of CYP induction. One of the unsolved issues is the wide interindividual variability in CYP induction. There are a large number of factors that could contribute to the variability, including genetic and environmental variables. The relative contribution between the genetic and environmental variables is not readily assessed due in part to the complexity of CYP induction and insufficiency of proper experimental tools. Another critical issue that may complicate the prediction and interpretation of CYP induction is the interplay between efflux transporters and CYP enzymes. Although there is now increasing evidence to suggest important interplay between CYPs and efflux transporters because of a broad overlap in substrates and inducers with CYP enzymes and efflux transporters, our knowledge about this interplay is very limited (111–114). This uncertainty remains because it is very difficult to accurately estimate the relative contribution of CYP enzymes and transporters to drug absorption and disposition. This is particularly true for drug interactions that are caused by CYP and transporter induction.

Because of the complexity of the contributing factors, quantitative prediction of CYP induction is very difficult, if not impossible. Although numerous *in vitro* systems have been developed to assess CYP induction, these systems are only useful for qualitative assessment whether a new drug candidate has a potential for CYP induction. Information obtained from *in vitro* induction studies is still limited in its ability to predict whether there is a "probability" of induction-based drug interactions. If the *in vitro* data suggest that a drug candidate could have a potential for CYP induction, clinical studies should be conducted earlier to assess the degree of induction. Therefore, it is of importance to emphasize that the data obtained from *in vitro* systems on CYP induction should not serve as a "no-go" decision for drug development without a proper clinical assessment. Finally, it should be noted that in many cases the CYP induction is manageable by adjusting dosage regimen.

REFERENCES

1. Fromm MF, Busse D, Kroemer HK, Eichelbaum M. Differential induction of prehepatic and hepatic metabolism of verapamil by rifampin. Hepatology 1996; 24:796–801.

2. Silva JM, Morin PE, Day SH, Kennedy BP, Payette P, Rushmore T, Yergey JA, Nicoll-Griffith DA. Refinement of an in vitro cell model for cytochrome P450 induction. Drug Metab Dispos 1998;26:490–496.
3. Kostrubsky VE, Ramachandran Venkataramanan VR, Dorko K, Esplen JE, Zhang S, Sinclair JF, Wrighton SA, Strom SC. The use of human hepatocyte cultures to study the induction of cytochrome P-450. Drug Metab Dispos 1999;27:887–894.
4. Maurel P. The use of adult human hepatocytes in primary culture and other in vitro systems to investigate drug metabolism in man. Adv Drug Deliv Rev 1996; 22:105–132.
5. LeCluyse EL. Human hepatocyte culture systems for the in vitro evaluation of cytochrome P450 expression and regulation. Eur J Pharm Sci 2001;13:343–368.
6. Kliewer SA, Moore JT, Wade L, Staudinger JL, Watson MA, Jones SA, McKee DD, Olicer BB, Wilson TM, Zetterström RH, Perlmann T, Lehmann JM. An orphan nuclear receptor activated by pregnanes defines a novel steroid signaling pathway. Cell 1998;92:73–82.
7. Sueyoshi T, Negishi M. Phenobarbital response elements of cytochrome P450 genes and nuclear receptors. Annu Rev Pharmacol Toxicol 2001;41:123–143.
8. Whitlock JP Jr. Induction of cytochrome P4501A1. Annu Rev Pharmacol Toxicol 1999;39:102–125.
9. Poland AP, Glover E, Kende AS. Stereospecific, high affinity binding of 2,3,7,8-tetrachlorodibenzo-p-dioxin by hepatic cytosol: evidence that the binding species is the receptor for induction of aryl hydrocarbon hydroxylase. J Biol Chem 1976;251:4936–4946.
10. Poland AP, Knutson JC. 2,3,7,8- tetrachlorodibenzo-p-dioxin and related aromatic hydrocarbons: Examination of the mechanism of toxicity. Annu Rev Pharmacol Toxicol 1982;22:517–554.
11. Denison MS, Fisher JM, Whitlock JP Jr. The DNA recognition site for the dioxin-Ah receptor complex. Nucleotide sequence and functional analysis. J Biol Chem 1988;33:17221–17224.
12. Jones KW, Whitlock JP Jr. Functional analysis of the transcriptional promoter for the CYP 1A1 gene. Mol Cell Biol 1990;10:5098–5105.
13. Baes M, Gulick T, Choi H-S, Martinoli MG, Simha D, Moore DD. A new orphan nuclear member of the nuclear hormone receptor superfamily that interacts with a subset of retinoic acid response elements. Mol Cell Biol 1994;14:1544–1551.
14. Wei P, Zhang J, Egan-Hafley M, Liang S, Moore DD. The nuclear receptor CAR mediates specific xenobiotic induction of drug metabolism. Nature 2000;407:920–923.
15. Waxman DJ. P450 gene induction by structurally diverse xenochemicals: central role of nuclear receptors CAR, PXR, and PPAR. Arch Biochem Biophys 1999;369:11–23.
16. Forman BM, Tzameli I, Choi HS, Chen J, Simha D, Seol W, Evans RM, Moore DD. Androstane metabolites bind to and deactivate the nuclear receptor CARß. Nature 1998;395:612–615.
17. Trottier E, Belzil A, Stoltz C, Anderson A. Localization of a phenobarbital-responsive element (PBRE) in the 5′ flanking region of the rat CYP2B2 gene. Gene 1995;158:263–268.

18. Kawamoto T, Sueyoshi T, Zelko I, Moore R, Washburn K, Negishi M. Phenobarbital-responsive nuclear translocation of the receptor CAR in induction of the CYP2B gene. Mol Cell Biol 1999;19:6318–6322.

19. Honkakoski P, Negishi M. Protein serine/threonine phosphatase inhibitors suppress phenobarbital-induced Cyp2b10 gene transcription in mouse primary hepatocytes. Biochem J 1998;330:889–895.

20. Zelko I, Sueyoshi T, Kawamoto T, Moore R, Negishi M. The peptide near the C terminus regulates receptor CAR nuclear translocation induced by xenochemicals in mouse liver. Mol Cell Biol 2001;21:2838–2846.

21. Xie W, Barwick JL, Downes M, Blumberg B, Simon CM, Nelson MC, Neuschwander-Tetri BA, Brunt EM, Guzelian PS, Evans RM. Humanized xenobiotic response in mice expressing nuclear receptor SXR. Nature 2000;406:435–439.

22. Bertilsson G, Heidrich J, Svensson K, Asman M, Jendeberg L, Sydow-Bäckman M, Ohlsson R, Postlind H, Blomquist P, Berkenstam A. Identification of a human nuclear receptor defines a new signaling pathway for CYP3A induction. Proc Natl Acad Sci USA 1998;95:12208–12213.

23. Squires EJ, Sueyoshi T, Negishi M. Cytoplasmic localization of pregnane X receptor and ligand-dependent nuclear translocation in mouse liver. J Biol Chem 2004;279:49307–49314.

24. Wang H, LeCluyse EL. Role of orphan nuclear receptors in the regulation of drug-metabolizing enzymes. Clin Pharmacokinet 2003;42:1331–1357.

25. Watkins RE, Wisely GB, Moore LB, Collins JL, Lambert MH, Williams SP, Willson TM, Kliewer SA, Redinbo MR. The human nuclear xenobiotic receptor PXR: structural determinants of directed promiscuity. Science 2001;292:2329–2333.

26. Goodwin B, Redinbo MR, Kliewer SA. Regulation of CYP3A gene transcription by the pregnane X receptor. Annu Rev Pharmacol Toxicol 2002;42:1–23.

27. Diaz D, Fabre I, Daujat M, Saint Aubbert B, Bories P, Michel H, Maurel P. Omeprazole is an aryl hydrocarbon-like inducer of human hepatic cytochrome P-450. Gastroenterology 1990;99:737–747.

28. McDonnell WM, Scheiman JM, Traber PG. Induction of cytochrome P450 1A genes (CYP1A) by omeprazole in human alimentary tract. Gastroenterology 1992;103:1509–1516.

29. Kocarek TA, Schuetz EG, Strom SC, Fisher RA, Guzelian PS. Comparative analysis of cytochrome P4503A induction in primary cultures of rat, rabbit and human hepatocytes. Drug Metab Dispos 1995;23:415–421.

30. Nebert DW, Gonzalez FJ. The P450 gene superfamily. In: Ruckpaul K, Rein H, editors. Principles, mechanisms and biological consequences of induction. London: Taylor & Francis; 1990. pp. 35–61.

31. Degawa M, Kobayashi K, Miura S, Arai H, Esumi H, Sugimura T, Hashimoto Y. Species difference among experimental rodents in induction of P450IA family enzymes by 2-amino-1-methyl-6-phenylimidazo[4,5-b]pyridine. Jpn J Cancer Res 1992;83:1047–1051.

32. White IN, Davies A, Smith LL, Dawson S, Matteis F. Induction of CYP2B1 and CYP3A1, and associated monooxygenase activities by tamoxifen and certain analogues in the livers of female rats and mice. Biochem Pharmacol 1993;7:21–30.

33. Fontaine SM, Hoyer PB, Halpert JR, Sipes IG. Role of induction of specific hepatic cytochrome P450 isoforms in expoxidation of 4-vinylcyclohexene. Drug Metab Dispos 2001;29:1236–1242.
34. Barwick JL, Quattrochi LC, Mills AS, Potenza C, Tukey RH, Guzelian PS. Transspecies gene transfer for analysis of glucocorticoid-inducible transcriptional activation of transiently expressed human CYP3A4 and rabbit CYP3A6 in primary cultures of adult rat and rabbit hepatocytes. Mol Pharmacol 1996;50:10–16.
35. Jones SA, Moore LB, Shenk JL, Wisely GB, Hamilton GA, McKee DD, Tomkinson NC, LeCluyse EL, Lambert MH, Willson TM, Kliewer SA, Moore JT. The pregnane X receptor, a promiscuous xenobiotic receptor that has diverged during evolution. Mol Endocrinol 2000;14:27–39.
36. Choi H-S, Chung M, Tzameli I, Simha D, Lee Y-K, Seol W, Moore DD. Differential transactivation by two isoforms of the orphan nuclear hormone receptor CAR. J Biol Chem 1997;272:23565–23571.
37. Diliberto JJ, Akubue PI, Luebke RW, Birnbaum LS. Dose-response relationships of tissue distribution and induction of CYP1A1 and CYP1A2 enzymatic activities following acute exposure to 2,3,7,8-tetrachlorodibenzo-p-dioxin (TCDD) in mice. Toxicol Appl Pharmacol 1995;130:197–208.
38. Lin JH, Lu AYH. Interindividual variability in inhibition and induction of cytochrome P450 enzymes. Annu Rev Pharmacol Toxicol 2001;41:535–567.
39. Alvan G, Piafsky K, Linda M, von Bahr C. Effect of pentobarbital on the disposition of alprenolol. Clin Pharmacol Ther 1977;22:316–321.
40. Holtbecker N, Fromm MF, Kroemer HK, Ohnhaus EE, Heidemann H. The nifedipin-rifampin interaction: evidence for induction of gut wall metabolism. Drug Metab Dispos 1996;24:1121–1123.
41. Kharasch ED, Hoffer C, Whittington D, Sheffels P. Role of hepatic and intestinal cytochrome P450 3A and 2B6 in the metabolism, disposition, and miotic effects of methadone. Clin Pharmacol Ther 2004;76:250–269.
42. Schmider J, Brockmoller J, Arold G, Bauer S, Roots I. Simultaneous assessment of CYP3A4 and CYP1A2 activity in vivo with alprazolam and caffeine. Pharmacogenetics 1999;9:725–734.
43. Ohnhaus EE, Brockmeyer N, Dylewiez P, Habicht H. The effect of antipyrine and rifampin on the metabolism of diazepam. Clin Pharmacol Ther 1987;42:148–156.
44. Villikka K, Kivisto K, Luurila H, Neuvonen PJ. Rifampin reduces plasma concentrations and effects of zolpidem. Clin Pharmacol Ther 1997;62:629–634.
45. Villikka K, Kivisto K, Lamberg TS, Kantola T, Neuvonen PJ. Concentrations and effects of zopiclone are greatly reduced by rifampicin. Br J Clin Pharmacol 1997;43:471–474.
46. Backman JT, Olkkola KT, Neuvonen PJ. Rifampin drastically reduces plasma concentrations and effects of oral midazolam. Clin Pharmacol Ther 1996;59:7–13.
47. Villikka K, Kivisto K, Backman JT, Olkkola KT, Neuvonen PJ. Triazolam is ineffective in patients taking rifampin. Clin Pharmacol Ther 1997;61:8–14.
48. Hebert MF, Fisher RM, Prueksaritanont T, Benet LZ. Bioavailability of cyclosporine with concomitant rifampin administration is markedly less than predicted by hepatic enzyme induction. Clin Pharmacol Ther 1992;52:453–457.
49. Herbert MF, Fisher RM, Marsh CL, Dressler D, Bekersky I. Effects of rifampin on tacrolimus in healthy volunteers. J Clin Pharmacol 1999;39:91–96.

50. Damkier P, Hansen LL, Brosen K. Rifampicin treatment greatly increases the apparent oral clearance of quinidine. Pharmacol Toxicol 1999;85:257–262.
51. Crixivan® package insert. Pennsylvania: Merck & Co; 1997.
52. Synold TW, Dussault I, Forman BM. The orphan nuclear receptor SXR coordinately regulates drug metabolism and efflux. Nat Med 2001;7:584–590.
53. Gerbal-Chaloin S, Pascussi J-M, Pichard-Garcia L, Daujat M, Waechter F, Fabre JM, Carrére N, Maurel P. Induction of CYP2C genes in human hepatocytes in primary cultures. Drug Metab Dispos 2001;29:242–251.
54. Heimark LD, Gibaldi M, Tragger WF, O'Reilly RA, Goulart DA. The mechanism of the warfarin-rifampin drug interaction in humans. Clin Pharmacol Ther 1987;42:388–394.
55. Baldwin SJ, Clarke SE, Chenery RJ. Characterization of the cytochrome P450 enzymes involved in the in vitro metabolism of rosiglitazone. Br J Clin Pharmacol 1999;48:424–432.
56. Park J-Y, Kim K-A, Kang M-H, Kim S-L, Shin J-G. Effect of rifampin on the pharmacokinetics of rosiglitazone in healthy subjects. Clin Pharmcol Ther 2004;75:157–162.
57. Langtry HD, Balfour JA. Glimepiride. A review of its use in the management of type-2 diabetes mellitus. Drugs 1998;55:563–584.
58. Niemi M, Kivisto KT, Backman JT, Neuvonen PJ. Effect of rifampicin on the pharmacokinetics and pharmacodynamics of glimepiride. Br J Clin Pharmacol 2000;50:591–595.
59. Park J-Y, Kim K-A, Park P-W, Park C-W, Shin J-G. Effect of rifampin on the pharmacokinetics and pharmacodynamics of gliclazide. Clin Pharmacol Ther 2003;74:334–340.
60. Niemi M, Backman JT, Neuvonen M, Neuvonen PJ, Kivisto KT. Effects of rifampin on the pharmacokinetics and pharmacodynamics of glyburide and glipizide. Clin Pharmacol Ther 2001;69:400–406.
61. Dvorak Z, Modriansky M, Pichard-Garcia L, Balaguer P, Vilarem M-J, Ulrichová J, Maurel P, Pascussi JM. Colchicine down-regulates cytochrome P450 2B6, 2C8, 2C9, and 3A4 in human hepatocytes by affecting their glucocorticoid receptor-mediated regulation. Mol Pharmacol 2003;64:160–169.
62. Madan A, Graham RA, Carroll KM, Mudra DR, Burton LA, Krueger LA, Downey AD, Czerwinski M, Forster J, Ribadeneira MD, Gan LS, LeCluyse EL, Zech K, Robertson P Jr., Koch P, Antonian L, Wagner G, Yu L, Parkinson A. Effects of prototypical microsomal enzyme inducers on cytochrome P450 expression in cultured human hepatocytes. Drug Metab Dispos 2003;31:421–431.
63. Sahi J, Stern RH, Milad MA, Rose KA, Gibson G, Zheng X, Stilgenbauer L, Sadagopan N, Jolley S, Gilbert D, LeCluyse EL. Effects of avasimibe on cytochrome P450 2C9 expression in vitro and in vivo. Drug Metab Dispos 2004;32:1370–1376.
64. Pascussi JM, Gerbal-Chaloin S, Drocourt L, Maurel P, Vilarem MJ. The expression of CYP2B6, CYP2C9 and CYP3A4 genes: a tangle of networks of nuclear and steroid receptors. Biochim Biophys Acta 2003;1619:243–253.
65. Drocourt L, Ourlin J-C, Pascussi J-M, Maurel P. Expression of CYP3A4, CYP2B6, and CYP2C9 is regulated by vitamine D receptor pathway in primary human hepatocytes. J Biol Chem 2002;277:25125–25132.

66. Drocourt L, Pascussi J-M, Assenat E, Fabre J-M, Maurel P, Vilarem M-J. Calcium channel modulators of the dihydrophyridine family are human pregnane X receptor activators and inducers of CYP3A, CYP2B, and CYP2C in human hepatocytes. Drug Metab Dispos 2001;29:1325–1331.

67. Jeong HG, Yang K-H. Expression of 2,3,7,8-tetrachlorodibenzo-*p*-dioxin-inducible cytochrome P4501A1 in human splenic lymphocyte cultures. Cancer Lett 1996;98: 193–198.

68. Li W, Harper PA, Tang B-K, Okey AB. Regulation of cytochrome P450 enzymes by aryl hydrocarbon receptor in human cells: CYP1A2 expression in the LS180 colon carcinoma cell line after treatment with 2,3,7,8-tetrachlorodibenzo-*p*-dioxin or 3-methylcholanthrene. Biochem Pharmacol 1998;56:599–612.

69. Santostefano MJ, Ross DG, Savas U, Jefcoate CR, Birnbaum LS. Differential time-course and dose-response relationships of TCDD-induced CYP1B1, CYP1A1, and CYP1A2 proteins in rats. Biochem Biophys Res Commun 1997;233:20–24.

70. Kedderis LB, Mills JJ, Andersen ME, Birnbaum LS. A physiologically based pharmacokinetic model of 2,3,7,8-tetrabromodibenzo-*p*-dioxin (TBDD) in the rat: tissue distribution and CYP1A induction. Toxicol Appl Pharmacol 1993;121:87–98.

71. Leung H-W, Ku RH, Paustenback DJ, Andersen ME. A physiologic-based pharmacokinetic model for 2,3,7,8-tetrachlorodibenzo-*p*-dioxin in C57BL/6J and DBA/2J mice. Toxicol Lett 1988;42:15–28.

72. Leung H-W, Paustenback DJ, Murray FJ, Andersen ME. A physiological pharmacokinetic model of tissue distribution and enzyme-inducing properties of 2,3,7,8-tetracholordibenzo-*p*-dioxin in the rat. Toxicol Appl Pharmacol 1990;103:399–410.

73. Rost KL, Brosick H, Heinemeyer G, Roots I. Specific and dose-dependent enzyme induction by omeprazole in human beings. Hepatology 1994;20:1204–1212.

74. Rost KL, Brosick H, Brockmoller J, Heinemeyer G, Roots I. Increase of P450IA2 activity by omeprazole: evidence by the 13-C[N3-methyl]caffeine breath test in poor and extensive metabolizers of S-mephenytoin. Clin Pharmacol Ther 1992;52:170–180.

75. Faber MS, Fuhr U. Time response of cytochrome P450 1A2 activity on cessation of heavy smoking. Clin Pharmacol Ther 2004;76:178–184.

76. Kostrubsky VE, Lewis LD, Strom SC, Wood SC, Schuetz EG, Sinclair PR, Wrighton SA, Sinclair JF. Induction of cytochrome P4503A by taxol in primary cultures of human hepatocytes. Arch Biochem Biophys 1998;355:131–136.

77. Nallami SC, Glauser TA, Hariparsad N, Setchell K, Buckley DJ, Buckley AR, Desai PB. Dose-dependent induction of cytochrome P450 (CYP) 3A4 and activation of pregnane X receptor by topiramate. Epilipsia 2003;44:1521–1428.

78. Sahi J, Hamilton G, Sinz M, Barros S, Huang S-M, Lesko LJ, LeCluyse EL. Effects of troglitazone on cytochrome P450 enzymes in primary cultures of human and rat hepatocytes. Xenobiotica 2000;30:273–284.

79. Herman RJ, Nakamura K, Wilkinson GR, Wood AJ. Induction of propranolol metabolism by rifampicin. Br J Clin Pharmacol 1983;16:565–569.

80. Ndanusa BU, Mustapha A, Abdu-Aguye I. The effect of single dose of rifampicin on the pharmacokinetics of oral nifedipine. J Pharm Biomed Anal 1997;15:1571–1575.

81. Wu CY, Benet LZ, Hebert MF, Gupta SK, Rowland M, Gomez DY, Wacher VJ. Differential of absorption and first-pass gut and hepatic metabolism in humans: Studies with cyclosporine. Clin Pharmacol Ther 1995;58:492–497.

82. Gorski JG, Vannaprasaht S, Hamman MA, Ambrosius WT, Bruce MA, Haehner-Daniels B, Hall SD. The effect of age, sex, and rifampin administration on intestinal and hepatic cytochrome P450 3A activity. Clin Pharmacol Ther 2003;74:275–287.
83. Lin JH, Chiba M, Baillie TA. Is the role of the small intestine in first-pass metabolism overemphasized? Pharmacol Rev 1999;51:135–157.
84. Paine MF, Khalighi M, Fisher JM, Shen DD, Kunze KL, Marsh CL, Perkins JD, Thummel KE. Characterization of interintestinal and intraintestinal variations in human CYP3A-dependent metabolism. J Pharmacol Exp Ther 1997;283:1552–1562.
85. Läpple F, van Richter O, Fromm MF, Richter T, Thon KP, Wisser H, Griese EU, Eichelbaum M, Kivistö KT. Differential expression and function of CYP2C isoforms in human intestine and liver. Pharmacogenetics 2003;13:565–575.
86. de Waziers I, Cugnene PH, Yang CS, Leroux JP, Beaune PH. Cytochrome P450 isozymes, epoxide hydrolase and glutathione transferase in rat and human hepatic and extrahepatic tissues. J Pharmacol Exp Ther 1990;253:387–394.
87. Lin JH, Chiba M, Nishime JA, DeLuna FA, Yamazaki M, Lin Y.. Effect of dexamethasone on the intestinal first-pass metabolism of indinavir in rats: evidence of cytochrome P-450 3A and P-glycoprotein induction. Drug Metab Dispos 1999;27:1187–1193.
88. Sesardic D, Cole KJ, Edwards RJ, Daveis DS, Thomas PE, Levin W, Boobis AR. The inducibility and catalytic activity of cytochromes P450c (P450 1A1) and P450d (P450 1A2) in rat tissues. Biochem Pharmacol 1990;39:499–506.
89. Kleinbloesem CH, van Hartan J, Wilson JPH, Danhof M, van Brummek P, Breimer DD. Nifedipine: kinetics and hemodynamic effects in patients with liver cirrhosis after intravenous and oral administration. Clin Pharmacol Ther 1986;40:21–28.
90. Eichelbaum M, Albrecht M, Kliems G, Schafer K, Somogyi A. Influence of mesocaval shunt surgery on verapamil kinetics, bioavailability and response. Br J Clin Pharmacol 1980;10:527–530.
91. Ged C, Rouillon JM, Pichard L, Combalbert J, Bressot N. The increase in urinary excretion of 6-hydoxycortisol as a maker of human hepatic cytochrome P450IIIA induction. Br J Clin Pharmacol 1989;28:373–387.
92. Glaeser H, Drescher S, Eichelbaum M, From MF. Influence of rifampicin on the expression and function of human intestinal cytochrome P450 enzymes. Br J Clin Pharmacol 2005;59:199–206.
93. Kolars JC, Schmiedlin-Ren P, Scheutz JD, Fang C, Watkins PB. Identification of rifampicin-inducible P450IIIA4 (CYP3A4) in human small bowel enterocytes. J Clin Inves 1992;90:1871–1878..
94. Petty KJ, Vega JM. Drug-drug interactions: marketing perspectives. In: Rodrigues AD, editor. Drug-Drug Interactions. New York: Marcel Dekker; 2002. pp. 633–643.
95. Smith DA. Induction and drug development. Eur J Pharm Sci 2000;11:185–189.
96. Lin JH Challenges in drug discovery: lead optimization and prediction of human pharmacokinetics. In: Borchardt RT, Kerns EH, Lipinski CA, Thakker DR, Wang B, editors. Pharmaceutical profiling in drug discovery for lead selection. Arlington (VA): AAPS Press; 2004. pp. 293–325.
97. Whitlock JP, Dension MS Induction of cytochrome P450 enzymes that metabolize xenobiotics. In: Ortiz de Montellano PR, editor. Cytochrome P450 structure, mechanism and biochemistry. 2nd ed. New York: Plenum Press; 1995. pp. 367–390.

98. Clarke SE, Jones BC Human cytochromes P450 and their role in metabolism-based drug-drug interactions. In: Rodrigues AD, editor. Drug-drug interactions. New York: Marcel Dekker; 2002. pp. 55–88.

99. Luo G, Cunningham M, Kim S, Burn T, Lin J, Sinz M, Hamilton G, Rizzo C, Jolley S, Gilbert D, Downey A, Mudra D, Graham R, Carroll K, Xie J, Madan A, Parkinson A, Christ D, Selling B, LeCluyse EL, Gan LS. CYP3A4 induction by drugs: correlation between a pregnane X receptor reporter gene assay and CYP3A4 expression in human hepatocytes. Drug Metab Dispos 2002;30:795–804.

100. Roins MJJ, Ingelman-Sunderg M. Induction of human drug-metabolizing enzymes: Mechanisms and implications. In: Woolf TF, editor. Handbook of drug metabolism. New York: Marcel Dekker; 1999. pp. 239–262.

101. Hsu A, Grannemain GR, Bertz RJ. Ritonavir, clinical pharmacokinetics and interactions with other HIV agents. Clin Pharmacokinet 1998;35:275–291.

102. Chang TK, Yu L, Maurel P, Waxman DJ. Enhanced cyclosphosphamide and ifosfamide activation in primary human hepatocyte cultures: response of cytochrome P-450 inducers and autoinduction by oxazaphosphorines. Cancer Res 1997;57:1946–1954.

103. Lin JH. Sense and nonsense in the prediction of drug-drug interaction. Curr Drug Metab 2000;1:305–331.

104. Ito K, Iwatsubo T, Kanamitsu S, Ueda K, Suzuki H, Sugiyama Y. Prediction of pharmacokinetic alterations caused by drug-drug interactions: metabolic interaction in the liver. Pharmacol Rev 1998;50:387–411.

105. Takahashi H, Kashima T, Kimura S, Murata N, Takaba T, Iwade K, Abe T, Tainaka H, Yasumori T, Echizen H. Pharmacokinetic interaction between warfarin and a uricosuric agent, bucolome: application of in vitro approaches to predicting in vivo reduction of (S)-warfarin clearance. Drug Metab Dispos 1999;27:1179–1186.

106. Tang C, Lin JH, Lu AYH. Metabolism-based drug-drug interactions: what determines individual variability in cytochrome P450 induction? Drug Metab Dispos 2005;33:603–613.

107. Madan A, Dehaan R, Mudra D, Carroll K, LeClyuyse EL, Parkinson A. Effect of cryopreservation on cytochrome P450 enzyme induction in cultured rat hepatocytes. Drug Metab Dispos 1999;27:327–335.

108. Lu C, Li AP. Species comparison in P450 induction: effects of dexamethasone, omeprazole, and rifampin on P450 isoforms 1A and 3A in primary cultures from man, Sprague-Dawley rat, minpig, and beagle dog. Chem Biol Interact 2001;134:271–281.

109. Raucy JL, Mueller L, Duan K, Allen SW, Strom S, Lasker JM. Expression and induction of CYP2C P450 enzymes in primary cultures of human hepatocytes. J Pharmacol Exp Ther 2002;302:475–482.

110. Benet LZ, Williams RL Design and optimization of dosage regimens: pharmacokinetic data. In: Gilman AF, Rall TW, Nies AS, Taylor P, editors. The pharmacological basis of therapeutics. New York: McGraw-Hill; 1990. pp. 1650–1735.

111. Lin JH, Yamazaki M. Role of P-glycoprotein in pharmacokinetics: clinical implications. Clin Pharmacokinet 2003;42:59–98.

112. Lin JH. Drug-drug interaction mediated by inhibition and induction of glycoprotein. Adv Drug Deliv Rev 2003;55:53–81.

113. Benet LZ, Cummins CL, Wu CY. Transporter-enzyme interactions: implications for predicting drug-drug interactions from in vitro data. Curr Drug Metab 2003;4:393–398.

114. Benet LZ, Cummins CL, Wu CY. Unmasking the dynamic interplay between efflux transporters and metabolic enzymes. Int J Pharm 2004;277:3–9.

INDEX

Predictive Approaches in Drug Discovery and Development: Biomarkers and In Vitro/In Vivo Correlations, First Edition. Edited by J. Andrew Williams, Jeffrey R. Koup, Richard Lalonde, and David D. Christ.

Wiley Series on Technologies for the Pharmaceutical Industry

Sean Ekins, Series Editor

Computational Toxicology: Risk Assessment for Pharmaceutical and Environmental Chemicals
Edited by Sean Ekins

Pharmaceutical Applications of Raman Spectroscopy
Edited by Slobodan Šašić

Pathway Analysis for Drug Discovery: Computational Infrastructure and Applications
Edited by Anton Yuryev

Drug Efficacy, Safety, and Biologics Discovery: Emerging Technologies and Tools
Edited by Sean Ekins and Jinghai J. Xu

The Engines of Hippocrates: From the Dawn of Medicine to Medical and Pharmaceutical Informatics
Barry Robson and O.K. Baek

Pharmaceutical Data Mining: Approaches and Applications for Drug Discovery
Edited by Konstantin V. Balakin

The Agile Approach to Adaptive Research: Optimizing Efficiency in Clinical Development
Michael J. Rosenberg

Pharmaceutical and Biomedical Project Management in a Changing Global Environment
Scott D. Babler

Predictive Approaches In Drug Discovery and Development: Biomarkers and In Vitro/In Vivo Correlations
Edited by J. Andrew Williams, Richard Lalonde, Jeffrey Koup, and David D. Christ